Management for Professionals

For further volumes:
http://www.springer.com/series/10101

Georg Weiers

Innovation Through Cooperation

The Emergence of an Idea Economy

Georg Weiers
African Development Bank
Carthage
Tunisia

ISSN 2192-8096　　　　　　　　　ISSN 2192-810X (electronic)
ISBN 978-3-319-00094-7　　　　　　ISBN 978-3-319-00095-4 (eBook)
DOI 10.1007/978-3-319-00095-4
Springer Cham Heidelberg New York Dordrecht London

Library of Congress Control Number: 2013948755

© Springer International Publishing Switzerland 2014
This work is subject to copyright. All rights are reserved by the Publisher, whether the whole or part of the material is concerned, specifically the rights of translation, reprinting, reuse of illustrations, recitation, broadcasting, reproduction on microfilms or in any other physical way, and transmission or information storage and retrieval, electronic adaptation, computer software, or by similar or dissimilar methodology now known or hereafter developed. Exempted from this legal reservation are brief excerpts in connection with reviews or scholarly analysis or material supplied specifically for the purpose of being entered and executed on a computer system, for exclusive use by the purchaser of the work. Duplication of this publication or parts thereof is permitted only under the provisions of the Copyright Law of the Publisher's location, in its current version, and permission for use must always be obtained from Springer. Permissions for use may be obtained through RightsLink at the Copyright Clearance Center. Violations are liable to prosecution under the respective Copyright Law.
The use of general descriptive names, registered names, trademarks, service marks, etc. in this publication does not imply, even in the absence of a specific statement, that such names are exempt from the relevant protective laws and regulations and therefore free for general use.
While the advice and information in this book are believed to be true and accurate at the date of publication, neither the authors nor the editors nor the publisher can accept any legal responsibility for any errors or omissions that may be made. The publisher makes no warranty, express or implied, with respect to the material contained herein.

Printed on acid-free paper

Springer is a part of Springer Science+Business Media (www.springer.com)

Foreword

The way firms innovate is changing. Top-line growth is increasingly driven by innovation. Research and Development (R&D) budgets are growing, but also much more is demanded of them to churn out new inventions quicker and more efficiently. A lot of focus is on how to make such innovation processes leaner and more effective. And as the old model of brick and mortar R&D infrastructure is under pressure, new avenues are being explored. 'Open innovation' is becoming the new normal.

Start with two observations: (1) Competition is increasingly driven by innovation. Product cycles have become shorter. Competition has become more global. New products and new production methods drive success. (2) Inventions increasingly come from outside the firm. Companies increasingly understand this and are starting to adapt their business models. Connecting to inventors from outside the firm, companies can speed their innovation and time to market. They take on inventions that work. The risk of discovery and initial development is with the inventors. This reduces the risk of in-house research as well as the administrative burden of running R&D units and raises the return on the company's core business of producing and marketing products.

The world is a big pond with many sources of ideas. Any one company does not have all the ideas in the world. New ideas come from unanticipated places, many adjacent to, or farther away from, a specific company or sector. This is a huge opportunity for companies to leverage the capabilities of the world, not just their own R&D. No one can do everything themselves. No one expects in-house researchers to know all. For most challenges a company may face, there are ready-to-go solutions out there used somewhere else. To make the best use of research capacity, good networks are needed to link up with others outside to discover promising inventions and bring these ideas from the outside inside the firm.

Achieving such a more open innovation model is challenging. It is not easy to tap into this wealth of inventions from the outside, connect to the inventors around the world and jointly develop a product. Many firms developing extensive networks with other companies, research institutions or suppliers, to be tuned into new developments, have early access and co-create new products. Another approach, one that has developed into a multibillion-dollar industry, is using innovation prizes. Firms publish an R&D need to attract effective solutions that already exist or can easily be found instead of trying to develop a solution themselves, often from scratch. This is changing the business approach to product development by starting from the product and finding appropriate solutions, not from the technology identifying a market for it.

Procter and Gamble (P&G) has been a pioneer in opening up the innovation process. It was one of the first to realize this huge potential and take appropriate measures to realize it. Connect and Develop was launched to move from a traditional R&D model to a more open one: connecting to inventors outside and developing products jointly–connect and develop. We realized that for any one of our 9,000 researchers inside the various units of P&G, there were 200 just as qualified outside. Instead of relying on the 9,000 inside, tapping into a network of 1.809 million offers a quite different potential. It multiplies the number of ideas to work with. It can leverage the best ideas outside and combine them with expertise inside the company. R&D productivity went up by 60 %. More than 250 products resulted from it. Insourced innovation went up from 15 % to 50 %.

Many companies are embracing similar strategies. But not everyone has yet realized its immense potential of looking beyond the company's walls. Those that have realized the potential of connecting to the world often have a hard time making the necessary changes. Turning a company's innovation model on its head from an inward- to an outward-looking approach requires a whole new strategy.

From the perspective of the inventors, the picture is a different one. Firms were closed to them and not willing or able to connect to and work together with them. Now they are opening up, welcoming solutions and actively looking for inventions across the globe. Managers recognize that inventions are developed throughout the globe that can help companies innovate and grow. This offers new prospects for inventions to be scaled up and profited from.

This book pulls together many of these aspects, laying out how innovation has changed and is changing, opening up to inventions from the outside. It captures the increasingly collaborative nature of innovation from both the inventor and the firm perspective. It goes a step further even by proposing collaboration at an earlier stage than is currently the case where intellectual property is not fully defined. If it were possible to connect not only with inventors who have a ready product safeguarded by intellectual property but earlier on in the invention process, many new opportunities would open up. Both firms and inventors could develop the invention together. This has been a crucial constraint of open innovation. Going beyond it raises a number of new challenges that intellectual property could not solve. An alternative solution would hold tremendous promise for both inventors and firms. It gives firms access to even more inventions and makes it easier for inventors to work together with firms to develop products jointly. The book is a pioneering attempt to identify the underlying principle that can make such joint development work. It will still have to prove itself in practice. Already it offers a promising way forward that deserves to be explored further, discussed, refined and developed.

Larry Huston

Preface

Imagine: you have an idea, a brilliant idea that, you are certain, could make you rich. Alas, you do not have the means or skills to realize it. Now imagine this: you have an idea and you can sell this idea to a big firm that can realize it easier and more profitable than you ever could. They profit, as do you. They pay you handsomely for your idea. Your idea, as an idea, already is valuable.

Nonsense, many would say, this does not work. And they would be right. Unfortunately, we have no ready solution to make such trading in ideas work in practice. This means countless good ideas get lost. Innovation does not live up to its full potential. Ideas may pop up in every imaginable corner – only to be forgotten there. And while invention is becoming more and more democratic, innovation is not.

This unfortunate enigma has been subject to long and extensive debates. Many a tree has fallen victim to academic controversies on it, politicians have littered speeches with it and perfectly good dinner conversations have been spoiled over it. Such disputes span far and wide. They have trickled down into many debates, political, scholarly, colloquial and otherwise. Often disparate, many never overlap. This book tries to pull many of them together. This has two implications: it prescribes the structure and sets the tone.

(i) *A simplified structure:* Though the proposed argument may sound simple on the outset, it is highly complex once you look behind the pretty facade. Much thinking has gone into these topics by legions of academics and practitioners. For each and every point, you can probably link to entire lifeworks of research and discussions, and for each you can find some specialist able to get into it in much, much more detail. No such attempt can ever be complete, only reasonably comprehensive. Balancing the need to venture into details, while maintaining a more general storyline, is challenging. It can be done. This is such an attempt. To make it easier to read, it is split into two parts. A deliberate choice to separate two broad perspectives: the macro and the micro view. The big picture first – for anyone more interested in the framework than the nitty-gritty of making it work. The argument extrapolates the current debate. It follows in the footsteps of Drucker, Audretsch, Chesbrough and von Hippel and extends their argument. It is a bird's-eye view, showing broad strokes, rough trends of how innovation has and will continue to evolve. The second part looks at the details – for anyone interested in how to make the vision laid out in Part I a reality. It translates various academic discussions into a coherent approach to demonstrate what the problem is (Arrow's Fundamental Paradox of Information) and what new processes need to be developed to overcome it and make trading ideas possible. It is about the practical details of how cooperation can work.

(ii) *An accessible tone:* As much of the discussion is sparsely connected, straddling this divide, spanning across disparate fields and various groups with mostly siloed edifices of literature, means translating one into the other, or better yet, making it accessible to all. The argument draws on diverse backgrounds, from economic literature to business cases to very theoretical discussions. Even within these broad fields, the lines are drawn, with members using their own genre, their own language, their own denotation. This book tries to translate it into a common and accessible format that anyone can read. It will be light to the expert in their respective field, yet hopefully accessible to the unacquainted.

Acknowledgments

Because it is so broad, because it is quite bold, it has been a lengthy, yet fascinating, journey from early sketches to final drafts, a journey that alone would surely have been most boring, most tedious and utterly unsuccessful. The manuscript has benefited from innumerable discussions with colleagues and friends, too many to thank all of them in person. Some should be mentioned here, nonetheless, for their tireless efforts and for graciously donating their time to provide detailed comments and suggestions to help improve and refine the various drafts. I am especially grateful for this to Christoph Meineke, Gabriele Fattorelli, Günther Tschabuschnig, Johannes Kiess, Mariana Todorova and Vasco Steltenkamp. David Audretsch, Joshua Gans, Nikolaus Franke, Larry Huston and Kenneth Arrow kindly agreed to provide feedback and guidance. Also, I am grateful to the World Bank, African Development Bank and George Washington University for providing the much-needed access and stimulating intellectual environment to pursue this work. Of course, a significant gratitude is owed to my editor. Publishing is a little like trading ideas. Without a ready product, you need to convince a stranger that it is worth their time and effort to team up. Prashanth Mahagaonkar, Springer's Business and Economics Editor, took a chance on an outsider, far afield the academic centres, submitting a proposal from Africa. Most importantly still, I would like to thank my wife *Ursula* for her inspiration, motivation, patience and tireless support in finishing this book.

Contents

Part I Introduction

1 Introduction   3
   References   14

Part II The Emergence of an Idea Economy

2 Economic Evolution   19
   References   32

3 The Entrepreneurial Society   35
   References   51

4 Open Innovation   59
   References   68

5 The Idea Economy   73
   References   85

6 More and Better Innovation   87
   References   95

Part III Cooperative Innovation

7 Setting the Stage   99
   References 110

8 The Missing Link 115
   References 123

9 Narrowing the Gap 125
   References 139

10	**Bridging the Gap**		143
	10.1	Trust	148
	10.2	Fairness	160
	References		166
11	**Cooperative Innovation**		171
	11.1	Matching	172
	11.2	Feedback Mechanism	177
	11.3	Code of Conduct	183
	11.4	Guidance	189
	References		193
12	**Cooperative Innovation and Its Future**		195
	References		200
13	**Concluding Remarks**		201
	References		205
References			207
Index			235

Part I
Introduction

Introduction 1

In 1879, in a laboratory not far from New York, Thomas Edison invented the incandescent light bulb.[1] After many failed attempts, finally it worked using a carbonized cotton filament. His first bulb lasted only around 15 h, soon after, using different materials, to be expanded to some 1,200 h. By 1882, electric lamps illuminated the mansions of the first Manhattan customers. It was an important leap towards the modern electrical and lighting industry. This was by far not Edison's only, perhaps not even his most important invention. Edison started off inventing devices for the nascent telegraph industry. The inventor of the first successful apparatus for recorded music and motion pictures many consider him the father of the movie and the music recording industry. He developed the carbon microphone, and even an electric pen, a vote recorder, and a storage battery intended for an electric car can be found among his inventions. As one of the most prolific inventors, he accumulated a record 1,093 US patents in his lifetime.[2] His most important contribution is often considered another still: Inventing the method of invention – inventing invention. In Menlo Park and West Orange he set up the first industrial research laboratory devoted entirely to invention and benefitting from its output – the first 'invention factory'. Many different experts gathered there, working on a broad range of inventions in various industries, without any affiliation to a specific company. Yet he aspired to be more than being a great inventor. He aspired to create whole industries from his inventions and become a huge commercial success. He did, but much less than you might expect. As gifted an inventor that he was, as unfortunate as a businessman he fared. He "had no real feel for what the public wanted or how to convert his discoveries into products that might sell. It took him years to understand that the phonograph was not a business tool but an entertainment breakthrough. He helped create movies but resisted the idea that

---

[1] See especially Smil (2005, pp. 37–49). More generally, see Israel (1998) and Millard (1990).
[2] For an exhaustive list and descriptions, see edison.rutgers.edu/inventions.htm (02.03.2013).

people might want to go into a theatre to watch one."[3] Henry Ford attested Edison to be "the world's greatest inventor and worst businessman."[4] Management guru Peter Ducker's judgement on him is unequivocally devastating: "Edison, the nineteenth century's most successful inventor, converted invention into the discipline we now call research. His real ambition, however, was to be a business builder and to become a tycoon. Yet he so totally mismanaged the businesses he started that he had to be removed from every one of them to save it."[5]

Many other great inventors did not fare any better. Consider a contemporary of Edison: the ingenious Nikola Tesla. He is often considered the more brilliant inventor of the two, but also the one who commercially fared even worse. His work includes amongst many others, pioneering work on wireless communication, the induction motor, x-rays, radar, energy weapons, weather control, and especially long distance and wireless energy transmission. His many brilliant and often visionary inventions, as well as his profound scientific work are said to have helped 'usher in the second industrial revolution'. A contemporary praised him: "were we to seize and eliminate from our industrial world the result of Mr. Tesla's work, the wheels of industry would cease to turn, our electric cars and trains would stop, our towns would be dark, our mills would be idle and dead. Yes, so far reaching is his work, that is has become the warp and woof of industry. [...] His name marks an epoch in the advance of electrical science. From that work has sprung a revolution in the electrical art."[6] Yet, he never became a successful businessman himself. His bookkeeper attested: "This genius was totally lacking in business ability".[7] As a businessman, Tesla failed dismally. From his own Tesla Electric Light & Manufacturing his investors soon removed him. The money he made from his inventions he lost to many commercially fruitless endeavours often refusing to pursue commercially more viable options.[8]

---

[3] Time Magazine (2010).

[4] Millard (1990, p. 49).

[5] Drucker (1986, pp. 12–13). This surely is a harsh and often undeserved verdict. Many consider it a widely held myth (see Pretzer 1989, p. 136, Millard p. 50). Ample evidence exists to take a more nuanced view. Especially the work by Millard restores Edison as a businessman to some extent "The history of Edison's business enterprise is not one of uninterrupted success, but it is also not a catalogue of failure [...] His business associates knew a different Edison, a shrewd, calculating man who exercised fine judgement of the marketplace." (Millard 1990, p. 50). Still, it cannot be fully dispelled, and the 'myth', true or not, may still prove an important point (Taussig 1915, p. 17, recalls an anecdote placing less emphasis on Edison's ability to manage then his attitude towards money versus invention challenge).

[6] Bernard Behrend, Vice President of the Institute of Electrical Engineers, in 1917 lauding Tesla in the prize ceremony where he received the Edison Medal from the American Institute of Electrical Engineers, quoted in O'Neill (2007, p. 236). Similarly, grand claims are sometimes made about Edison (compare with, for example, Millard 1990).

[7] George Scherff, Tesla's bookkeeper and secretary, quoted in O'Neill (2007, p. 21).

[8] See O'Neill (2007). Also see Tesla's (1919) autobiography.

William Shockley's story is even more remarkable.[9] He was a co-inventor of the transistor, which gave birth to modern electronics and earned him the Nobel Prize in Physics in 1956. As the director of Shockley Semiconductor, he could have created the worldwide dominant chipmaker. But apparently, he managed it so badly as a leader and businessman that a key group of people left: among them Robert Noyce and Gordon Moore who soon after set up Intel – the now behemoth and quasi-monopoly of computer processors currently worth some $100 billion.[10] Shockley is sometimes seen as the father of all of Silicon Valley, not least precisely because so many of his employees left to set up Silicon Valley's initial computer companies. "In the late 1970s, the American Electronics Association published a genealogy of Silicon Valley. The table showed that virtually every company in the valley could show a line leading directly to someone who worked at and eventually left Fairchild Semiconductor. [...] Everyone from Fairchild originally came from Shockley Semiconductor. Shockley's company was the seed of Silicon Valley."[11] But unlike so many in Silicon Valley, creating vast fortunes and industry defining companies, Shockley never made a profit. Though a brilliant inventor, as a businessman he utterly failed. "In all of the history of business, the failure of Shockley Semiconductors is a class by itself."[12] Shockley went back to teaching.

Not everyone has to be a Nobel laureate or timeless genius. Anyone can come up with a brilliant idea – in general or in any area of experience and expertise: the coffee cup sleeve, the pop-up toaster, electric toothbrush, Nutella, the ATM, PageRank, or the idea for Groupon or Facebook, and countless more. Anyone can have ideas, but only some are entrepreneurs. Jack Dorsey the inventor of Twitter was soon forced out of his company, as he "was not the greatest manager"[13]; Reed Hastings the later founder of Netflix even asked to be removed from his first company Pure Software as he lost faith in his own management skills.[14] History is full of such examples. Many prominent ones exemplify this widespread discrepancy: ingenious and bright inventors, with great ideas, but often bad or unfortunate businessman. Many people have great ideas. Most are not also able and skilled at realizing it in the market.

Think of an idea you had. Recall that enthralling jolt of excitement when first you thought of it, the flow of creativity, and the sense of accomplishment. What happened to it? In your mind, it all played out well. It sounded so easy. In reality, it appeared all the more challenging. You put it on paper; you dwelled on it for a while; you dreamed about its possible future, envisioned its success. Maybe you even developed it further, spent time on it, tried to refine it; gave it more shape.

---

[9] For a riveting tale of the rise and fall of Shockley, see Shurkin (2006).
[10] According to Bloomberg (bloomberg.com/quote/INTC:US 02.03.2013).
[11] Shurkin (2006, p. 187).
[12] Shurkin (2006, p. 164).
[13] See Vanity Fair (2011).
[14] USA Today (2006).

You may have done some search on the web, maybe set up shop in your garage or basement, tinkered and fiddled about with your idea, even sketched out a business plan. Yet, somewhere along the way, somewhere in your mind doubt festered and grew. A sense of overwhelming challenges, of needed commitment, of responsibilities, of required knowledge and expertise emerged; a reality check; unease; anguish. Hoping one day will be more suited, one day you will be better prepared, one day the situation will be more opportune, likely, you pushed it back, back into your mind, your drawer, your garage; you got side-tracked; you backed off. It became one of those things-to-do. Likely, it will remain just that. At some point, it will become redundant. Someone else will do it, someone else will benefit from it, someone else will profit. Regret for you. Profit for others.

This is nothing unusual. Possibly, this is the norm. It is the fate of all too many ideas. If only you could realise it, see a realistic chance of success, not an outlier or a distant wish, detached from what is reasonably possible. Realising ideas is not only challenging, it is daunting. Few have the knowledge of what to do, fewer the means and skills to do it. You may know of an ideal candidate, an established firm that could easily realise it and make a profit. But you have no way in. The company is like a fortress. Literally. You may get to the gate, you may get to the reception, but meeting with the right person who might be willing to listen to your idea is impossible. And even if you were to find such competent contact, why would she listen, or if she did, and liked the idea, she could simply use it and make a profit for the company, leaving you with nothing. Mostly therefore the only reasonable option that comes to mind to realise your idea is to become an entrepreneur: develop the idea to a marketable product, start a company, market your product. Apart from the fact that this only works for a narrow segment of ideas – many are of a different nature or cannot be realised without the support of others (e.g. you have a brilliant idea to improve the efficiency of a production line. Without the owners of the production line, this does not help you much. You need them to realise your idea.) – entrepreneurship is not as simple as the often romantic picture media and urban myth convey. You don't just set up a firm in your garage, your great idea will produce and sell itself, and your company will grow and prosper. It requires means, skills, and knowledge, all of which may not be at your disposal, which you may not have ready access to, or which presents a risk, a choice of lifestyle a lot of people are not willing to engage in.

What this means is that many people have brilliant ideas. Most never get to realize them. Thus, many ideas are never or only much later realised, mostly by others. This is inefficient, and to some extent unfair. All should be able to realise their ideas more freely and be able to profit from their ingenuity. You would be better off as would anyone who benefits from your inventions.

On the other hand, you can find an array of companies that stumbled or failed due to a lack of ideas – even ably managed, well-established incumbents, failing to stay ahead. Remember Commodore Business Machines? In the 1980s, Commodore swept the market with the C64 and later the Amiga. "Its Commodore 64 was the Volkswagen of computers, a low-cost, dependable model that became the best- selling machine in the business."[15] Between 1982 and 1993, more than 22 million units were sold – making it "the greatest selling single computer model of all time",[16] warranting an entry in the Guinness Book of World Records. Yet, Commodore failed to mature in the PC market and missed the move to consoles. It failed to keep up with innovations made by IBM PCs or Apple's Macintosh as well as the emerging Nintendo or Sega consoles. In 1994, Commodore declared bankruptcy.[17]

Or remember the time you used a film role to place in a camera to take a photo. It was limited, expensive, and took ages to be developed. Kodak does – reminiscing about the good old times. Founded during the advent of photography in 1880 by George Eastman, Kodak basically established the market for film; it pioneered photography. By the middle of the century, it was a household name. "By 1976 Kodak accounted for 90 % of film and 85 % of camera sales in America. Until the 1990s it was regularly rated one of the world's five most valuable brands."[18] In January 2012, Eastman Kodak filed for bankruptcy protection. Kodak was a corporate giant for 100 years, but since the 1990s, it lost 90 % of its value, and reduced 90 % of its staff. Kodak missed out on the move to digital photography, missed out to transform itself into a service provider or find a lucrative niche. It lacked good ideas. Fujifilm, the eternal rival managed. So did innumerable other companies, grabbing the now uncontested market where once Kodak ruled. While Facebook is doling out a billion dollars for Instagram, a company of just 13 employees, barely more than a year old, producing neither film nor cameras, but a software to edit digital pictures,[19] Kodak, a proud corporation of once 145,000 employees, now has a market capitalization of a mere $50 million.[20]

Countless such examples exist: Radio Corporation of America (RCA), Digital Equipment Corporation (DEC), America Online (AOL), Xerox, Borders, Blockbuster Video, even Apple for a while in-between Steve Jobs and the return of Steve Jobs, and many, many more. Though often key players for a while many giants were 'simply' out-innovated. Certainly not due to a lack of resources. They did not have enough, or the right ideas. Typically, they command vast financial prowess, competent staff, and exhibit cutting edge organizational structures. Still they stumbled. "Innovate or die". Those with the better ideas prosper – those without lose. But how to come by great ideas? How to get a hold of them; how to get access to them?

Incumbents are just the obvious ones. Think of all the potential ones, those that never show. Think of all those ambitious business students attending entrepreneurship

---

[15] Time Magazine (1986).
[16] computerhistory.org/timeline/?category = cmptr (02.03.2013).
[17] See Bagnall (2006). He traces the cause to management mistakes and personal agendas. Essentially it remains one of being out innovated. Also commodore.ca/history/company/chronology_portcommodore.htm (02.03.2013).
[18] The Economist (2012a).
[19] See BBC News (2012).
[20] According to Bloomberg (bloomberg.com/quote/EKDKQ:US 02.03.2013).

courses, all those aspiring entrepreneurs, those experienced and gifted serial entrepreneurs – desperate for a good idea.[21] What wouldn't they give to have a great idea? Even the best firms, the brightest, most gifted and pragmatic doers need good ideas to truly prosper in the market. Their potential often lays dormant or dwindles. The ability to implement needs a purpose and needs ideas to be implemented. Coming up with ideas is not that easy. Ideas cannot be prompted or simply requisitioned.

Companies, entrepreneurs, investors, experts all are searching for ideas, longing for more and better innovations to grow, prosper and profit. Say you run a successful firm. Competition is fierce. Pressure to innovate is enormous. You spend vast sums on R&D. New firms keep disrupting the market. You are besieged from all sides, under constant pressure to keep up or stay ahead, constantly on your toes looking for ideas. You are told: "only the paranoid survive"[22] Indeed, you see many clever start-ups, emerging technologies, methods, ideas, closing in on you, wishing you had had access to them first. You scramble to merge with or acquire start-ups, to buy vetted concepts and ideas at vast sums. Often it is too late, or ridiculously expensive. You survive, but do not profit. Mostly it is inefficient. You wish you could have gotten hold of them much sooner. But how to get hold of ideas earlier? How to get hold of them first?

Or you are an entrepreneur, or are aspiring to become one. Better yet, you are a serial entrepreneur. You are very good at realizing ideas. You have what it takes, the skills, the means, the access, the entrepreneurial talent. But you do not have a striking idea. At least not, right now. You may not be the content expert, but you could make it work. You are a doer. But where to get ideas?

You could also be an expert, a gifted specialist, a highly trained researcher, skilled tinkerer, experienced pragmatist, making ideas work in practice. Making products out of ideas. Though you are very good, you need inspiration, a cue, a solid hunch. You are sure you could achieve much. You see others around you, less talented, less knowledgeable, less pragmatic that have come far with their ideas. You could have done better. If only they would have come to you, if only they were your roommates, were in your club, your relatives, your neighbours, your friends. Together, you could have made the most of it. But how to get them to come to you?

Many have the needed skills and means to realise them, if only they had the ideas. Alas, all too often, others do. How to access them? How to tap into this vast potential source of ideas? The mere internal approach of dedicated in-house R&D, or desperate tinkering in your garage, and esoteric ideation attempts, and the hope that you will come up with a brilliant idea, that you can instigate the generation of great ideas is limited and inefficient. Many ideas are spontaneous and cannot be prompted. More importantly, many more, many disruptive and highly competitive ideas are increasingly generated outside of your company, by competitors, by other

---

[21] For a sense of scope, see, for example, Kelley et al. (2012).
[22] Grove (1999).

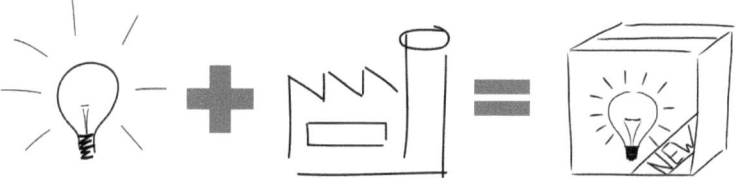

Good ideas require necessary skills and means to turn into successful products

Fig. 1.1 Innovation through cooperation

people, by people who, unlike you possibly, often do not have the means to realise them – but also no inclination to share them with you.

Innovation needs both, bright ideas and the skills to realize them. It is very hard to come up with great ideas. But it is also highly challenging to realize them. To inventors ideas often have strong appeal, but "the other side is usually an afterthought. It is humdrum. It is behind the scene. It is dirty work." And yet, "Ideas are only beginnings".[23] What comes next is challenging. It requires different skills. Realizing ideas is just as hard as having them, if not more so. For innovation to really succeed both are required, ideas and the skills to succeed in the marketplace. The same holds in reverse. Though ready and willing to work hard to realize ideas, ideas first need to be available to be realized. Without them the best capabilities, and implementation potential remains without purpose, without use. As the Harvard Business Review put it: "Profitable survival requires an edge derived from some combination of a creative idea and a superior capacity for execution."[24] (See Fig. 1.1)

Alas, only very few have both – the ideas, as well as the skills and means to realize them. Those fortunate few often become hugely successful: think of Walt Disney; of Elon Musk the co-founder among others of PayPal, SpaceX and Tesla Motors; Sam Walton the founder of Walmart, or Ingvar Kamprad of IKEA; Pierre Omidyar of eBay, Jeff Bezos of Amazon, or Andrew Mason of Groupon; of Frederick Smith of FedEx, just as James Dyson of Dyson Ltd.

Some have it all. Others are fortunate to find the right mix of partners, combining great ideas with implementation capacity: for example Steve Wozniak together with Steve Jobs of Apple, Herbert Boyer with Robert Swanson of Genentec, Larry Page and Sergey Brin together Eric Schmidt of Google. Without such opportune combination of able partners, none would likely have been as successful. Without Silicon Valley veteran Eric Schmidt it is safe to say, Google would not be where it is today – worth over $250 billion.[25] Without his years of experience, knowledge, his network and sheer business acumen any of the many contenders of the likes of AltaVista, Lycos, MSN/Bing, Ask, etc. could just as well have come to dominate the search market. When he

---

[23] Govindarajan and Trimble (2010, pp. 2–3).
[24] Bhide 1994, p. 151.
[25] According to Bloomberg (bloomberg.com/quote/GOOG:US 02.03.2013).

stepped down in 2011 after 10 years of growing the company, he twittered tellingly: "Day-to-day adult supervision is no longer needed."[26] Good for Google.

Some have both. Others are lucky to find the right partners. What about all the rest – what about all those with ideas but neither the skills nor means to realize their ideas? What about all those without a talented friend with a knack for business or a chance encounter with a gifted entrepreneur, willing and able to implement the idea together; those without access to promising partners? No outlet for their ideas exists, no real opportunity to profit from them. They have no real chance to succeed. Many ideas remain dormant, underused, and under-realized. What about the many gifted entrepreneurs and incumbent firms desperately looking for new ideas? Without the serendipity of an idea, or fortuitous encounter with a bright and willing inventor, their potential too remains dormant, underused, and underappreciated. If only the two – those with ideas looking for a partner, and those able to realize ideas, looking for ideas – could work together. If only it were possible to connect the two, to connect those able to realize ideas, with those with ideas. Together they could achieve much. Both would benefit tremendously.

What if it became a lot easier, much more realistic to realize your idea? What if the idea itself would already be valuable? What if you could get help to realise it, could team up with others to realise it with, or trade it to a firm or eager entrepreneur to even realise it for you? You invent, and together with an expert, entrepreneur, or firm you realize your idea. You have a more realistic chance to succeed. You need not be an entrepreneur, need not bear all the risk, need not start from scratch. You collaborate with others, entrepreneurs, firms, those best able to realize the idea with you. Both gain, both profit from it. This is indeed in the process of becoming reality – slowly, but we are getting there. Ideas are becoming increasingly valuable; ideas are becoming increasingly commoditized, becoming the commodity of the future. Soon you will be able to realise your idea much more easily. You will be able to collaborate with experts to develop and realise your idea. You will be able to trade your idea with companies or entrepreneurs to see it realised. You will be able to fairly profit from your idea.

This is just as appealing to companies and entrepreneurs. What if they could have ready access to such ideas? What if people were able and willing to collaborate, to trade ideas with you? Imagine you could freely trade ideas, have access and compete within the entire spectrum of ideas. This is what is increasingly happening. And it will intensify. Companies are increasingly opening up their innovation process, actively searching for bright ideas, actively looking to gain access to the latest ideas. Increasingly they are open to ideas from outside – and willing to collaborate and compensate the ones with ideas for their ingenuity. The competition for ideas and the need to attract the best and most promising ideas will increase. Ideas will be competed for, ideas will be traded, ideas will be increasingly viewed

---

[26] Chicago Tribune (2011).

as the fundamental source of profit. Be prepared. Be the first to move; be the most appealing; be the one that inventors seek out. Compete. Collaborate. Profit.

All this sounds alluring. Imagine the possibilities! Trading ideas; cooperation between inventors and implementers; ideas as valuable commodities. The question is why they are not already working together. Why is it so hard? Why has no such market for ideas been established? Why can you not yet really trade your idea to someone very able and willing to realize it? These are critical questions. The short answer is trust. We guard our ideas. We do not share them. And for good reasons. What is to keep others from taking the idea and realize it themselves? What is to keep them from stealing the idea? We cannot walk up to someone, least of all a firm, an expert, or established entrepreneur who could with ease realise it, and tell them our idea. They would gladly take it. They may throw you a bone, pat your back, and send you a fruit basket. But why would they compensate you for your idea? Trust is essential here. You may trust your family, some close friends. But how can you trust strangers or big companies to adequately reward your idea? Legal provisions such as patens typically do not apply. They hold for a fraction of ideas, and typically apply much later after much money is spent, and much expertise is already needed. Most ideas are not protected.

This notion has become part of popular culture: If you watched the movie about the Facebook story, 'The Social Network': Mark Zuckerberg may not have stolen the idea per se from HarvardConnection (Winklevoss Twins and Divya Narendra), but it spawned another, a better one, albeit one that precluded the HarvardConnection idea to prosper. He made billions with this idea, an idea he probably would not have had otherwise. HarvardConnection did not – likely because of Facebook.[27] Similarly prominent is the case of Robert Kearns popularized in the movie 'Flash of Genius', a film about his legal battle with Ford Motor Company about stealing his ideas for the intermittent windshield wiper.[28] He offered an early version of it to Ford. They rebuilt it and offered it in their cars – without acknowledging or compensating Kearns. Such a notion of valuable ideas easily being stolen once revealed has become widespread – for good reason. No wonder most people safeguard their ideas. They do not share them. Least of all with able experts and large corporations that could easily realise and appropriate them. It seems, the inventors mantra, repeated again and again: "never ever share your ideas – with anyone". This is a problem. This is a reason why so many never get to realize their ideas. Many cannot realize them on their own, or only very inefficiently. But they also cannot safely share them with the ones that are best able to realize them.

What is lacking is trust – trust that the others will not steal the idea, trust that they will not cheat you, and that you will be rewarded. Yet, the very nature of ideas makes it impossible to share without the risk of others stealing them. You cannot take them back. Ideas are unlike a car or a computer where you show what you have. People can kick the tires; there is typically some reference value; there is a market for this exact

---

[27] See Fincher (2010). Based on Mezrich (2010).
[28] See The New Yorker (1993) and Abraham (2008).

or at least very similar product. For ideas, there is no such thing. With ideas, this is generally impossible. They may tremendously vary is scope and value. By their very nature, they are typically unique. In order to offer your idea to someone you have to tell them what it is about. You reveal it. What then is there to keep them from taking it up and realizing it without giving you any credit or reward? On the other hand, without telling them your idea, how would they know what you are trying to sell them? How can they make up their mind? How could they commit to any reward up front? This intangible, fluid as well as its often ephemeral nature makes ideas special – and is the reason why no such market for ideas exists. It is unlike other goods. It has special properties that create special challenges to trading ideas.

This is a fundamental problem of ideas. Yet, it can be overcome. It may even be surprisingly simple. Admittedly, it is not *that* simple. If it were easy and straightforward, it would have been solved. A market for ideas would have sprung up. Yet, maybe it has, we just have not identified it as such. Also, the world has changed. We have gained more insights; we have made huge leaps in our understanding of markets, of cooperation, of trust, and ideas. New possibilities have arisen in a globalised and more interconnected world. New opportunities have arisen that were not available before. Exploring these possibilities in more detail and developing the particulars of what a solution could look like and what implications this will have on innovation and the economy is the subject of this book.

Part I puts forward the vision of an *Idea Economy*. It paints a picture of how innovation is evolving, where we are today, and what will happen once trading ideas becomes possible. Such a vision, and the analysis of its underlying dynamics, is important to understand how the *Idea Economy* is emerging, how it affects you, how you can partake, how you can prepare, engage, and benefit from it. Imagine what will happen when indeed it becomes possible to trade ideas. It will have profound implications; will have a very real impact. Both, the ones with ideas and those looking for ideas will benefit from such a development. The economy as a whole will benefit. Bringing the two together, enabling such cooperative innovation will have a tremendous socio-economic impact. Ideas will become the key to prosperity. The search for ideas will take priority – implementing them will become a service. The economy will change. What was once the era of big business, an era of large corporate research labs, an era of Detroit, Bell Labs, Xerox PARC, and RCA Laboratories, has already been undermined by a vibrant culture of entrepreneurialism. Behemoths are frequently challenged and brought to their knees by bold start-ups. The entrepreneur now symbolizes innovation. Bill Gates, Steve Jobs, Richard Branson, Carlos Slim, Philip Knight, Francois Pinault, Leonardo Del Vecchio, Bernie Ecclestone. Still only, a fortunate few succeed and profit from their ideas – the fortunate few that personify both ideas and entrepreneurialism.[29] In the future all inventors can succeed, even without becoming entrepreneurs. Inventors will replace the entrepreneurs as heroes;

---

[29] Typically, it seems the entrepreneur receives most of the credit. Even in a partnership of inventors and entrepreneurs, the more visible business leader seems to receive most of the attention.

firms will become implementers; vertical integration will dissolve; research and development will decouple; new services and business models will be introduced. Such a shift will have far-reaching and profound implications. It will upset the status quo. Imagine what this will mean for you. You can trade your ideas to firms, you can cooperate with entrepreneurs. You need to be less and less the entrepreneur. You need ideas. Light bulb moments are the key. The inventor again is the hero, without having to be the businessman. At the same time, this places more responsibility on those able to realize ideas. They need to be more cooperative to gain access to ideas, to compete for ideas. Together, through cooperative innovation, many more ideas will come to fruition; together, both will prosper.

But we are not there yet. The path may be set. The pace is not. Though underway, the *Idea Economy* has not yet materialized. Realising ideas is still very much confined to accustomed processes and common innovation paths, to internal corporate innovation or independent entrepreneurship. In practice, direct cooperation between those with ideas and those able to realize them is mostly limited to friends and family. Getting access to ideas is still a secondary search process outside of corporations. A true division of labour between ideas and its realisation, a true cooperation between the two has not progressed much beyond rudimentary and fragmented approaches. Some clever and promising examples of progress exist, but not on a broad scale, not accessible to most. This has good reasons that will be addressed in Part II. Ideas are special, making trading them difficult. How can cooperative innovation succeed? Solving this impasse is key to fostering innovation. Only then can the *Idea Economy* truly unfold and with it the full evolution towards a socio-economic framework where everyone has the chance to realize and profit from their ideas. Many challenges exist, but solutions are slowly emerging. Analysing them, identifying their underlying mechanisms can help design better and broader solutions; it can help find pragmatic ways to bridging the gap between the innovators and implementers. Enabling a true division of labour in the innovation process will allow for more and better innovation, it will help make the most of ideas, will help innovation to flourish, you to prosper, and the economy to thrive.

As *The Economist* pointed out: "Today there is no hotter topic in management theory than "sperm in the air". How do companies generate new ideas? And how do they turn those ideas into products? Hardly a week passes without someone publishing a book on the subject. Most are rubbish."[30] This is no book on management theory. It is one on ideas. Still it has to live up to the standard, and live up to its claim. Any claim of singlehandedly enabling cooperative innovation must seem preposterous. It is. This is merely a contribution to this evolution towards a practical solution. Still, even such a claim sounds radical. It is not. It is not as farfetched as first it may appear. The development towards an *Idea Economy* is but a logical extrapolation of current developments. The path is set. Conceptualising this development from an analytical perspective is, to narrow it down to the most prominent contributors, a melange of the

---

[30] The Economist (2010e).

inspiring works of Eric von Hippel, particularly on the democratization of innovation, Peter Drucker, especially his vision of the entrepreneurial society, and Henry Chesbrough, and his concept of open innovation. This mingled with some common sense and academic curiosity, drawing on a wide range of telling examples and relevant literature, provides for a robust analysis and a glimpse of things to come.[31] Going the extra step of claiming a solution to the age-old challenge of trading ideas is what makes it pretentious. But again, this too is not that farfetched. The problem has long been identified and framed as 'Arrow's Fundamental Paradox of Information'. Many efforts have been made to address and overcome it. Better yet, in practice, such solutions already exist. In several niches such as venture capital or the movie industry trading ideas is common. The trick is to identify the underlying mechanisms that allow for it and to identify the challenges that hinder it. Together a more general solution to enable cooperative innovation can be designed. This in turn is only possible because the environment has progressed and the tools have become readily available. Economic theory has advanced dramatically, offering a range of tools to better analyse social and economic interaction; the world has changed and people have become more aware of the prospects cooperation may offer as well as the challenges it still faces, and are more open to solutions – a necessary requirement; advances and ubiquitous availability of modern technologies allow for a completely different environment to propose solutions in. This fortunate confluence of developments is what makes it possible. It simply is the right time. Solutions are ripe. They do not have to be conjured. They are low hanging and 'merely' need to be identified. It is more a synthesis than a revolution. Never may such solutions be perfect, but this still can be a huge leap forward. If robust enough, and sufficiently pragmatic, they may just work. Ideas can be traded and the *Idea Economy* can unfold. Imagine what this would mean.

References

Abraham, M. (2008). *The flash of genius* [Movie]. Directed by Abraham, M., screenplay by Railsaback, P., Universal Pictures/Spyglass Entertainment/Strike Entertainment.
Audretsch, D. B. (2003). *The entrepreneurial society*. Oxford: Oxford University Press.
Bagnall, B. (2006). *On the edge: The spectacular rise and fall of commodore*. Winnipeg: Variant Press.
BBC News (2012). *Facebook's instagram deal: Can one app be worth $1bn?* By Weber, T., 10 April. www.bbc.co.uk/news/business-17666032
Bhide, A. V. (1994). How entrepreneurs craft strategies that work. *Harvard Business Review, 72* (2), 150–161.
Chesbrough, H. W. (2003a). *Open innovation: The new imperative for creating and profiting from technology*. Cambridge: Harvard Business School Press.
Chesbrough, H. W. (2006a). *Open business models: How to thrive in the new innovation landscape*. Cambridge: Harvard Business School Press.

---

[31] See von Hippel (2005) and Drucker (1986). The modern version of this entrepreneurial society is best described by David Audretsch (2003) and propagated by Carl Schramm (2006). For Open Innovation, see Chesbrough (2003a, 2006a) and Chesbrough et al. (2006).

References

Chesbrough, H. W., Vanhaverbeke, W., & West, J. (2006). *Open innovation: Researching a new paradigm.* New York: Oxford University Press.
Chicago Tribune (2011). *Google shake-up returns page to CEO post surprise move comes as challenges mount for internet search giant.* 21 January, by Guynn, J. www.chicagotribune.com/business/ct-biz-0121-google-20110121,0,2955817.story. Accessed 02 March 2013.
Drucker, P. F. (1986). *Innovation and entrepreneurship.* New York: HarperCollins. Reprint 2006.
Fincher, D. (2010). *The social network.* [Movie] Directed by D Fincher, screenplay by Sorkin, A. Columbia Pictures/Sony Pictures International.
Govindarajan, V., & Trimble, C. (2010). *The other side of innovation: Solving the execution challenge.* Boston: Harvard Business School Publishing.
Grove, A. S. (1999). *Only the paranoid survive.* New York: Doubleday.
Israel, P. (1998). *Edison: A life of invention.* New York: Wiley.
Kelley, D. J., Singer, S., & Herrington, M. (2012). *The global entrepreneurship monitor.* Babson College, Wellesley. www.gemconsortium.org/docs/download/2409. Accessed 02 March 2013.
Mezrich, B. (2010). *The accidental billionaires: The founding of facebook: A tale of sex, money, genius and betrayal.* New York: Doubleday.
Millard, A. J. (1990). *Edison and the business of innovation.* Baltimore: Johns Hopkins University Press.
O'Neill, J. J. (2007). *Prodigal genius: The life of Nikola Tesla.* San Diego: The Book Tree.
Pretzer, W. S. (1989). *Working at inventing: Thomas A. Edison and the Menlo Park experience.* Dearborn: Henry Ford Museum & Greenfield Village.
Schramm, C. J. (2006). *The entrepreneurial imperative – how America's economic miracle will reshape the world (and change your life).* New York: Collins.
Shurkin, J. N. (2006). *Broken genius: The rise and fall of William Shockley, creator of the electronic age.* New York: Palgrave Macmillan.
Smil, V. (2005). *Creating the twentieth century: Technical innovations of 1867–1914 and their lasting impact (Technical revolutions and their lasting impact).* New York: Oxford University Press.
Taussig, F. W. (1915). *Inventors and money-makers.* New York: Macmillan.
Tesla, N. (1919). *My inventions.* Originally published in the electrical experimenter magazine, available at www.teslaplay.com/auto.htm. Accessed 02 March 2013.
The Economist (2010e). *Schumpeter: The innovation machine.* 26 Aug 2010.
The Economist (2012a). *Technological change. The last Kodak moment? Kodak is at death's door; Fujifilm, its old rival, is thriving. Why?* 14 Jan 2012.
The New Yorker (1993). *The flash of genius: Bob Kearns and his patented windshield wiper have been winning millions of dollars in settlements from the auto industry, and forcing the issue of who owns an idea.* 11 Jan, by Seabrook, J. www.newyorker.com/archive/1993/01/11/1993_01_11_038_TNY_CARDS_000363341. Accessed 02 March 2013.
Time Magazine (1986). *Adios, Amiga?* 24 Feb, by Henry, G. M., Hackman, W. & McCarroll, T. www.time.com/time/magazine/article/0,9171,960694,00.html. Accessed 02 March 2013.
Time Magazine (2010). *The making of America: Thomas Edison.* The inventor. 23 Jun, by Stengel, R. www.time.com/time/specials/packages/article/0,28804,1999143_1999496_1999498,00.html. Accessed 02 March 2013.
USA Today (2006). *'Charismatic' founder keeps netflix adapting.* 23 April, by Hopkins, J. http://usatoday30.usatoday.com/tech/products/services/2006-04-23-netflix-ceo_x.htm. Accessed 02 March 2013.
VanityFair (2011). *The accidental activist: Part I – twitter was act one.* April, by Kirkpatrick, D. www.vanityfair.com/business/features/2011/04/jack-dorsey-201104.
von Hippel, E. (2005). *Democratizing innovation.* Cambridge: MIT Press.

Part II

The Emergence of an Idea Economy

Economic Evolution 2

Why is innovation important? Innovation is essentially, what has made us so well off. We can achieve more with less. Innovation is the fundamental driver of prosperity. But what are the fundamentals of innovation? It seems we know little of how to foster innovation. It is critical to understand this better. If it is possible to nurture more innovation we can achieve so much more.

Innovation is important. Case and point! Everyone would agree. We hear it constantly, repeat it, use it. It has become an integral part of our vocabulary, conversation, and commentary. Just how important, though, and why, is often underappreciated or shrouded in myth. Innovation is the key to prosperity. It is not a convenient aspect of gadgetry, industrial competition, and academic vanities. In daily life, we encounter the wording in media, policy, casual conversation, as economic dogma. It has a familiar hollowness to it. Often we hold it as a highly abstract notion of continuous 'progresses'. We take it for granted. We expect and anticipate the arrival of new, better, and cheaper things without knowing what and how. It is almost absurd. We embrace Moore's law that computers double in speed ever so often,[1] new devices will be offered, products will become cheaper, and so on, thinking innovation is a given, a continuous stream. We defer purchases or even sometimes adopt legislation based on such innovation expectation, in full confidence that the necessary innovation will occur – a new iPad to arrive soon, the new generation of electric cares to be launched.[2] At the same time, in many cases we

---

[1] This, of course, is only a popular proxy of Moore's law actually stating that the number of circuits double every circa 2 years. The implication still holds. People expect the speed of computers to increase constantly without understanding the severe challenges and the need for cutting edge research and remarkable new ideas to make this possible. It is expected. It often drives consumer behaviour.

[2] For an example of such supply push legislation, see the California legislation on the energy efficiency of televisions, expecting to have a broad availability of more efficient television sets by 2013 (see California Energy Commission 2009). Other broader scale initiatives include Climate Change legislation and carbon trading schemes. Also see Norberg-Bohm (2000) and van den Ende and Dolfsma (2005).

hardly notice the profound innovations. It occurs at the backend, it seems often expected or incremental. Sometimes even a burden. A Blackberry, though offering some convenience is often perceived a restraint more than a glorious innovative leap and productivity gain. That it offers colour, becomes ever faster is expected. A touchscreen is a given. That prices fall is a must; that a new version will come soon an obvious fact. Still, sometimes we are interrupted and in awe of innovations, then quickly absorb them into daily life. The iPhone – shock and awe, now a standard. Other innovations, much more fundamental, often occur much less noticeable: the printing press, power loom, or less tangible ones such as matrix organizations and franchising, and so on. Its overall impact can hardly be overstated. At its inception, it often seemed marginal, expected, or unwelcome. Some innovations seem trivial, though with game changing impact (e.g. money, containerized shipping).[3] Other, more prominent achievements, but with less direct innovative impact at the time (e.g. the lunar landing, the compass) we praise as mankind's ingenuity, though do not necessarily connect it to personal affluence. Many things have changed and are constantly changing. We accept, even expect it. It is easy to attribute importance to innovation. But it is more than just a happenstance and matter of convenience. It is hard work with often extraordinary consequences.

That innovation exists is agreed. Its true impact is not. Think of innovation not as petty increments, but consider leaps, and the impact of such ingenuity becomes strikingly obvious. Increments add up. Look at it over a longer period. Pick two points in time. Spot the difference. Think of a world with no mobile telephony, without internet, without aviation, cars, etc. We have come a long way. We are moving forward with rapid pace. Tomorrow will not be the same as today. Who knows in 10 years? This is what science fiction tries to grasp – over, but often underpredicting the future prospects of yet unimaginable possibilities. Looking backward it is all too obvious – an astounding path of ingenuity. Innovation drives prosperity. It defines our limits; it defines what we can do and how we can live. It drives economic growth, it pushes the limits of what we can achieve – and it has, remarkably so.

To see this impressive account of human achievement, consider a long, a very long run perspective. Clearly, we have become richer, a lot richer and more affluent than anyone before. The conveniences of life have become more readily available, have become cheaper and have made a substantial impact on the quality of life. To better demonstrate this progress, this remarkable contribution of ideas, consider the light bulb. It epitomizes the sparking thought, the flash of genius, the brilliant idea. It symbolizes invention per se. The light bulb is now the ultimate cliché of a bright idea and invention. It is synonymous with invention. What better then to demonstrate the magnificent advances of human ingenuity that the light bulb, more general: lighting. We take it for granted. Who really thinks about the intriguing history and evolution of such common a thing as lighting when buying a light bulb or reading a book in the evening? We rather complain about the 'high' costs of lighting than relish this

---

[3] For example see Bernhofen et al. (2013) on containerized shipping or Bughin et al. (2011) as well as Chen et al. (2013) on internet search.

impressive feat of human ingenuity that has made it possible in the first place. Just imagine you had to sit there with a torch, a candle, or an oil lamp. The manner in which we produce light has progressed significantly – without us really appreciating it. The first approaches of our early ancestors are sure to have been a controlled use of fire, such as simple sticks used as torches. Fairly simple oil lamps were used by the Babylonians and ancient Greeks and Romans. This was already a significant improvement. Further innovation made lighting continuously better and cheaper to obtain. In the Middle Ages tallow candles were used, later replaced by sophisticated oil (e.g. Argand oil lamp), later still with gas lamps. Edison's incandescent light bulb marked the start of the electrical lamps. They have steadily been improved. Fluorescent light bulbs are increasingly used, and promising results of modern light emitting diodes (LEDs) seem to point a way forward. From the use of fire to the use of electrical lamps has been an impressive evolution. It made the production of light several hundred times more efficient. To generate light for your nighttime read, what you can achieve with a simple light bulb today would have required a truckload of candles then. This from a technical perspective. It may not tell you much. A more interesting way of looking at it may be another: costs. How much would you have had to work to afford a certain amount of lighting? Prominent Yale economist William Nordhaus brought together this information demonstrating the power of such innovation (Fig. 2.1[4]).[5]

The price represents the amount of labour time that would be required to purchase a certain quantity of light (in hours worked to attain 1,000 lm-h. Note the logarithmic scale!). This is to say how long you would have to work to earn the money to afford an hour of light from a normal light bulb.[6] The hours of work required to earn this money fell only slowly until the mid-nineteenth century. Since then it has fallen drastically, by almost a factor of 10,000. Over an entire year, using a 100 W standard incandescent light bulb for 3 h each evening to read, today would costs you a total of a few dollars. In the early 1800s, generating this much light would have required 17,000 candles for which at the time "the average worker would have had to toil almost 1,000 h to earn the dollars to buy the candles."[7] Or in another way: where the Babylonians still had to work 40 h to buy enough oil to generate an equivalent of an hour of light generated from a light bulb, Beethoven's contemporaries would have had to work about an hour, while today it will take you

---

[4] Based on calculation by Nordhaus (1996). Note the horizontal scale intervals. The intervals become shorter, thus shifting the data outwards, making it appear even less steep.

[5] See Nordhaus (1996, 1997). Actually, the papers have a different objective: the challenges and far-reaching implication of calculating adequate price indexes. The history of Lighting is used to demonstrate a potentially significant underestimation of economic progress. This, in the current context, would mean the achievements of innovation would be even greater with an even higher impact. For an equally impressive example demonstrating the power of invention, see Smil (1994).

[6] "1,000 lm-h are approximately the light cast by a modern 75-watt incandescent light bulb in one hour, or the light from burning a standard candle for about 60 h" Nordhaus (1997, p. 1552).

[7] Nordhaus (1996, p. 50). "The Age of Invention shows a dramatic improvement in lighting efficiency, with an increase of efficiency from 1800 to 1992 by a factor of 900, representing an annual rate of 3.6 % per year" Nordhaus (1997, p. 1553).

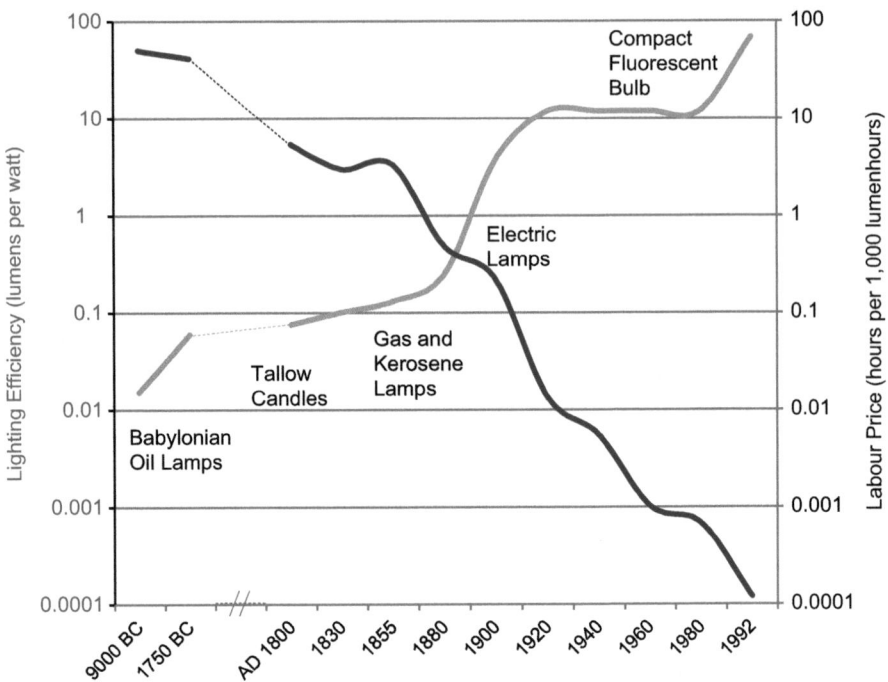

Fig. 2.1 Innovation has made lighting cheaper and better

a few seconds the most. Of course, these are but estimates, using averages and standardized objectives, but you get the idea. Producing light has become ridiculously cheap compared to anytime in history. Innovation made this possible.[8]

Lighting is just one example. It hides a more important truth. We have not only progressed in the manner in which we produce light, but have progressed in many other areas as well. We can produce cheaper light because we have gotten better at building lamps, in generating electricity, producing the materials needed, and in general, have become better at many things.[9] We have become richer so that the relative portion we spend on lighting is getting ever lower as we get more and more productive in doing other things as well. Think of such prominent things as computers (they are constantly becoming faster and/or cheaper), music devices (from tapes, to CD, to mp3), or transportations (carriages, cars, hybrids). In almost all aspects of life, you can trace magnificent advances and progress. The growth in

---

[8] It probably could have been more impressive still if not for the Phoebus cartel, where its members – the major light bulb producers among others Phillips, Tungsram, Osram, General Electric – agreed to make light bulbs less long-lasting to increase replacement rates and thus turnover and profit (see United States v. General. Elec. Co., 82 F. Supp. 753, 890–91 (D.N.J. 1949)).

[9] For example, between 1880 and 1883 alone the price of producing an incandescent lamp fell from $1.12 to 30 cents (see Millard 1990, p. 89).

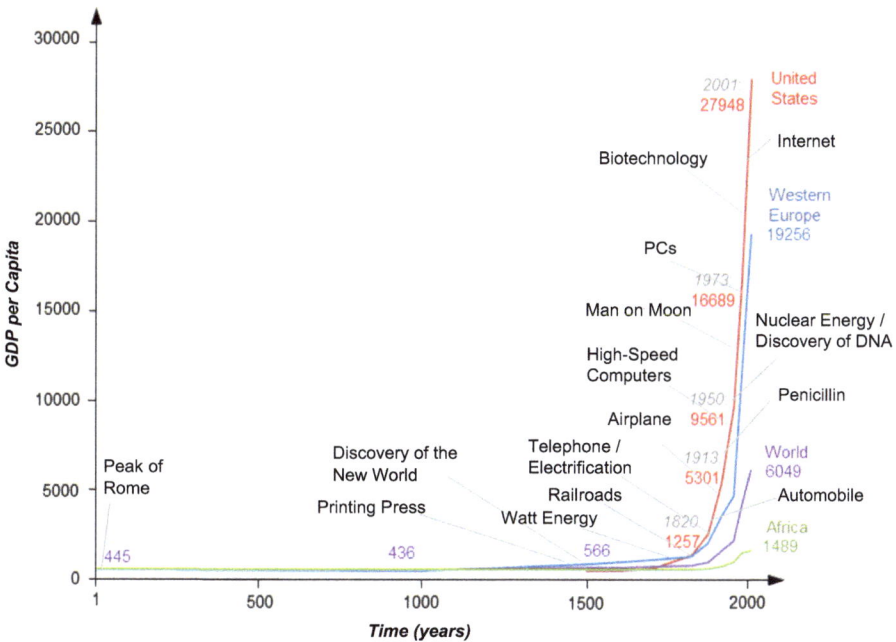

Fig. 2.2 Economic growth and innovation in the very long run

GDP may reflect this. It may at least be a good proxy. It provides for a similarly impressive account (see Fig. 2.2[10]). The tireless work of some clever economic historians has made possible to trace this progress.

We have prospered tremendously.[11] Population rose 22-fold. World GDP almost 300 times. The first 1,000 years almost so no progress. "From the year 1000–1820 the advance in per capita income was a slow crawl–the world average rose about 50 %. Most of the growth went to accommodate a fourfold increase in population. Since 1820, world development has been much more dynamic. Per capita income rose more than eightfold, population more than fivefold."[12] Many factors contributed to these

---

[10] This is a mélange of the brilliant work by Robert Fogel (1999) and the magnificent efforts by Angus Maddison (2001). Two aspects should be noted: the invention is always assigned to the most prosperous GDP line. It is but indicative.

For a more extensive timeline of innovations, see, for example, greatachievements.org (02.03.2013), or en.wikipedia.org/wiki/Timeline_of_historic_inventions (02.03.2013). For a more detailed account of the more recent history, see Smil (2005, 2006). Similarly, even more ambitious data can be found also in Kremer (1993). Also see Maddison's Historical Statistics dataset for disaggregated data (ggdc.net/maddison/ 02.03.2013) and Maddison (2007).

[11] Such progress is very hard to measure. Different measures may reveal vastly different results. We possibly even prospered far more than the data reveals (see Nordhaus 1997, 1996).

[12] Maddison (2001, p. 17).

staggering dynamics.[13] "Advances in population and income over the past millennium have been sustained by three interactive processes: (a) Conquest or settlement of relatively empty areas which had fertile land, new biological resources, or a potential to accommodate transfers of population, crops and livestock; (b) international trade and capital movements; (c) technological and institutional innovation."[14] The latter, innovation, seems the only sustainable over a longer term, even enabling and determining the former two to a large extent (though surely the reverse could also be said to some extent). At least for now, conquest and settlement seems to be reaching some geographic, economic and often environmental limitation. Gains from international trade are still incurred, but likely to be diminishing, as more and more such gains are being exploited in an increasingly globalized world. Future gains from trade are likely driven by innovation driven differentiation and specialization rather than the expansion of transport and capital movements alone. Essentially innovation drives economic growth. As the OECD put it: "Innovation is inextricably linked to past and future economic performance and to societal wellbeing. Most of the rise in material standards of living since the industrial revolution has been the consequence of innovation. New or improved products and services, and new and improved ways of producing them, have for a long time been the main motor of economic growth. This trend is expected to continue."[15]

By any standard, this is an extraordinary account. Many countries have advanced and grown immensely rich compared to anyone before. We have become used to such progress; we have become used to an ever – advancing economy. To us such numbers, ½ %, 1 %, 2 % of economic growth, etc. may seem sometimes insignificant when we hear about China, Brazil, Botswana, and so on, with sometimes two-digit growth figures. But think about it using some simple numbers[16]: Imagine one of your forefathers 2,000 years ago had one dollar. He/she invested it for you. Its real value grew by 1 % a year. Today[17] it would be $480 million! Had it grown by

---

[13] Decomposing growth, disentangling the different factors is far from easy. See, for example, Solow (1957) and Mankiw et al. (1992). Also Jorgenson and Griliches (1967), Aghion and Howitt (2007), Fernald and Neiman (2011). The question why then, and why in the places that it did, mainly Europe, still is subject of intense discussion. This will remain an on-going discussion that has produced many explanations. Whether it is institutions, the rise of intellectual property, enclosure movement, trade, colonial expansion, hereditary laws, climate, Protestant Ethics, a confluence of all, etc. (see, for example, North and Thomas 1973; Landes 1998; Bauer and Matis 1988; Rosen 2010; Weber 1950), or for a recent and highly intriguing perspective see, for example, McCloskey (2010). Many more exist.

[14] Maddison (2001, p. 18).

[15] OECD (2009b, p. 7).

[16] Following a similar illustrative example Barro and Sala-i-Martin (1995, p. 1) use to demonstrate the importance of growth. The calculations here are based on Maddison (see ggdc.net/maddison/ 02.03.2013).

[17] For the sake of simplicity, the year 2000 is chosen. The original calculations go until then. Also it provides a better time horizon: 2000 years. This may be inexact, and the additional 10 years to 2010 do make a difference, but this is merely a thought experiment. The intuition still holds just the same.

Table 2.1 The impact of growth in the very long run

World	Annual growth rate (%)	Income per capita year 1 (USD^a)	Income per capita in 2000 (USD^b)
Actual	0.13	450	6,029
Scenario 1: slightly higher growth	0.20	450	24,569
Scenario 2: higher growth	1.00	450	218,324,337,934
Scenario 3: same as USA after 1820	1.73	450	510,374,292,028,365,000

^{a}1990 International Geary-Khamis dollars. Data from Maddison 2008
^{b}Calculated as (450) · e^ (2000 *interest rate)

only 0.5 % you would end up with 'only' $22,000. At 1.5 % with $10 trillion. In reality, compounded growth was far, far less than that. Crude estimates would put it at 0.13 %. You would end up with a little more than $13. That's it. If growth would have been slightly higher, unbelievably huge effects on prosperity would have resulted over such a long time (see Table 2.1). In year one anno domini, average world GDP per capita is estimated at $450. With a compounded growth rate of 0.13 %, this has grown to an average of some $6,029 per person today. Imagine it would have been a little bit higher. Instead of 0.13 % and 0.2 %. This small difference would have had a big impact. World GDP per capita would now be at $24,569 instead of some $6029. Would growth have been a 'meagre' 1 % over this time it would result in some $218 billion per person. At 1.73 %, the rate the US grew since 1820, it would have resulted in $510,374,292 billion per person. Per person! A noticeable difference. You would be quite a lot richer.[18]

What such flimsy 'what-if' numeric exercises show is that slight differences in growth can come a long, long way over time. And one, if not the most important determinant of growth is innovation. There are possibly good reasons why growth was 'only' at the rate it was. But try to grasp the impact if indeed it were possible to nurture growth, possible to nurture innovation, even only slightly, to continuously make better use of the innovation potential. Its economic impact would be enormous.[19]

---

[18] Similar would hold for the US economy. Take a shorter time horizon: 'only' 180 years (but an important 180 years of rapid industrialization and growth). Compounded it grew at 1.73 % annually. Imagine what a percentage more or less each year would have resulted in – in 2000 GDP per capita would have been at $172,216; that is six times as much. Or even if it had grown only slightly faster, at a compounded annual rate of 1.91 % as Japan did over the past 180 years (though from a lower initial starting point), per capita income per person in the US would be significantly higher, at close to $40,000, almost 50 % higher.

[19] In the short run, innovation may cause considerable adverse effects: structural unemployment, displacements, even conflicts (see, for example, Heertje 1973, or Vivarelli 1995) for an overview of the classical debate, including James Steuart, Ricardo, Marx, etc. Also, see other discussions, such as Neisser (1942), and most recently, the controversy surrounding Rifkin (1995). Also Freeman et al. (1982), Acemoglu (1997, 2002), Postel-Vinay (2002), Standing (1984), Vivarelli and Pianta (2000), Woirol (1996). This is an on-going debate. Empirically, no conclusive evidence has yet been found Pianta (2006, p. 577). Also, growth differs vastly across the globe. Catching up is hugely different

The history of economic growth is essentially a history of ideas. Ideas have been the single most defining element in determining and enabling prosperity. Surely not a linear one. At times we have diverged, faltered, regressed, we have at times forgotten ideas, gone wrong. The history of science is littered with marvellous advances, curious developments and embarrassing missteps.[20] Yet, all in all, we have progressed – and magnificently at that. Ideas made it possible. Such claim builds on two essential features of ideas: (i) ideas are – more or less – cumulative, and (ii) ideas prevail. We constantly add new insights to the stock of knowledge. It expands. We have the opportunity of standing on the shoulders of giants, making the giant grow by adding new knowledge. This provides us with a fascinating view of the world, an increasingly sophisticated understanding of the universes – and how to benefit from it.[21] New and exciting opportunities open up to improve our lives through ideas. This also stresses a second appealing feature of ideas. While structures decay, resources wane, capital depreciates, riches are squandered or destroyed, ideas remain. With knowledge, at least the potential to achieve prevails. We may not be able to live in a Roman villa or Viking hut, but the knowledge that was generated to build it remains and not only offers us the chance to easily repeat it, but has enabled us to build better and/or cheaper living quarters using our time and resources more effectively. Capital accumulation may determine the wealth of nations. Knowledge and ideas define its prosperity. In the long run ideas prevail and ideas dominate. They determine what we can achieve. They define prosperities envelop. As economist Paul Romer illustrated this intuition: "The raw materials that we use have not changed, but as a result of trial and error, experimentation, refinement, and scientific investigation, the instructions that we follow for combining raw materials have become vastly more sophisticated. One hundred years ago,

---

than pushing the envelope (see, for example, Abramowitz 1986 or Mankiw et al. 1992). Many other factors determine growth, and many correlate with innovation. Incremental differs to radical innovation. Some are disruptive; others complements. Many act through externalities; others are not diffused. Some cluster, others are highly independent. Some occur in waves; others randomly. Sometimes they are forgotten; rediscovered or belatedly employed. Many are borne out of need; others come from affluence. And so on. These are all important aspects. The general claim though remains intact: over the long run innovation is the single most important contributor to economic growth and prosperity. As an indicative measure, its track record is sound.

[20] The most obvious setback, on a large scale, is surely the dark ages and the subsequent renaissance. Some more particular examples are the Harvey-Descartes Controversy (see Gorham 1994), the dispute between Huygens and Newton, or the "wounded dog theory" with the powder of sympathy to determine longitude (see Sobel 1995, Chap. 5). An entertaining example may also be the changing attitudes to hygiene over time (see The Economist 2009d). Such notions of non-linear progress in science have long been subject to extensive debate (most prominently the Popper-Kuhn-Feyerabend-Lakatos' debate (e.g. Popper 1989; Kuhn 1970; Feyerabend 1975; Lakatos and Feyerabend 1974; also Blaug 1976).

[21] Despite the discouraging philosophical debate on truth, we often apply the pragmatic 'no-miracles-argument' – 'according to which the success of science would be miraculous if scientific theories were not at least approximately true descriptions of the world' (see Smart 1963 and Putnam 1975). However, it too faces severe constraints and methodological problems (e.g. Matheson 1998 or Ghins 2002).

all we could do to get visual stimulation from iron oxide was to use it as a pigment. Now we put it on plastic tape and use it to make videocassette recordings."[22] In short: Ideas allow us to achieve more with less – not only repeating the same with more resources, but crafting new and better outcomes with the same or less inputs.[23] And, it is progressive. Building on what we know, adding ideas, we can achieve even more. Thus, utilizing the existing stock of knowledge, better yet expanding this envelop through new ideas can vastly improve lasting prosperity.

This provides a legitimate basis for the dull and numbing litany of the importance of 'innovation'. It must seem trivial. In retrospect, it often looks so simple. Easy to put in some nonchalant metaphor: "economic growth springs from better recipes, not just from more cooking".[24] Clearly, innovation is the main driver of prosperity. Thanks to the pioneering work of many luminaries, serendipitous circumstances and profound historic developments, we have fully embraced innovation as the fundamental driving force of prosperity. Technological progress, and more generally innovation, is now considered *the* most important driver of prosperity. Ideas are a key resource in a modern economy, "not just another resource alongside the traditional factors of production – labor, capital, and land – but the only meaningful resource today".[25] This notion has found its way deep into public conscience. Media has popularized and romanticized innovation, has idealised inventors and entrepreneurs, stressed the gains of innovation for the individual and society. Innovation is now widely accepted as the source of prosperity. Innovation is mainstreamed. This is reflected also in policy. Political debates are shaped around this topic. It has spawned an edifice of innovation policy, and an exuberance of hope and belief in its power of salvation.[26] In Europe 2009 was declared the 'European Year of Creativity and Innovation.'[27] India declared the entire next decade as the

---

[22] Romer (1990). Also see Romer (1992).

[23] This is a populist version of what the diligent denotative economist would consider an to produce the same with less or more with the same. Both are analytically and equally valid propositions. The resulting economic effects on growth and employment, however, depend on the assumptions being made, whether the inputs are held constant or the output. Clearly, if either one holds true, for any increment, however small, within such increment more with less is a valid statement – populist, agreed, true none the less. This captures an important point still: The choice of perspective yields different results. Either emphasizes too strongly or neglects entirely a possible negative effect innovation might have in the short run. The fear: the same with less – less has often been seen as less labour, hence unemployment, stirring reluctance, even luddite sentiment. This is a valid concern to be addressed in the text below. But it should be acknowledged, though not overstressed, nonetheless. Venturing into such debate is daunting given the dogmatic nature of it. This short reference to it here may suffice.

[24] Romer (2007).

[25] Nonaka and Takeuchi (1995, p. 6).

[26] "There is growing awareness among policymakers that innovative activity is the main driver of economic progress and well-being as well as a potential factor in meeting global challenges in domains such as the environment and health" OECD (2007, p. 3). Also see OECD (2005).

[27] create2009.europa.eu (02.03.2013).

Indian 'Decade of Innovation'.[28] In the U.S. President Obama discovered "innovation is what America has always been about"[29] and the National Innovation Initiative proclaimed: "Innovation will be the single most important factor in determining America's success through the twenty-first century." with the affix: "Innovate or Abdicate"[30] The Chinese 'Medium- to Long-Term Strategic Plan for the Development of Science and Technology' envisions China by 2020 to be one of the world's leading 'innovation economies': "By 2020, the progress of science and technology will contribute at least 60 % to the country's development."[31] The government envisions "to build China into a country of innovation."[32] Similar could be found in most countries. This is typically backed by large funding volume, legislation, and large-scale initiatives. The EU for example earmarked some 60 billion Euros in research and innovation promotion funding; the US has the Creating Opportunities to Meaningfully Promote Excellence in Technology, Education, and Science Act (COMPETES); Germany is the *'land of ideas'*.[33]

In short: innovation matters gravely. We know that now. And with confidence, we can now proclaim: Ideas define prosperity. Innovation creates opportunities, creates affluence. The fundamental importance of innovation is well understood (if not fully appreciated). Policy is besotted with innovation. Grand claims are made. Many reports are written. Incredible amounts of money are spent on it. Innovation is fundamental!

But what are the fundamentals of innovation? It is suspiciously curious how little we really know about innovation – the inner working of what drives it, how it comes about. We just accept it. We have identified its importance, but not how it comes about. Indeed, "windy talk about innovation is mind-numbingly abundant."[34] Despite all the research, despite all theorizing we are still a long way from a practical approach. Many disciplines seem to experience a recent revival in search of the principle and practices of innovation, as our understanding is unmasked as rudimentary at best, guesswork if you will, theorizing on the surface, with little depth and fundamental insights. It still seems an 'intellectual onion': 'You peel it back layer by

---

[28] Pratibha Devisingh Patil, President of India (Patil 2009).

[29] President Obama (2012).

[30] Council on Competitiveness (2004, p. 7 and p. 9). Also see President Obama (2009).

[31] english.gov.cn/2006-02/09/content_183426.htm (02.03.2013) Also OECD (2007).

[32] China's Peaceful Development: What China Aims to Achieve by Pursuing Peaceful Development. September 2011, Information Office of the State Council, Beijing, english.gov.cn/official/2011-09/06/content_1941354.htm (02.03.2013).

[33] (i) Combining FP7 (cordis.europa.eu/fp7/ 02.03.2013) and CIP (ec.europa.eu/cip/ 02.03.2013), both running from 2007 until 2013. In addition, substantial amounts within the Structural Funds (ec.europa.eu/regional_policy/thefunds/index_en.cfm 02.03.2013) aim at this too; (ii) thomas.loc.gov/cgi-bin/query/z?c110:H.R.2272: (02.03.2013); (iii) land-der-ideen.de (02.03.2013).

[34] The Economist (2010d). See especially the feisty article in the Wall Street Journal May 23 2012 quoting the Securities and Exchange Commission figure that, in 2011, the word innovation was used 33,428 times in quarterly and annual reports.

layer and when you get to the centre, there is nothing there, but you are crying.'[35] Devoid of a clear view, the most sophisticated approach still, it seems, is to lavishly throw money at it. In principle, this can be a good idea as innovation is in essence often a question of resources, risk and incentive. But how? Current practices seem a shameful example of how not to proceed. Many are not eligible or neglected by doting politicians attempting of pick winners in information and communication technology (ICT), biotechnology and high and green technology sectors. Ideas can pop up anywhere, and often stem from less voguish endeavours. The second popular approach: institutional provisions for entrepreneurship, whatever that may mean to most (e.g. 'Doing Business', Intellectual Property, liability laws, access to finance, etc.). Grander approaches still: Dwell on the notion of education, culture, incentives per se. Alas, despite its prominence, innovation remains elusive. No wonder then innovation is often mused as the 'secret sauce' of business success, or a 'mysterious process', and 'magic formula'.[36] It seems hard to grasp, even harder to nurture. Attempts to foster, increase innovation are often platitudinal interpolations, are best guesses.[37] Some work, some do not. Most only work in particular circumstances. No one seems sure how to truly bring about innovation. To be fair, such harsh appraisal should be placed into a more elaborate context. In principle, these are all good points it seems. Many valuable insights do indeed exist. But, on the one hand, are they typically very vague, fragmented and often incoherent, and in practice fail to be effective as the conceptual linkages are poorly understood. On the other hand, and this is more intriguing, underlying all this is typically already a very distinct concept of innovation, characterizing our perception of innovation processes, clichés of how ideas are realised, ought to be realised. This determines the existing approaches – and often their failure.

How do you think innovation comes about? A flash of genius? Hard work? And the process? Large corporations and high-tech, gadget-laden laboratories? Some tinkerer in a shabby garage? A clever entrepreneur making his way in the world? Serendipitous circumstances where the stars align? Innovation has many facets, has surely many possibilities and stories to it. But when it comes to the question of what to do, how to foster it, how to nurture innovation, a more general understanding is needed to address the challenges head-on. We all have some preconceptions, some notion of how innovation manifests. These determine how we approach innovation, how we perceive our chances of success, our possibilities and prospects in participating in the innovation process, in shaping policy. Taking a step back, allowing for an unspoiled perspective can help to better identify such underlying concepts, can help connect some of the dots, to provide for a more comprehensive

---

[35] To paraphrase the delightful image used by Stevenson (2004). Admittedly, Stevenson, a Harvard Business School Professor for entrepreneurship, portrays a more positive picture of entrepreneurship in this article. This remains to be argued.

[36] See Dyer et al. (2009), Hargadon and Sutton (2000), The Economist (2009b), Burnham (2009), Berkun (2007); etc.

[37] See, for example, OECD (2009b).

or at least more dynamic picture of innovation processes. It can provide new perspectives, new insights, and help to keep an open mind to appreciate the things to come. It can help identify challenges and obstacles to overcome. Imagine the prospects if only it were possible to nurture innovation, allow easy and more successful participation in the innovation process.

Such an approach needs to be broad enough to be inclusive, granular enough to be applied, and pragmatic enough to work. Thus the plan: First, capture innovation as an evolving concept in a larger socioeconomic framework (in the following, Part I). This can help identify more profound developments, a more dynamic perspective, not just, where we are but also expose the trends of where we are heading, and why. Then, as a next step, understand how, what obstacles, practical challenges, and possible pragmatic solutions exist to foster innovation in such a setting (Part II).

To capture innovation, broad strokes can help provide an organizing framework. Stereotypes can help crystallize important evolutionary steps. This is crude, but necessary. Such stereotypes may help paraphrase characteristics of an era, of a process, of socio-economic conditions. They expose our implicit preconceptions of innovation, what we mean by and how we handle it. Certainly, there are always many aspects that may not be fully captured, and any categorization falls short of the complexity it represents. But it is helpful. It helps conceptualise it, to think of it in terms of trends. None is exclusive, and they often apply overlapping or simultaneously. But they describe some major indicative characteristic of the socio-economic environment. We do it all the time, assigning prototypical traits to people, institutions, nations, etc. such as talking about 'The Germans', 'The Investment Bankers', 'The Youth Today', and so on. Surely not all Germans are orderly, surely not all investment bankers are overpaid neurotics, obsessed with business card quality as in the movie American Psycho,[38] and most certainly not all youth is Wii-crazed, uses strange lingo and is drunk. Yet they are often pictured and defamed in such terms, are stereotyped. Though often helpful, they bear a risk. We act on them. Such stereotypes often determine our actions, our policies and approaches, the manner we engage. They should not be thrown around lightly. Two things follow: first, stereotypes are helpful. For them to stick they need to be catchy, short, and precise. More importantly the second: they need to be sound. Anecdotes and tales of vague conjecture are not useful (e.g. broken society and youth). They need to be argued, rationalized, and supported by sound underpinnings. Thus better examples include characteristics that hold more closely and can be supported much better. Surely, we are moving increasingly from primary agricultural production to secondary manufacturing to a more tertiary, a 'service economy'. Data supports that. We surely not all work in services, but it may still well describe the overall current economic composition and its impact. We live in an era where we have to take and are exposed to more and more manmade risks – a 'Risk Society'. Logical

---

[38] Harron (2000).

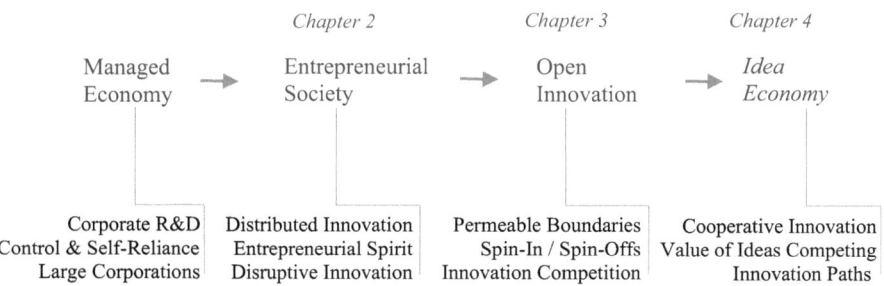

Fig. 2.3 Evolution of innovation (From the Managed to the Idea Economy)

argument supports that. The world is getting smaller – not literally – but globalized. With ease we travel, trade, communicated globally. We live in a 'globalized world'. Daily experience supports that. These are stereotypical descriptions of a situation, an economy, a society. Surely they do not affect all in the same way, overstress certain aspects, and clearly cannot be seen the only aspect. Often they coexist depending on the context. But, they may capture important developments, trends, changes and complex concepts; they may convey important developments, and illustrate important features. They are helpful. We indeed do the same when describing innovation. More granularity is needed. What kind of innovation? How does it manifest? What are the conditions, the enabling environment? What are the underlying dynamics? In discussing this, stereotypes can help. Exposing different facets of innovation in terms of stereotypes can help better understand them; crafting new ones may help provide perspective and vision. Such vision, such broader picture is useful. It can be assessed and discussed, and it can help prepare for it, even foster its arrival. To ruin the suspense: the innovation framework has progressed from the 'managed economy' to an 'entrepreneurial society' and will, via 'open innovation', move towards an '*Idea Economy*' (see Fig. 2.3).

The managed economy, where innovation was largely a function of corporate R&D and implementation by the incumbent firm, has been replaced with the entrepreneurial society. Today entrepreneurship exemplifies innovation, and with it defines our attitudes and policy approaches towards innovation. The arrival of the entrepreneurial society has in turn set in motion other developments. 'Open innovation' is gaining ground – as corporate response to the entrepreneurial threat. Corporations are increasingly pressured, their markets contested by start-ups realizing better ideas. Firms are adjusting, opening up to gain access to such ideas in order to compete and not to be left behind. Together this dynamic will culminate in yet another socio-economic stereotype: the *Idea Economy*. Ideas will be traded; a division of labour between invention and implementation will drive innovation. With its arrival, much will change. New opportunities will open up. Understanding the evolution towards it, the implications and challenges may help to better prepare, may help to accelerate it and may help to bring it about. It may help to foster innovation, and with it, enhance growth and prosperity.

References

Abramovitz, M. (1986). Catching up, forging ahead, and falling behind. *The Journal of Economic History, 46*(2), 385–406.
Acemoglu, D. (1997). Technology, unemployment and efficiency. *European Economic Review, 41*(3), 525–533.
Acemoglu, D. (2002). Technical change, inequality, and the labor market. *Journal of Economic Literature, 40*(1), 7–72.
Aghion, P., & Howitt, P. (2007). Capital, innovation, and growth accounting. *Oxford Review of Economic Policy, 23*(1), 79–93.
Barro, R., & Sala-i-Martin, X. (1995). *Economic growth*. New York: McGraw-Hill.
Bauer, L., & Matis, H. (1988). *Geburt der Neuzeit: Vom Feudalsystem zur Marktgesellschaft*. Munich: DTV Deutscher Taschenbuch.
Berkun, S. (2007). *The myths of innovation*. Sebastopol: O'Reilly.
Bernhofen, D. M., El-Sahli, Z., & Kneller, R. (2013). *Estimating the effects of the container revolution on world trade*. CESifo working paper series no. 4136. http://www.cesifo-group.de/DocDL/cesifo1_wp4133.pdf. Accessed 31 March 2013.
Blaug, M. (1976). Kuhn versus Lakatos or Paradigms versus research programmes in the history of economics. Reprinted In Hausman, D. M. (1994). *The philosophy of economics – an anthology* (2nd ed., pp. 348–375). Cambridge: Cambridge University Press.
Bughin, J. et al. (2011). *The impact of internet technologies: Search. McKinsey &Company*. http://ssl.gstatic.com/think/docs/the-impact-of-internet-technologies-search_research-studies.pdf. Accessed 31 March 2013.
Burnham, J. B. (2009). Economic growth, entrepreneurship, and the deployment of technology. In N. Aydogan (Ed.), *Innovation policies, business creation and economic development* (pp. 13–35). New York: Springer.
California Energy Commission (2009). Appliance efficiency regulations. Docket Number 09-AAER-1C, California Code of Regulators, Sections pp.1601–1608.
Chen, Y., Jeon, G. Y., & Kim, Y. M. (2013). *A day without a search engine: An experimental study of online and offline searches*. http://yanchen.people.si.umich.edu/papers/VOS_2013_03.pdf. Accessed 31 March 2013.
Council on Competitiveness (2004). *Innovate America: National innovation initiative summit and report*. Council on competitiveness, Washington, DC. www.compete.org/images/uploads/File/PDF%20Files/NII_Innovate_America.pdf. Accessed 02 March 2013.
Dyer, J. H., Gregersen, H. B., & Christensen, C. M. (2009). The innovator's DNA. *Harvard Business Review, 87*(12), 61–67.
Fernald, J. G., & Neiman, B. (2011). Growth accounting with misallocation: Or, doing less with more in Singapore. *American Economic Journal Macroeconomics, 3*(2), 29–74.
Feyerabend, P. K. (1975). *Against method*. London: New Left Book.
Fogel, R. W. (1999). Catching up with the economy. *The American Economic Review, 89*(1), 1–21.
Freeman, C., Clark, J., & Soete, L. (1982). *Unemployment and technical innovation: A study of long waves and economic development*. Westport: Greenwood Press.
Ghins, M. (2002). Putnam's no-miracle argument: A critique. In S. Clarke & T. D. Lyons (Eds.), *Recent themes in the philosophy of science: Scientific realism and commonsense* (pp. 121–137). Dordrecht: Kluwer Academic.
Gorham, G. (1994). Mind-body dualism and the Harvey-Descartes controversy. *Journal of the History of Ideas, 55*(2), 211–234.
Hargadon, A., & Sutton, R. I. (2000). Building an innovation factory. *Harvard Business Review, 78*(3), 157–166.
Harron, M. (2000). *American psycho*. [Movie] Directed by Harron, M., screenplay by M Harron amd G Turner, based on a novel by BE Ellis, Lions Gate Films.
Heertje, A. (1973). *Economics and technical change*. London: Wiley.

References

Jorgenson, D. W., & Griliches, Z. (1967). The explanation of productivity change. *The Review of Economic Studies, 34*(3), 249–283.

Kremer, M. (1993). Population growth and technological change: One million B.C. to 1990. *Quarterly Journal of Economics, 108*(3), 681–716.

Kuhn, T. S. (1970). *The structure of scientific revolutions*. Chicago: University of Chicago Press.

Lakatos, I., & Feyerabend, P. K. (1974). *Kritik und Erkenntnisfortschritt*. Braunschweig: Vieweg.

Landes, D. L. (1998). *The wealth and poverty of nations: Why some are so rich and others so poor*. New York: Norton.

Maddison, A. (2001). *The world economy: A millennial perspective*. Paris: OECD Development Centre. doi:10.1787/9789264189980-en.

Maddison, A. (2007). *Contours of the world economy 1–2030 AD: Essays in macro-economic history*. Oxford: Oxford University Press.

Maddison, A. (2008). Statistics on world population, GDP and per capita GDP, 1-2006 AD. http://www.ggdc.net/maddison/Maddison.htm. Accessed 6 July 2009.

Mankiw, N., Romer, D., & Weil, D. (1992). A contribution to the empirics of economic growth. *Quarterly Journal of Economics, 107*(2), 407–438.

Matheson, C. (1998). Why the no miracles argument fails. *International Studies in the Philosophy of Science, 12*(3), 263–279.

McCloskey, D. (2010). *Bourgeois dignity: Why economics can't explain the modern world*. Chicago: University of Chicago Press.

Millard, A. J. (1990). *Edison and the business of innovation*. Baltimore: Johns Hopkins University Press.

Neisser, H. P. (1942). "Permanent" technological unemployment: "Demand for commodities is not demand for labor". *The American Economic Review, 32*(1), 50–71.

Nonaka, I., & Takeuchi, T. H. (1995). *The knowledge-creating company: How Japanese companies create the dynamics of innovation*. New York: Oxford University Press.

Norberg-Bohm, V. (2000). Creating incentives for environmentally enhancing technological change: Lessons from 30 years of U.S. energy technology policy. *Technological Forecasting and Social Change, 65*(2), 125–148.

Nordhaus, W. D. (1996). *Do real output and real wage measures capture reality? The history of lighting suggests not*. Cowls foundation working paper no 957, http://cowles.econ.yale.edu/P/cp/p09b/p0957.pdf. Accessed 02 March 2013.

Nordhaus, W. D. (1997). Traditional productivity estimates are asleep at the (technological) switch. *The Economic Journal, 107*(444), 1548–1559.

North, D. C., & Thomas, R. P. (1973). *The rise of the western world: A new economic history*. Cambridge: Cambridge University Press.

Obama, B. (2009). *Remarks by the president: On a new beginning*, Cairo University, Cairo, Egypt, June 4, The white house, office of the press secretary. www.whitehouse.gov/the-press-office/remarks-president-cairo-university-6-04-09. Accessed 02 March 2013.

Obama, B. (2012). *State of the Union 2012*. Remarks by the president in State of the union address, United States Capitol, Washington, DC, 24 Jan, The White House, Office of the Press Secretary. www.whitehouse.gov/the-press-office/2012/01/24/remarks-president-state-union-address. Accessed 02 March 2013.

OECD. (2005). *Oslo manual* (3rd ed.). Paris: OECD. doi:10.1787/9789264013100-en.

OECD. (2007). *OECD Reviews of innovation policy China*. Paris: OECD. doi:10.1787/9789264039643-en.

Patil, P. D. (2009) *President of India, Smt. Pratibha Devisingh Patil's address to the joint session of 15th Lok Sabha in New Delhi*. 4th June, AKT/AD/HS/LV (Release ID :49043). http://pib.nic.in/newsite/erelease.aspx?relid=49043. Accessed 02 March 2013.

Pianta, M. (2006). Innovation and employment. In J. Fagerberg, D. C. Mowery, & R. R. Nelson (Eds.), *The Oxford handbook of innovation* (pp. 568–598). Oxford: Oxford University Press.

Popper, K. R. (1989). *Logik der Forschung* (9th ed.). Tübingen: Mohr.

Postel-Vinay, F. (2002). The dynamics of technological unemployment. *International Economic Review, 43*(3), 737–760.
Putnam, H. (1975). *Mathematics, matter and method (Philosophical papers I)*. London: Cambridge University Press.
Rifkin, J. (1995). *The end of work: The decline of the global labor force and the dawn of the post-market era*. New York: Putnam.
Romer, P. M. (1990). Endogenous technological change. *Journal of Political Economy, 98*(5), 71–102.
Romer, P. M. (1992). Two strategies for economic development: Using ideas and producing ideas. *Proceedings of the World Bank Annual Conference on Development Economics, 1992*, 63–115.
Romer, P. M. (2007). Economic growth. In: D. R. Henderson (Ed.), *The concise encyclopedia of economics* (2nd ed.). Indianapolis: Liberty Fund. www.econlib.org/library/Enc/Economic Growth.html. Accessed 02 March 2013.
Rosen, W. (2010). *The most powerful idea in the world: A story of steam, industry, and invention*. New York: Random House.
Smart, J. J. C. (1963). *Philosophy and scientific realism*. New York: Humanities Press.
Smil, V. (1994). *Energy in world history*. Boulder: Westview Press.
Smil, V. (2005). *Creating the twentieth century: Technical innovations of 1867–1914 and their lasting impact (Technical revolutions and their lasting impact)*. New York: Oxford University Press.
Smil, V. (2006). *Transforming the twentieth century: Technical innovations and their consequences*. New York: Oxford University Press.
Sobel, D. (1995). *Longitude: The true story of a lone genius who solved the greatest scientific problem of His time*. New York: Walker and Company.
Solow, R. M. (1957). Technical change and the aggregate production function. *The Review of Economics and Statistics, 39*(3), 312–320.
Standing, G. (1984). The notion of technological unemployment. *International Labour Review, 123*(2), 127–147.
Stevenson, H. H. (2004). Intellectual foundations of entrepreneurship. In H. P. Welsch (Ed.), *Entrepreneurship: The way ahead* (pp. 3–1). New York: Routledge.
The Economist (2009b). *Special report on entrepreneurship: Magic formula. The secrets of entrepreneurial success*. 12 March 2009.
The Economist (2009d). *Filth. The joy of dirt: Why cleanliness may be going out of fashion*. 17 Dec 2009.
The Economist (2010d). *Promoting innovation. Growth on the cheap: The OECD tells governments how to unleash business's creative potential*. 27 May 2010.
van den Ende, J., & Dolfsma, W. (2005). Technology-push, demand-pull and the shaping of technological paradigms: Patterns in the development of computing technology. *Journal of Evolutionary Economics, 15*(1), 83–99.
Vivarelli, M. (1995). *The economics of technology and employment: Theory and empirical evidence*. Cheltenham: Elgar.
Vivarelli, M., & Pianta, M. (Eds.). (2000). *The employment impact of innovation: Evidence and policy*. London: Routledge.
Weber, M. (1950). *The protestant ethic and the spirit of capitalism*. (trans: Talcott, P.). London: Scribner. http://ia700306.us.archive.org/5/items/protestantethics00webe/protestantethics00webe.pdf. Accessed 02 March 2013.
Woirol, G. R. (1996). *The technological unemployment and structural unemployment debates*. Westport: Greenwood Press.

The Entrepreneurial Society 3

How has innovation happened so far? After the Second World War, innovation was mainly a corporate process. In-house researchers were tasked to innovate for firms. Towards the end of the 20th century, entrepreneurship became more and more important where anyone with a bright idea and the skills and means to set up a firm could innovate. An Entrepreneurial Society arose, 'an economy full of innovators and entrepreneurs, with entrepreneurial vision and entrepreneurial values, with access to venture capital, and filled with entrepreneurial vigor'. Now innovation is essentially equated with entrepreneurship. Innovation is entrepreneurship. Entrepreneurs are hailed the modern economic heroes, the bringers of growth and prosperity.

The nature of innovation, and with it the socioeconomic environment that characterizes the economy, has evolved. What used to seem like mostly a closed process in laboratories of large corporation, has increasingly become a more public phenomenon – that of entrepreneurship. Ideas are increasingly realised by ambitious entrepreneurs rather than lethargic incumbents; young companies are challenging and replacing existing structures, opening up new markets, defining new fields. They create massive economic effects and are hailed a key driver of the modern economy. At the same time, they are equally sources of impressive socioeconomic change.

It is not that entrepreneurship has not always been an integral part of economic evolution.[1] Entrepreneurship has always been an important aspect throughout

---

[1] Despite its critical importance and historic prominence, little consensus exists on defining 'entrepreneurship' (see the Delphi exercise conducted by Gartner 1990). "The entrepreneur is at the same time one of the most intriguing and one of the most elusive characters in the cast that constitutes the subject of economic analysis" (Baumol 1968, p. 64. Also see Baumol 2002, p. 58). The entrepreneur is still left without a workable, commonly acceptable concept (see Shane and Venkataraman 2000, p. 217). Also Davidsson 2004, Chap. 1; Bull and Willard 1993; Hebert and Link 1988. Also see discussion in Baumol 1993; Iversen et al. 2008, Chap. 2; Carland et al. 1984 (For an early attempt and review also see Cole 1946). Many definitions exist, and different concepts of entrepreneurship are in use, often based on long traditions, including Richard

history and in most economic systems. Consider many Roman traders, Dutch merchants, British industrialists, or American railroad tycoons. But the magnitude and perspective that accompany entrepreneurship today are fundamentally different. Entrepreneurship has changed, changed dramatically. It is the results and cause of socio-economic evolution.[2]

One way to conceptualize the emergence of entrepreneurship as a defining characteristic of innovation is by contrasting it to what is known as 'closed' or 'managed' innovation.[3] The predominant concept of innovation for a long time was that of large incumbent corporations investing in research and development. Engineers, scientists, designers, managers, etc. focus on creating new or better products or services, cheaper production, opening up new markets, and selling goods, through creativity, analysis, and experiments. Such self-reliance is still widespread practice. Some employees are explicitly tasked with generating ideas; others are encouraged to propose new ideas. Research and development departments are explicitly mandated to invent, to come up with new products. Sometimes also other parts of organizations have similar tasks of generating ideas (process innovations, marketing, etc.). Decisions on the applicability, and subsequent initiation of the realisation process, are taken by management. Highly evolved decision structures and incentive schemes have been designed to optimise this process. A member of the organization proposes an idea, a designated manager decides to fund or reject its realisation or at least on further advancing the proposal to appropriate decision-making bodies. Often elaborate review and vetting procedures are in place to decide on the viability, the appropriateness and future potential of such proposals for the corporation (see Fig. 3.1).

Cantillon, Jean Baptiste Say, John Stuart Mill, Joseph Schumpeter, Frank Knight, Israel Kirzner, Peter Drucker and many more. This has left a diverse field of inquiry. Several broad conceptual approaches run in parallel. These include: (*i*) *Bearing Risk (see* especially Knight 1921); (*ii*) *Starting a new firm* (for a nuanced argument for and against see Reynolds et al. 2005; also Gartner 1985, 1988; Gartner et al. 2004); (*iii*) *Realising ideas* (see, for example, Schumpeter 1947, p. 153. Also see Schumpeter 1942, p. 132, or Baumol 1968, p. 65). Neither one approach seems distinct and specific enough to capture what one typically considers entrepreneurship. "There is no generally accepted definition of entrepreneurship"(Grilo and Thurik 2006, p. 4. Also Audretsch 2005, p. 22; Carland et al. 1984; Peneder 2009; Hebert and Link 1989; Gartner 1988; Shane and Venkataraman 2000). Many official definitions mostly add to the confusion, rather than making it easier. (OECD 1998, p. 11. Also see Ahmad and Hoffman 2008; OECD 2010a, pp. 32–33; Wennekers and Thurik 1999, pp. 46–47; Peneder 2009; Gartner 1990; Davidsson 2004; Gartner et al. 2004). The best approach, therefore, it seems is to keep whatever conception you may have, vague and undogmatic.

[2] For a fascinating overview into the richness of entrepreneurial activity over the ages, see Landes et al. (2010). Also see Carlsson et al. (2009), Wingham (2004). Also Chandler (1977) and Blackford (1991).

[3] For a vivid account, see Audretsch (2003). For a different perspective. also see Chesbrough's description of the 'closed' economy (Chesbrough 2003a). For more structured description, see Audretsch and Thurik (2000, 2001a, 2004), Thurik (2008); also see Schramm (2006). Note, however: "Contrasting the managed economy with the entrepreneurial economy is not, however, symmetric." Audretsch and Thurik 2001a, p. 270; also Thurik 2008; Harrison 1997.

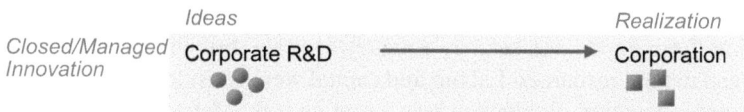

Fig. 3.1 The managed/closed innovation path

Clearly not all ideas are suited for realization within the corporation. Some are filtered out. The design of such filters is important. What to do with ideas that are deemed not promising. What to do with ideas that possibly undermine or disrupt the corporate strategy. What to do with ideas that might be useful somehow, but are outside the purview of the organization, yet valuable in another sector, a different market. Typically, if ever they were put forward, these ideas were filtered and shelved. Only few were realized with other means such as licensed out or realized outside the corporation. Innovation, in short, was perceived as a corporate process. Big corporations hired the best and the brightest thinkers to conduct research for them (e.g. Bell Labs researchers won seven Nobel Prizes, Xerox PARC and IBM Watson Scientific Computing Laboratory were the envy of any researcher, among them several winners of Charles Stark Draper Prizes and Turing Awards). With huge hiring potential and attractive wages, they absorbed able and willing researchers. Research and development, from basic research to developing final products, often became fully integrated.[4] Innovation became streamlined, somewhat path dependent. What did not fit in the line of the business, was shelved, with little in or outflow. Innovation was mostly a closed, a managed process. As Schumpeter described it: "innovation itself is being reduced to routine. Technological progress is increasingly becoming the business of teams of trained specialists who turn out what is required and make it work in predictable ways. The romance of earlier commercial adventure is rapidly wearing away, because so many *things can be strictly calculated* that had of old to be visualized in a flash of genius"[5]

To better understand this closed innovation model it helps to place it within the larger, corporation centric socio-economic environment of its time. This was characterized by "hierarchical and bureaucratic organisations that were in the business of making long runs of standardised products. They introduced "new and improved" varieties with predictable regularity; they provided their workers with life-time employment; and they enjoyed fairly good relations with the giant trade unions."[6] Vertical integration was the norm. Large corporations dominated.[7] Innovation was believed to be fairly predictable, continuous and linear. What to

---

[4] This was in line with a general perception of innovation as a specialized process. Other sources of innovation were ignored. For an early critique, see Schmookler (1957).

[5] Schumpeter (1942, p. 132).

[6] The Economist 2001 (as pointed out in Thurik 2008).

[7] In the US, for example, big businesses like steel, automobile, chemicals, food industry and heavy manufacturing, dominated the view of the economy. This included names like GM, Ford, IBM, US Steel, AT&T, RCA. Also see especially Chandler (1977).

produce seemed clear. Consumer preferences were believed to be obvious, and stable. Producing these goods was seen as a matter of scale, bringing together machinery, labour, and natural resources. Labour and capital were considered the main drivers of growth, not innovation. Production was a routine task. Management was a scientific process. Organizations were bureaucracies. Workers seemed loyal, homogenous, and to identify with their employer, in exchange for seeming lifetime employment. The stylized lifestyle was suburban family life, community oriented. Stability, continuity, and homogeneity characterized the managed economy. Static concerns with scale and scope were more important than flexibility and inventiveness. Small-scale production was considered a waste. Policy reflected this. Public efforts were directed at supporting the managed economy. Large corporations were seen as the driving force of the economy, creating growth and jobs. Schumpeter concluded: "What we have got to accept is that the large-scale establishment or unit of control has come to be the most powerful engine of progress and in particular of the long-run expansion of output."[8] Car companies of the 1960s with its presumptuous claims such as 'What is good for General Motors is good for America' epitomized this.[9]

These were the stereotypes to characterize the managed economy with its closed innovation system. It reflected the post-war period to the 1980s, not just in the US, but in most other developed courtiers as well. Despite the negative connotation many of these attributes have today, this economic model performed well, indeed, provided "the engine of jobs, growth, stability and security."[10] But it evolved into something different. And with it has innovation.

Several problems arise in such a system, problems that have increasingly undermined the managed economy. Start with the innovation process: Typically significant internal knowledge filters exists. Such 'knowledge filters' within corporations are often too narrow to allow ideas to be realised within organisations, least to incorporate ideas from the outside.[11] This rests on the view that corporations often lack the vision, are risk averse, are victim to 'not-invented-here' and 'not-sold-here' syndromes, and

---

[8] Schumpeter (1942 (2006), p. 106).

[9] This expression (though the actual quote may differ) typically referenced to Charlie Wilson, Chairman and CEO of General Motors in 1955 is often used to exemplify this era. Several important books reflect this spirit, stressing a variety of aspects: Chandler 1962, 1977, 1990, 1994; Galbraith 1952, 1967; Servan-Schreiber 1968; Schumpeter 1942; Also: Whyte 1956; Riesman et al. 1953; Mills 1951; Even Wilson 1955; Kerr et al. 1960; For a more critical account, also see Langlois (2004), as well as and Mowery (1983).

As Thurik noted: "The alleged success of the communist, centrally-led economies plays a huge role in the prevailing way of thinking of that era. These economies thrived on uniform, stable mass production. It is straightforward that entrepreneurship is viewed as behaviour hostile to the communist system and declared criminal" (Thurik 2008, p. 4).

[10] Audretsch and Thurik (2001b, p. 18).

[11] See Audretsch (2003) and Chesbrough (2003a) especially p. 70. Also see especially Hamel (1999) and Henderson and Clark (1990). Related also Prahalad and Bettis (1986, 1995), Daft and Lewin (1990). Connected to this, also see the discussion on the 'not sold here virus' in Chesbrough (2006a).

are too focussed on maintaining the status quo, on processes, on daily management. Many ideas are rejected, neglected and cannot flourish. They are filtered out. It represents a mind-set, an organizational design, a certain kind of work process. These ideas are not realized. Realizing such ideas requires entrepreneurial endeavour, involving risk, vision, leadership and other distinct skills that are typically found outside big corporations. This holds especially for more radical and disruptive innovations.[12] They all too often are neglected. Examples of failure in such traditional corporate innovation framework are numerous. Among the most cited and best researched are surely companies like IBM, Phillips, Kodak, Lucent or General Motors.[13] Xerox is probably the most infamous. It seemed to have excelled at this. Having 'invented' the computer mouse, the Graphical User Interface (GUI), spreadsheet, and many of the features that came to define the computer age were neglected, shelved or abandoned, only to be taken up by others – to the huge detriment of Xerox. Apple was the first commercialize the GUI – using images to interface with the computer rather than text command-line – in the Lisa and Macintosh operating systems in the early 1980s, soon followed by Microsoft. Xerox, who built the Alto and the Star, the first such computers, failed, mainly due to bureaucracy and internal neglect.[14]

Many ideas fall victim to such knowledge filters. They are either shelved or discarded. In effect, it may even often be the case that, aware of the rigidity of such filters, ideas are never put forward. People become discouraged, mainstreamed in their search for ideas, focussed on the most obvious, on incremental, and those ideas easiest to convey. More radical, but potentially also more valuable ideas are not pursued. Ideas falling outside the often rigid framework of corporate needs and acceptable range of ideas are neglected.

The second, possibly more challenging feature of internal innovation is the absence of appropriate incentives. Most reward structures within organizations do not or only very narrowly account for ideas. Promotion, future career, income, peer recognition, etc. often do not begin to compare to the economic gain ideas may generate. Thus even if the corporation were able and potentially willing to realize such ideas, many are not proposed due to a lack of adequate remuneration. This in turn may have two aspects to it. Either people do not propose their ideas to the respective organization because they perceive the reward as inadequate and unfair, or, and this is important for the further argument, they perceive it as less appealing than the potential value of pursuing alternative realization path, i.e. of trying to

---

[12] See, for example, Christensen (1997).

[13] See, for example, Chesbrough (2003a, 2006a).

[14] See Smith and Alexander (1999). For a different version of Apple stealing the GUI, see The New Yorker 2011. Such filters can also appear not only in large companies, but even in start-ups. A telling example is that of Dr. Gelernter. He developed a proposal for what now is used as the Facebook timeline "from the moment of your birth to the day of your death, containing every document, photo, message or web page you have ever interacted with all in a single, searchable stream, and held safely online." (The Economist 2011c). His plan, similar to Facebook, was to grow gradually by first attracting college students as early adopters starting at Yale, only to grow. Investors decided to focus on corporate clients instead. It never took off. The Economist 2011c.

realise the idea on their own. Say you work in a big corporation and you have a promising idea. Why would you share it with your employer? True in some cases this is part of your job (e.g. developer, marketing expert, etc.) and in such cases there are often some incentive structures in place (income, career, non-compete agreements, intellectual ownership, etc.). Many ideas however you are not paid to generate. The corporation may choose to reward you for it. In many cases, however this does not hold true. You may have ideas that lie outside of your core competence, or are more valuable than the reward your employer would offer. Think of an assembly line worker coming up with a process improvement to save valuable time and money, think of an engineer coming up with an entirely new product, though he is really paid to improve an existing one, and so on. The very nature of ideas makes them un-monitorable. It is a voluntary act of sharing – one that requires proper incentives to reveal the idea. If the reward offered by the corporation is too low, either in comparison to other alternatives, especially such as entrepreneurial realisation on your own, or because the offered incentive in comparison to its value is considered as unfair (e.g. you save the company millions, and they offer you a bronze suit pin or pad on the shoulder), you are unlikely to reveal your idea. To the corporation, it is lost. Especially in the presence of increasing ease of realising ideas outside the corporation, the easier entrepreneurship becomes, the less appealing it becomes to leave it to the corporation to benefit from it instead of the potential gain of you realising it.

So far, this covers only internal idea generation. It gets far more problematic when it comes to external ideas. Ideas generated by people outside the organization have next to no incentive to share it with a corporation. They might be willing to offer it to the corporation in circumstances where no alternative exists to realising the idea and/or some at least rudimentary reward structures exist (e.g. getting a job in the corporation, receiving some voluntary reward for proposing the idea, having some benefit from having the desired product or service available to themselves, etc.), but in the presence of alternative realisation routes it is unlikely they are willing to share it. If entrepreneurship is easily attainable, and the potential economic benefit large enough, it seems highly unlikely they would share it with an anonymous corporation, but rather instead pursue its realisation on their own.

Many such ideas, both internally and externally generated, may have vast economic potential and realistic chance to be realised outside the corporation, and may thus benefit the entrepreneur rather than the corporation. In such cases, the alternative to corporate innovation to realising such ideas is to establish a new firm, to be entrepreneurial on your own. Thus, increasingly, in parallel to the 'closed' or 'managed' economy a second stream is opening up: the entrepreneurial realisation of innovation (see Fig. 3.2).

Again, this of course cannot be seen outside the socioeconomic context. It was no sudden epiphany, a game changing insight. Entrepreneurship has always been an option. Entrepreneurs have always existed. But not on such scale, with such forceful magnitude that it should be considered a characterizing element and to be

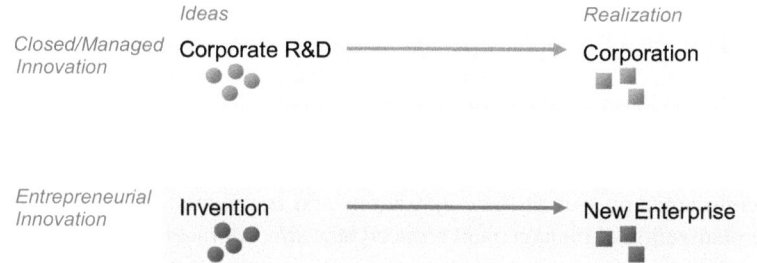

Fig. 3.2 The entrepreneurial innovation path

seemingly equated with innovation per se.[15] Why now and why with such force has entrepreneurship replaced the 'managed economy'? It has been a complex evolution with many facets and interacting factors that have made this evolution possible – and is far less straightforward then merely (though these are important factors) pointing to deregulation, and ease of doing business. It is an encompassing socio-economic development based on a range of facilitating factors. As Peter Drucker envisioned it in the 1980s: "Without doubt, the emergence of the entrepreneurial economy is as much a cultural and psychological, as it is an economic or technological, event"[16] A wide range of factors has contributed to this development. This includes, to cluster them broadly, societal, economic, cultural, as well as institutional and theoretical developments increasing both opportunity and willingness to engage in entrepreneurial activities.[17]

Social A range of *social* factors had a profound impact on the economy. Many of those factors lie far outside the economic realm. Many are historic accidents rather than deliberate or predictable evolutions. Together they form a critical range of factors that have contributed to the emergence of the entrepreneurial society: A social emancipation from conformity ('lonely crowd', 'organizational man') to individualism and the emergence of a post-industrial notion of self-fulfilment in work encouraged diversification and the pursuit of self-employment; a proliferation of consumer taste towards less standardized mass produced goods often required

---

[15] This is not quite true. In fact, Kenneth Galbraith (1967), the most vocal proponent of the managed economy, describes the replacement of the previous seemingly prevalent entrepreneurial society with the managed approach. Also, Schumpeter (1942) stresses such development (Also see Schramm 2006 for a more succinct version of a re-emergence, rather than the rise, of entrepreneurial activity in the US.). It could be argued that the entrepreneurial society did indeed exist. Thus, the analysis here could be limited to the post-war area.

[16] Drucker (1984, p. 63), also (1986, p. 14). Similarly also Audretsch and Thurik (2001b, p. 5): "The consequences of economic restructuring away from the managed economy to the entrepreneurial economy are enormous and encompass virtually every dimension of economic life. No field of economics alone is capable of capturing shifting economic systems."

[17] For more elaborate discussions and to get a better sense of the complexity of this evolution, see especially Audretsch (2003, 2009), Audretsch and Thurik (2000, 2001a, 2004), Carree and Thurik (2003); Also Brock and Evans (1989) and Drucker (1986).

more targeted production and allowed for more niche products; increasing labour mobility in terms of location, but also between firms and sectors lead to wider dispersion of skills, and more knowledge exchange; decreasing job security and loss of lifetime employment cultures, often said to have been triggered by waves of corporate downsizing due to economic fluctuations, disruptive innovations and tumbling corporate giants (e.g. IBM), in combination with a perceived lack of social conscience on part of the corporations and fragmentation of ownership and professionalization of management reduced labour/corporate loyalty and a different notion of economic security emerged; with rising prosperity and economic prospects, taking risk became acceptable and became more and more a part of culture, including a decreasing stigmatization of failure (even the reverse); demographic changes occurred, including a generational shift replacing the war and post-war generations bringing different skills and values to the labour markets; increasing participation of women placed pressure on the labour market, and added new skills and approaches to economic activity, encouraged the emergence of new services and work environments; political views changed with the post-war era, the cold war and the end of the cold war, elevating the achievements of entrepreneurial activity; the education base broadened considerably, dispersing knowledge much wider, and allowing for new skill sets to occur, and to be absorbed faster as well as leading to more labour flexibility; the success of some high profile companies such as Intel or Microsoft have had a significant impact on public perception, attitude and believe in the possibilities and potential of entrepreneurship; media contributed it part by instilling an entrepreneurial romanticism, crafting an image of lone genius, friendly geeks, idyllic garages, etc. spawning quasi-mythical anecdotes of phenomenal entrepreneurial success, seasoned with some David-versus-Goliath charm.

Economic Also *economic* aspects have highly influenced this shift: with ever smaller returns to accumulation and expansion innovation became an increasingly important competitive factor; opportunities through catching up and convergence diminished, necessitating new approaches, placing more emphasis on innovation; in a globalised competitive environment markets became larger, offering higher overall returns to new or better products; increasing division of labour often undermined vertical integration, fostering a more modular production process, offering opportunities to new entrants, and access to competitors; the availability of finance, and the willingness to bear risk to generate above market returns, especially in an environment of low interest, offered liquidity, access to loans, seed finance and venture capital – and the emergence and establishment of such industries; structural change, often driven by innovation, placed more emphasis on services, which exhibited smaller economies of scale and rather called for innovation driven competition; with prosperity, demand rose, markets expanded allowing for sufficiently large more specialised markets, and new market segments; new needs arose and possibilities of serving them opened up, often creating entirely new industries and thus a stomping ground for entrepreneurs; new production structures and services were introduced assisting market entrants and lowering risk levels through better market research, marketing, communication, etc. This also meant a higher division of labour and the possibility to partake in the market with more focussed skills, allowing for smaller firms and lower capital investment needs; new products to hedge and management techniques to manage risk further supported this; overstrained public finances together with new liberal dogma places increasing pressure on research institutions to become more applied and generate alternative sources of funding by cooperating with industry or better utilising their research, often leading to entrepreneurial university spin-offs spawning entire entrepreneurial clusters such as Silicon Valley, Route 128, or the Research Triangle.

Institutional Changes in the *institutional* framework also facilitated entrepreneurship: Markets became increasingly deregulated, opening up new opportunities, and making existing markets contestable and disruptable; competition regulation contributed to this, breaking up monopolies and monopsonies; tax incentives favouring vertical integration were reduced; entrepreneurship favourable taxation and regulatory requirements did their part; reduced trade restrictions, in terms of trade legislation, import/export taxes, but also the emergence of currency unions and internal markets, meant larger markets, less friction, but also the exposure to foreign competition and the need to catch up and/or stay ahead; strengthened Intellectual Property legislation offered better protection for smaller companies and enabled and encouraged research activity unaffiliated with existing production structures; the ease of doing business, such as setting up an enterprise, liability laws, transaction costs and time, etc. improved further facilitating entrepreneurial activity; the emergence of better and more extensive support structures allowed for better exchange of knowledge, gave it voice and offered supportive services.

Technological The arrival of new *technologies* themselves fostered innovation and entrepreneurship: while in many cases increasing productivity allowed for smaller organizations, elsewhere increasing scale and scope effects enabled a faster market penetration and return on innovation making it easier for new entrants; more flexible production structures allowed to be more responsive and reduced fixed costs investments; smaller scale research possibilities enabled independent research at lower costs and lower risk; information and communication technologies allowed for better diffusion of knowledge and ideas and thus often offered opportunities and better market overview, and at the same time provided access and better marketing possibilities to consumers; consumers also become more demanding as markets became more transparent and competition increased often allowing for more niches and individualisation of products; better and faster transport opened up new markets and entirely new production and distribution chains and with it opportunities for new and often smaller niche players.

Theory But also shifts in economic and business *theory* have highly fostered the rise of entrepreneurship, helping understand and shape the phenomena by enabling and supporting better guidance, services, products, and policies for promoting entrepreneurship. Considerable change was needed. "Even advocates of small business agreed that small firms were less efficient than big companies. These advocates were willing to sacrifice a modicum of efficiency, however, because of other contributions – moral, political, and otherwise – made by small business to society. Small business policy was thus "preservationist" in character"[18] This notion changed with a better understanding of innovation, and the emergence of better theory of conceptualizing the underlying economic and social shifts. Public sentiment shifted with new theory and increasing at least anecdotal evidence of entrepreneurial success. Innovation become more important, and identified as the driving force of growth. Entrepreneurship became better understood and identified as an important vehicle of innovation. But theory also directly influenced entrepreneurship: understanding of sources of innovation (e.g. lead users), the emergence of management science, of entrepreneurial studies; the understanding and subsequent design of risk products; the emergence of adequate valuation and accounting methods for start-ups; the academic debates driving and shaping intellectual property considerations; the mainstreaming of the notion of spill-overs leading to cluster policies; entrepreneurial research directly leading to the emergence of services and guidance tools (writing business plans, consulting services, etc.).[19]

These developments illustrate a complex underlying shift, from corporate innovation, to a more accessible form of innovation: entrepreneurship. More people gained access and became willing to invent, shifting innovation increasingly outside of corporations. Entrepreneurship opened up as an appealing and increasingly used alternative path to realising ideas. In effect, indeed, today realising, utilising on, and converting invention into innovation is often equated with entrepreneurship. As Peter Drucker opened his famous book on Innovation and Entrepreneurship where he envisions such development: "Entrepreneurs innovate. Innovation is the specific instrument of entrepreneurship. The act that endows resources with a new capacity to create wealth."[20] Such a notion that innovation is essentially an entrepreneurial activity may adequately characterise a broadly accepted notion of innovation today. To realise new and groundbreaking ideas people set up new enterprises.[21] Though this may not comprise all of the on-going innovation

---

[18] Audretsch (2009, p. 249). Note that small business does not need to coincide with entrepreneurship. The ambition can and often is to become a large firm. Even a large scale entrepreneurial entry is possible. Typically, however, the two overlap to a considerable extent. The quote reflects a general perception of small business as opposed to large incumbents.

[19] For an overview of the evolution of research in entrepreneurship, see Murphy et al. (2006). On the other hand, much research is still needed. Entrepreneurship research still has a long way to go (Brazeal and Herbert 1999; similarly Low 2001).

[20] Drucker (1986, p. 30). To be precise this argument does not hold in reverse, which might be the more telling logical proposition. It does not read: 'Entrepreneurship is the specific instrument (read path) of innovation.' Others have indeed gone further: "One view of entrepreneurship and innovation is that they are virtually synonymous [...] innovation and entrepreneurship are almost a tautology" (Acs and Audretsch 2005, p. 1; similarly Wennekers and Thurik 1999. Also see McDaniel 2000. A similar criticism is also in Bruyat and Julien 2001).

[21] Note the somewhat narrow version of entrepreneurship as enterprise creation adopted here. It corresponds best to innovative entrepreneurship. It stresses the process of non-corporate, external innovation processes.

3 The Entrepreneurial Society

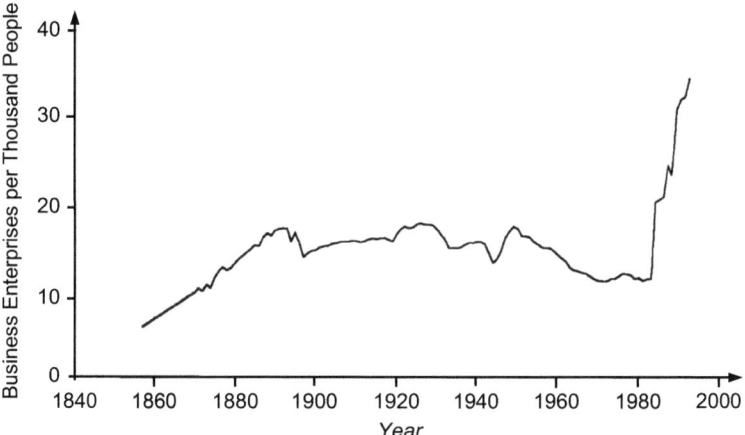

Fig. 3.3 The rise of entrepreneurship: business enterprises per 1,000 people

activities – a sizable portion is still being done by big corporations – entrepreneurship in this sense is becoming an increasingly defining characteristic. Next to the earlier typical 'managed innovation', entrepreneurship is firmly established as an alternative route. It is increasingly being used (see Fig. 3.3[22]).

---

[22] Gartner and Shane (1995), based on Dun & Bradstreet data, as well as US census data. For a critical discussion, see Harrison 1994; Shane 2008; Davis et al. 1996; see Stangler and Kedrosky 2010. Using different definitions may yield considerably different results (using different times frames, stock v flow, relative v absolute numbers, etc. (e.g. New Business Incorporations (see database and Statistical Abstracts (census.gov/history/www/reference/publications/ statistical_abstracts.html 02.03.2013); New Establishments (see US Census data: Business Dynamics Statistics (census.gov/ces/dataproducts/bds/ 02.03.2013) also see Haltiwanger et al. 2009). If an unambiguous surge in entrepreneurship cannot be found in this data, it questions the validity of the current policy regime including legislation and public expenditure on entrepreneurship. Two hypotheses may further support such claim: (i) the nature of entrepreneurship has changed (due to changes within the data (e.g. some sectors are declining, some are growing – see for example (see Mansel and Blackford 1991))) and (ii) entrepreneurship has become more innovative. The 'entrepreneurial' endeavours of past generations are not the same as today. Starting a business today in many industries requires more innovation than in the past. Innovation based competition is greater (See especially Baumol 2002). Research and development is increasingly conducted also in smaller firms. In this sense, entrepreneurship has become more innovative, and closer to a common notion of innovative entrepreneurship (see Data on R&D Expenditure (nsf.gov/statistics/iris/ and nsf.gov/statistics/industry/ 02.03.2013)). Small enterprises share of R&D spending has increased from 4 % in 1973 to 24 % in 2007. Venture Capital investments rose from a mere $100 million in 1970 to some $28.1 by 2008 (see National Venture Capital Association 2009, p. 10. Also see Kaplan and Lerner 2010). Angel investments rose even more steeply to around $20 billion in 2010 (see Sohl 2011. In general see Center for Venture Research at the University of New Hampshire (wsbe.unh.edu/cvr 02.03.2013)).

Academic interest in entrepreneurship has surged drastically in recent decades (see Katz 2003. Also see Katz 1991, 2004; Vesper and Gartner 1997; Solomon and Fernald 1991; Solomon et al. 1994; Cooper 2003. Also see Zahra 2006, or Kuratko 2005). The number of publications dealing

The rise of entrepreneurship has also somewhat altered the common perception of innovation. This is now often seen as mostly challenging existing firms, challenging incumbents or expanding markets beyond existing structures. The notion of the young high tech entrepreneur in some garage or dorm room comes to mind, setting up Microsoft, Apple, Amazon, or Facebook. And although sometimes viewed indeed as a myth, this does capture some essential features of the entrepreneurial society. Competition can come from anywhere. No longer do only immediate competitors pose a challenge. Any start-up can potentially upset, disrupt, and fundamentally change the market. Anyone able and willing to endeavour into entrepreneurship can challenge incumbents, can define new markets and can become rich from their ideas and efforts. What had previously fallen victim to the knowledge filter or lay dormant in some drawer or shoebox, is now realised externally, or in parallel to the corporate innovation path. No longer is the internal process the only or necessary, or even optimal manner to realise ideas. The broad availability and pursuit of entrepreneurship is challenging the closed innovation concept and has become an increasingly important path to release the innovation pressure. Ideas no longer need to linger for an opportune moment to be spotted and realised by some attentive manager. Self is the entrepreneur to realise it.[23]

This parallel process that redefined innovation by opening up an alternative realization path, thereby offering a promising and increasingly accessible outlet for ideas, is what is meant by the entrepreneurial society. Its impact goes beyond

---

with entrepreneurship has increased significantly. For example, JSTOR publications on 'Entrepreneurship' have surged from 845 in the 1950s to 6,061 in the 1990s. Google Scholar shows a similar increase from 659 to 142,000 (Google and JSTOR accessed 05.04.2010).

There is no unambiguous data showing a definite and significant rise entrepreneurship. The data may still be indicative. Proxies may help. Qualitative changes may support the argument. But perhaps "any one measure of entrepreneurship is insufficient to account for its various and diverse characteristics." "The message from this discussion of levels of analysis, rates and stocks, and time frames, is not that any one measure of entrepreneurship is superior to another, but that all measures have limitations and require trade-offs." (Gartner and Shane 1995, p. 296 and p. 192)

[23] This does not mean it has completely replaced the managed economy (see, for example, Thurik 2008). But entrepreneurship has become more prominent and ubiquitous. Much empirical work has been conducted trying to capture and quantify the rise of entrepreneurship, trying to support this claim of the arrival of the entrepreneurial society. Unfortunately, the data situation is poor, and the different definitions of what constitutes entrepreneurship, and what can reasonably be measured, complicate this (see Godin et al. 2008, p. 43. Similarly, Iversen et al. 2008. Also see Ahmad and Hoffman 2008). New indicators are emerging and much effort is invested in collecting better data (e.g. Global University Entrepreneurial Spirit Students Survey (GUESSS), Global Entrepreneurship Monitor (GEM), Kauffman Index of Entrepreneurial Activity, OECD Entrepreneurship Indicators Programme (EIP), World Bank Group Entrepreneurship Snapshots (WBGES)). Also see discussions, for example, in Ahmad and Seymour 2008; OECD 2009a; Parker 2008; Godin et al. 2008; Desai 2009; Acs et al. 2008; Katz 1992; Aldrich 1990; Dennis 1996; Armington 2004; Headd and Saade 2008; and especially Gartner, and Shane 1995. Also see discussion in Acs and Audretsch 1990. Most research has been conducted using data from the US, where the data situation is still the best. For an attempt to harmonize international data, see van Stel 2005; Also Freytag and Thurik 2007; Wennekers 2006; Caree et al. 2002; Audretsch and Thurik 2001b; somewhat also Blanchflower 2004.

a mere notion of economic gain and a mere change in the innovation processes. Entrepreneurship has become a socio-economic phenomenon. The entrepreneurial society is characterized as a "society in which innovation and entrepreneurship are normal, steady, and continuous", "an economy full of innovators and entrepreneurs, with entrepreneurial vision and entrepreneurial values, with access to venture capital, and filled with entrepreneurial vigor."[24] Entrepreneurship surely seems to be on its way of becoming, many would claim it already has become such a defining characteristic of the economy. It has found its way into politics and popular culture and has gone far beyond the ivory tower of academic description. Big became bad, small became beautiful.[25] The political emphasis is progressively shifting from a broad innovation imperative to an increasing focus on entrepreneurship. Entrepreneurship, rather than corporate R&D is considered the bringer of innovation, the source of growth and prosperity.[26] ""Entrepreneurship" is frequently advocated in the public debate as the solution to all (or at least *most* of) our trouble."[27] It is thus not uncommon to read statements like: "Only new companies bring forth new innovations and new jobs."[28] "Entrepreneurship is the engine fuelling innovation, employment generation and economic growth."[29] "They [entrepreneurs] are now bringing new ideas to the market".[30] Innovation as entrepreneurship is widely

---

[24] Drucker (1986, p. 254 and p. 257).

[25] As so eloquently put by Robert Kuttner in the Foreword to Harrison (1997). "Small was bountiful. Small was beautiful. Small was *in*." (Harrison 1994, p. 4).

[26] See Audretsch (2003, p. 5). This claim rests on much research and debate aiming to show positive impact (i.e. job creation, economic growth, poverty reduction, formalizing the informal sector. See Ahmad and Hoffman 2008). This is tricky. One is the transmission mechanism. The most obvious would be its mere existence. More enterprises, ceteris paribus, would mean more employment. Fast growing gazelles, innovation driven enterprises, etc. drive growth. But why should this be superior to any existing possibly large enterprises? Here a variety of arguments have been put forward ranging from the benefits of increased competition, larger knowledge spillovers, greater economic diversity, and most of all, more innovation through better realization possibilities, incentive structures and potential appropriability for the inventor (see, for example, Thurik 2009; and Audretsch and Thurik 2001a; Biggs 2002; More applied: Carlsson et al. 2009.)

Much empirical work has been conducted, by supporters and critics, to support their respective positions. The results are mixed. Many studies support a close link to growth, employment and innovation. Others have questioned this; have found ambiguous results or even negative effects. (For an overview, see, for example, van Praag and Versloot 2008; Wong et al. 2005; van Stel 2006; related also Sutton 1997; Also see Henrekson and Johansson 2009 and Haltiwanger et al. 2010.)

Almost more important is to note the prominence entrepreneurship has gained in the debate. This itself has an impact: The 'recognition' that entrepreneurship helps fostering growth led to the political mandate to promote entrepreneurship (Thurik 2009).

[27] Iversen et al. (2008, p. 43) (emphasis in original).

[28] Carl Schramm President of the Kauffmann Foundation (Schramm 2010). As the President of the influential Kaufmann Foundation it is his job almost to make such claims.

[29] Klaus Schwab, Founder and Executive Chairman of World Economic Forum in World Economic Forum 2009.

[30] OECD (2010b, p. 5).

accepted. Innovation is entrepreneurship.[31] Based on such simplistic equation policies are crafted to support this conviction.[32] The new political infatuation: entrepreneurship. "Western governments are obsessed with promoting small businesses and fostering creative ecosystems."[33] The European Commission has discovered: "We must promote the entrepreneur spirit of Europeans. Creation of jobs, innovation and competition are the keys to European success."[34] There are "two fundamental issues for Europe: 'How to produce more entrepreneurs' and 'How to get more firms to grow'?"[35] As has the U.S.: "a culture of entrepreneurship has been central to the economic success of the United States. [...] entrepreneurship drives economic growth [...] it fosters competition and dynamism [...] entrepreneurship facilitates the incorporation of the new technologies that fuel economic growth [...] Entrepreneurship provides opportunity and it supports freedom. [...] a strong entrepreneurial class makes a society freer. They provide choices for consumers. They provide options for those seeking jobs. They provide perspectives in the public sphere that do not come from the public sector. They provide for independence from large, hierarchical organizations."[36] Indeed, it could be argued: "It is abundantly clear that entrepreneurship is important for economic growth, productivity, innovation and employment, and many OECD countries have made entrepreneurship an explicit policy priority."[37] "Communities, cities, regions, and nations throughout the world have been turning to entrepreneurship as an engine of growth, jobs, and competitiveness."[38] With such focus on entrepreneurship as the bringer of all that is good, the innovation agenda becomes narrower. If you look at innovation policy more closely, you will easily find it is largely targeted towards entrepreneurial innovation – at least in its objective and justification. Now entrepreneurship weeks are organised, years of entrepreneurship are proclaimed,

---

[31] This equation often goes unnoticed. As the introduction to the immensely popular entrepreneurship week phrases it (gewusa.org/about 02.03.2013): "For one week, millions of young people around the world join a growing movement of entrepreneurial people, to generate new ideas and to seek better ways of doing things. Countries across six continents come together to celebrate Global Entrepreneurship Week, an initiative to inspire young people to embrace innovation, imagination and creativity. To think big. To turn their ideas into reality. To make their mark." Note how Innovation is equated to Entrepreneurship!

[32] For an overview of the different actors and initiatives, see, for example, OECD (2004). Also see OECD (2010a, b). Also see the overview of policy approaches in Audretsch (2005, p. 36).

[33] The Economist (2011d).

[34] Barroso (2008). In the same spirit, see new Europe 2020 strategy (ec.europa.eu/europe2020/ 02.03.2013).

[35] European Commission (2004, p. 4); See also the European Commission (2003, 2009).

[36] Director of the National Economic Council to President Obama, Larry Summers (Summers 2010). Also see Schramm (2006). Similarly the UK: "need entrepreneurial individuals with the vision to turn new ideas into winning products and processes. Entrepreneurship is the lifeblood of the new British economy." UK Department of Trade and Industry (1998, p. 7).

[37] OECD (2009a, p. 5).

[38] Acs et al. (2009, p. 2).

huge funding volumes dedicated to promote entrepreneurship, legislation adopted to facilitate, policies designed to enable, media frenzies encouraged to promote it.[39] This encompasses entrepreneurial training, simplifying regulation, providing access to finance, promoting entrepreneurial attitude, changing social perception, offering tools, advice, services, etc. By lowering the threshold, it is hoped that more people will take the entrepreneurial route, and through it achieve more innovation, foster growth and employment.

Similar applies to the general public domain. Entrepreneurship is now deeply rooted in popular culture and media. Entrepreneurship is portrayed as accessible to all, attainable for everyone. Go into any bookstore and pick up a copy of "The Young Entrepreneur's Guide to Starting and Running a Business", "Entrepreneurship for Dummies", "3 Weeks to Startup", or "The Art of the Start: The Time-Tested, Battle-Hardened Guide for Anyone Starting Anything" and simply "How to get rich".[40] Numerous prominent examples exist that have become recurring themes of public admiration, idealisation ... and envy. These include quasi-mythical accomplishments of now household names and national icons such as Bill Gates (Microsoft), Steve Jobs (Apple, Pixar, NeXT), Michael Dell (Dell), Larry Ellison (Oracle), Sergey Brin and Larry Page (Google), Hasso Plattner (SAP), Mark Zuckerberg (Facebook), Elon Musk (Paypal, SpaceX, Tesslar Motors), Richard Branson (Virgin), Martin Varsavsky (Jazztel, Viatel, Ya.com, FON), and many, many more that have inspired the public dream of 'anything goes', and 'anyone can'. Entrepreneurs are 'folk heroes'. They "personify freedom and creativity. They come up with the Big Ideas and build the organizations – the Big Machines – that turn them into reality. They take the initiative, come up with technological and organizational innovations, devise new solutions to old problems. They are the men and women who start vibrant new companies, turn around failing companies, and shake up staid ones. To all endeavours they apply daring and imagination."[41] From the garage, basement, dormroom, or kitchen to the boardroom, to fame and riches, became a popular impression, with "the image of the lone individual who relies primarily on his or her extraordinary efforts and talent to overcome the difficulties inherent in creating a new business".[42] "In 1989, the garage at 367 Addison Avenue was designated California Historic Landmark

---

[39] For example: now worldwide, there are Entrepreneurship Weeks (unleashingideas.org; gew.org.uk; unleashingideas.webjam.com/usa; gruenderwoche.de); The EU hopes to stimulate entrepreneurial initiative through, among others, a 3.6 billion Euro Competitiveness and Innovation Framework Programme (CIP) (see European Commission 2006); Germany is now "Gründerland Deutschland" with an ineptly named "Founder Generation 2.0" (bmwi.de/BMWi/Navigation/Presse/pressemitteilungen,did=327390.html 02.03.2013); (test.magazin-deutschland.de/en/business/creative-economy/article/article/die-gruendergeneration-20.html 02.03.2013).

[40] Mariotti et al. (2000), Allen (2001), Berry and Parsons (2008), Kawasaki (2004), Dennis (2008).

[41] Reich (1987, p. 78). Though dated, it perhaps still best captures this perspective of the entrepreneurial hero.

[42] Audia and Rider (2005, p. 7). See Audia and Rider (2005) for an overview and discussion of such belief. Also Purrington and Bettcher (2001).

Number 976 and a plaque declaring "Birthplace of Silicon Valley" was placed at the front of the garage."[43] Such heroic entrepreneurship is now so deeply rooted in the public mind it has almost become part of urban legend. As *The Economist* quipped about this 'cult of entrepreneurialism': "Entrepreneurialism has become cool". "In college, guys usually pretended they were in a band [...] now they pretend they are in a startup."[44]

Such entrepreneurial society is more than a fad. It seems here to stay. It is self-sustaining and self-promoting. As entrepreneurship becomes more widespread, its many factors are strengthened. New tools and services are emerging to support entrepreneurship. Many services to support new ventures have already sprung up such as support to writing business planes, more encompassing support from Angel networks where seasoned entrepreneurs and mangers support novices, incubators, often public, to nurture nascent entrepreneurial activity, and many more highly innovative support structures.[45] At the same time, the more promising and more active entrepreneurship becomes the more it will be appealing to bear risks and to invest in young companies, thus the volume and access to venture capital is expanding fast.[46] With entrepreneurship becoming fashionable, becoming an increasingly public and political focus area, it has spurred research on entrepreneurial activity itself, contributing to a better understanding and thus refinement and even better chances for successful entrepreneurship. Entrepreneurship education is becoming an increasingly widespread and appealing subject in the various education levels and formats.[47] Not only in management studies, but in many other subjects such as for example engineering, biology, informatics, are entrepreneurship courses taught and the entrepreneurial spirit instilled, producing increasing numbers of well-prepared entrepreneurs. Though already mainstreamed, entrepreneurship is more

---

[43] Audia and Rider (2005, p. 6). '367 Addison Ave' was the home of Dave Packerad where he and Bill Hewlett founded Hewlett-Packard (HP) in 1939.

[44] The Economist (2009a, p. 6). Such attitudes can also be supported by research findings: "The main results reveal most of university students consider desirable to create a new firm, although the perception of feasibility is not positive" (Guerrero et al. 2008, p. 35). "A recent Gallup poll reported that more than 90 % of Americans would approve if either a daughter or son attempted to start a small business." (Cooper 2003, p. 28). Also see the highly publicised 'Pop-events' celebrating Entrepreneurship. For example, the Global Entrepreneurship Week with more than 33,000 events in 123 countries, some 24,000 partners, engaging over seven million people (see unleashingideas.org/press-releases/global-entrepreneurship-week-coming-belgium 02.03.2013).

[45] For example, 'hackfwd', an innovative early-stage investment/pre-seed fund (hackfwd.com 02.03.2013). Also see emerging outfits, such as SecondMarket.com or SharesPost.com, facilitating secondary start-up trading.

[46] For Venture Capital, see National Venture Capital Association 2010, Kaplan and Lerner 2010. For Angel investments, see Sohl (2011).

[47] See, for example, EU policies for Entrepreneurship education (ec.europa.eu/enterprise/policies/sme/promoting-entrepreneurship/education-training-entrepreneurship/index_en.htm 02.03.2013). Also Kauffman Foundation (2007), World Economic Forum (2009). However, there is also an on-going debate whether it can be taught at all or at least to what extent it might be based on 'natural' ability.

and more becoming an integral part of culture and communication, further strengthening public appeal and belief in general attainability and desire to partake in the process. More and more people consider themselves entrepreneurs, or to have the potential for it.[48] More and more champions and promoters of entrepreneurship are appearing; further strengthening research, outreach and services for entrepreneurship. This includes private initiatives, most prominently the Kaufman Foundation, and ever more ambitious public programmes (for example the European Competitiveness and Innovation Framework Programme (CIP), or the US Small Business Association).[49] Most importantly however, the continuous success of entrepreneurs and the perception of broad public participation, that anyone can become one, and everyone can make a fortune with their idea through successful entrepreneurship constitutes the most important appeal, fostering hope and confidence, and strengthening the allure of entrepreneurship as a socio-economic phenomenon.

The emergence of widespread entrepreneurship has had a tremendous impact. Being driven by underlying changes in the socio-economic structure it is itself contributing to changing it and promoting and sustaining itself. But even though it may seem self-perpetuating, may to some even seem the attainment of a solution, it is still an on-going process and part of a broader economic evolution. It is a consequence, but also a cause of future change. In what direction and where this is going remains to be seen. But some aspects already are changing – and they may suggest a different future than the celebrated proclamation of universal entrepreneurial endeavour the proponents of the entrepreneurial society have proposed.

References

Acs, Z. J., & Audretsch, D. B. (1990). *Innovation and small firms*. Cambridge: MIT Press.
Acs, Z. J., & Audretsch, D. B. (2005). Entrepreneurship, innovation and technological change. *Foundations and Trends in Entrepreneurship, 1*(4), 1–65.
Acs, Z. J., Audretsch, D. B., & Strom, R. (2009). *Entrepreneurship, growth, and public policy*. Cambridge: Cambridge University Press.
Acs, Z. J., Desai, S., & Klapper, L. (2008). What does entrepreneurship data really show. *Small Business Economics, 31*(3), 265–281.
Ahmad, N., & Hoffman, A. (2008). *A framework for addressing and measuring entrepreneurship* (OECD statistics working paper 2008/2). Paris: OECD. doi:10.1787/243160627270.

---

[48] See Global University Entrepreneurial Spirit Students Survey (GUESSS) 2004, 2006, 2008 (guesssurvey.org), as well as Global Entrepreneurship Monitor 1999–2010 (gemconsortium.org) and Kauffman Index of Entrepreneurial Activity 1996–2010 and the YouthPuls 2010 (kauffman.org, Fairlie 2011). Also see discussion on Entrepreneurial self-efficacy (Bird 1988; Boyd and Vozikis 1994; Krueger and Brazeal 1994 and Koellinger et al. 2007; McGee et al. 2009). Also see Arenius and Minniti (2005).

[49] Much effort is put into exporting this idea and its associated impact to the developing world. (See Schramm (2004). Especially see remarks by Secretary Clinton (2010)).

Ahmad, N., & Seymour, R. G. (2008). *Defining entrepreneurial activity: Definitions supporting frameworks for data collection* (OECD statistics working papers 2008/1). Paris: OECD. doi:10.1787/243164686763.
Aldrich, H. E. (1990). Using an ecological perspective to study organizational founding rates. *Entrepreneurship Theory and Practice, 14*(3), 7–24.
Allen, K. (2001). *Entrepreneurship for dummies*. New York: Wiley Publishing.
Arenius, P., & Minniti, M. (2005). Perceptual variables and nascent entrepreneurship. *Small Business Economics, 24*(3), 233–247.
Armington, C. (2004). *Development of business data: Tracking firm counts, growth and turnover by size of firms*. U.S. Small Business Administration, Washington, DC. http://archive.sba.gov/advo/research/rs245tot.pdf. Accessed 02 Mar 2013.
Audia, P. G., & Rider, C. I. (2005). A garage and an idea: What more does an entrepreneur need? *California Management Review, 48*(1), 6–28.
Audretsch, D. B. (2003). *The entrepreneurial society*. Oxford: Oxford University Press.
Audretsch, D. B. (2005). The emergence of entrepreneurship policy. In D. Audretsch, H. Grimm, & C. W. Wessner (Eds.), *Local heroes in the global village: Globalization and the new entrepreneurship policies* (pp. 21–43). Ney York: Springer.
Audretsch, D. B. (2009). The entrepreneurial society. *The Journal of Technology Transfer, 34*(3), 245–254.
Audretsch, D. B., & Thurik, R. (2000). Capitalism and democracy in the 21st century: From the managed to the entrepreneurial economy. *Journal of Evolutionary Economics, 10*, 17–34.
Audretsch, D. B., & Thurik, R. (2001a). What's new about the new economy? Sources of growth in the managed and entrepreneurial economies. *Industrial and Corporate Change, 10*(1), 267–315.
Audretsch, D. B., & Thurik, R. (2001b). *Linking entrepreneurship to growth* (OECD science, technology and industry working papers 2001/2). Paris: OECD. doi:10.1787/736170038056.
Audretsch, D. B., & Thurik, R. (2004). A model of the entrepreneurial economy. *International Journal of Entrepreneurship Education, 2*(2), 143–166.
Barroso, J. M. (2008). *Presentation of the priorities of the Slovenian Presidency of the Council of the EU*. Strasbourg, 16 January 2008, SPEECH/08/17. http://europa.eu/rapid/press-release_SPEECH-08-17_en.htm?locale=en. Accessed 02 March 2013.
Baumol, W. J. (1968). Entrepreneurship in economic theory. *The American Economic Review, 58*(2), 64–71.
Baumol, W. J. (1993). Formal entrepreneurship theory in economics: Existence and bounds. *Journal of Business Venturing, 8*, 197–210.
Baumol, W. J. (2002). *The free-market innovation machine: Analyzing the growth miracle of capitalism*. Princeton: Princeton University Press.
Berry, T., & Parsons, S. (2008). *3 weeks to startup*. Irvine: Entrepreneurship Press.
Biggs, T. (2002). *Is small beautiful and worthy of subsidy?* http://rru.worldbank.org/documents/paperslinks/tylerspaperonsmes.pdf. Accessed 02 March 2013.
Bird, B. (1988). Implementing entrepreneurial ideas: The case for intention. *Academy of Management Review, 13*(3), 442–453.
Blackford, M. G. (1991). Small business in America: A historiographic survey. *The Business History Review, 65*(1), 1–26.
Blanchflower, D. G. (2004). Self-employment: More may not be better. *Swedish Economic Policy Review, 11*(2), 15–74.
Boyd, N., & Vozikis, G. (1994). The influence of self-efficacy on the development of entrepreneurial intentions and actions. *Entrepreneurship Theory and Practice, 18*(4), 63–77.
Brazeal, D., & Herbert, T. T. (1999). The genesis of entrepreneurship. *Entrepreneurship Theory and Practice, 23*(3), 29–45.
Brock, W. A., & Evans, D. S. (1989). Small business economics. *Small Business Economics, 1*, 7–20.
Bruyat, C., & Julien, P. A. (2001). Defining the field of research in entrepreneurship. *Journal of Business Venturing, 16*(2), 165–180.

Bull, I., & Willard, G. E. (1993). Towards a theory of entrepreneurship. *Journal of Business Venturing, 8,* 183–195.
Caree, M., Van Stel, A., Thurik, R., & Wennekers, S. (2002). Economic development and business ownership: An analysis using data of twenty-three OECD countries in the period 1976–1996. *Small Business Economics, 19,* 271–290.
Carland, J. W., Hoy, F., Boulton, W. R., & Carland, J. A. C. (1984). Differentiating entrepreneurs from small business owners: A conceptualization. *The Academy of Management Review, 9*(2), 354–359.
Carlsson, B., Acs, Z. J., Audretsch, D. B., & Braunerhjelm, P. (2009). Knowledge creation, entrepreneurship, and economic growth: A historical review. *Industrial and Corporate Change, 18*(6), 1193–1229.
Carree, M. A., & Thurik, A. R. (2003). The impact of entrepreneurship on economic growth. In Z. J. Acs & D. B. Audretsch (Eds.), *Handbook of entrepreneurship research – an interdisciplinary survey and introduction.* Boston: Kluwer.
Chandler, A. D. (1962). *Strategy and structure: Chapters in the history of the American Industrial Enterprise.* Cambridge: M.I.T. Press.
Chandler, A. D. (1977). *The visible hand: The managerial revolution in American Business.* Cambridge: Belknap Press.
Chandler, A. D. (1990). *Scale and scope: The dynamics of industrial capitalism.* Cambridge: Belknap Press.
Chandler, A. D. (1994). The competitive performance of U.S. industrial enterprises since the second world war. *The Business History Review, 68*(1), 1–72.
Chesbrough, H. W. (2003a). *Open innovation: The new imperative for creating and profiting from technology.* Cambridge: Harvard Business School Press.
Chesbrough, H. W. (2006a). *Open business models: How to thrive in the new innovation landscape.* Cambridge: Harvard Business School Press.
Christensen, C. M. (1997). *The innovator's dilemma: When new technologies cause great firms to fail.* Cambridge: Harvard Business School Press.
Clinton, H. R. (2010). *Closing remarks at the presidential summit on entrepreneurship.* Washington, DC 27 April , 2010, PRN:2010/522. www.state.gov/secretary/rm/2010/04/140968.htm. Accessed 02 March 2013.
Cole, A. H. (1946). An approach to the study of entrepreneurship: A Tribute to Edwin F. Gay. *The Journal of Economic History 6,* Supplement: The tasks of economic history, 1–15.
Cooper, A. C. (2003). Entrepreneurship – the past, the present. In Z. J. Acs & D. B. Audretsch (Eds.), *Handbook of entrepreneurship research- an interdisciplinary survey and introduction* (pp. 21–34). Boston: Kluwer.
Daft, R. L., & Lewin, A. Y. (1990). Can organization studies begin to break out of the normal science straitjacket? An editorial essay. *Organization Science, 1*(1), 1–9.
Davidsson, P. (2004). *Researching entrepreneurship.* New York: Springer.
Davis, S. J., Haltiwanger, J., & Schuh, S. (1996). Small business and job creation: Dissecting the myth and reassessing the facts. *Small Business Economics, 8*(4), 297–315.
Dennis, W. J. (1996). More than you think: An inclusive estimate of business entries. *Journal of Business Venturing, 12*(3), 175–196.
Dennis, F. (2008). *How to get rich: One of the world's greatest entrepreneurs shares his secrets.* London: Penguin Books.
Desai, S. (2009). *Measuring entrepreneurship in developing countries.* UNU-WIDER research paper, no 2009/10.
Drucker, P. F. (1984). Our entrepreneurial economy. *Harvard Business Review, 62*(1), 59–64.
Drucker, P. F. (1986). *Innovation and entrepreneurship.* New York: HarperCollins. Reprint 2006.
European Commission (2003). *Green paper entrepreneurship in Europe.* (COM(2003) 27 final).
European Commission (2004). *Action plan – the European agenda for entrepreneurship.* (COM (2004) 70 final).

European Commission (2006). Decision No 1639/2006/EC of the European parliament and of the council of 24 October 2006 establishing a competitiveness and innovation framework programme (2007 to 2013). *Official Journal L, 310*, 0015–0040.

European Commission (2009). *Reviewing community innovation policy in a changing world.* (COM(2009) 442 final).

Fairlie, R. W. (2011). *Kauffman index of entrepreneurial activity (1996–2010).* Kauffman Foundation, Kansas City. www.kauffman.org/uploadedFiles/KIEA_2011_report.pdf. Accessed 02 March 2013.

Freytag, A., & Thurik, A. R. (2007). Entrepreneurship and its determinants in a cross-country setting. *Journal of Evolutionary Economics, 17*(2), 117–131.

Galbraith, J. K. (1952). *American capitalism: The concept of countervailing power.* Cambridge: The Riverside Press.

Galbraith, J. K. (1967). *The new industrial state.* Boston: Houghton Mifflin Company.

Gartner, W. B. (1985). A conceptual framework for describing the phenomenon of new venture creation. *Academy of Management Review, 10*(4), 696–706.

Gartner, W. B. (1988). Who is an entrepreneur? Is the wrong question. *American Journal of Small Business, 12*(4), 11–32.

Gartner, W. B. (1990). What are we talking about when we talk about entrepreneurship? *Journal of Business Venturing, 5*(1), 15–28.

Gartner, W. B., Shaver, K. G., Carter, N. M., & Reynolds, P. D. (2004). *The handbook of entrepreneurial dynamics: The process of organization creation.* Thousand Oaks: Sage.

Gartner, W. B., & Shane, S. A. (1995). Measuring entrepreneurship over time. *Journal of Business Venturing, 10*(4), 283–301.

Godin, K., Clemens, J., & Veldhuis, N. (2008). *Measuring entrepreneurship: Conceptual frameworks and empirical indicators. Studies in entrepreneurship and markets 7.* Fraser Institute. www.fraserinstitute.org/research-news/display.aspx?id=13202. Accessed 02 March 2013.

Grilo, I., & Thurik, R. (2006). Latent and actual entrepreneurship in Europe and the US: Some recent developments. *Scientific analysis of entrepreneurship and SMEs, SCALES-paper N200514.* http://www.ondernemerschap.nl/pdf-ez/n200514.pdf. Accessed 02 March 2013.

Guerrero, M., Rialp, J., & Urbano, D. (2008). The impact of desirability and feasibility on entrepreneurial intentions: A structural equation model. *International Entrepreneurship and Management Journal, 4*(1), 35–50.

Haltiwanger, J. C., Jarmin, R. S., & Miranda, J. (2009). Business dynamics statistics: An overview. *Ewing Marion Kauffman Foundation.* doi:10.2139/ssrn.1456465.

Haltiwanger, J. C., Jarmin, R. S., & Miranda, J. (2010). *Who creates jobs? Small vs. Large vs. Young. NBER working paper no 16300.* www.nber.org/papers/w16300. Accessed 02 March 2013.

Hamel, G. (1999). Bringing silicon valley inside. *Harvard Business Review, 77*(5), 70–84.

Harrison, B. (1994). The myth of small firms as the predominant job generators. *Economic Development Quarterly, 8*(1), 3–18.

Harrison, B. (1997). *Lean and mean: The changing landscape of corporate power in the age of flexibility.* New York: Guilford Press.

Headd, B., & Saade, R. (2008). *Do business definition decisions distort small business research results?* U.S. Small Business Administration. www.sba.gov/advo/research/rs330tot.pdf. Accessed 02 March 2013.

Henderson, R. M., & Clark, K. B. (1990). Architectural innovation: The reconfiguration of existing product technologies and the failure of established firms. *Administrative Science Quarterly, 35*(1), 9–30.

Hebert, R. F., & Link, A. N. (1988). *The entrepreneur, mainstream views and radical critiques.* New York: Praeger.

Hebert, R. F., & Link, A. N. (1989). In search of the meaning of entrepreneurship. *Small Business Economics, 1*(1), 39–49.

Henrekson, M., & Johansson, D. (2009). Gazelles as job creators: A survey and interpretation of the evidence. *Journal of Small Business Economics, 35*(2), 227–244.

Iversen, J., Jorgensen, R., & Malchow-Moller, N. (2008). Defining and measuring entrepreneurship. *Foundations and Trends in Entrepreneurship, 4*(1), 1–63.
Kaplan, S. N., & Lerner, J. (2010). It ain't broke: The past, present, and future of venture capital. *Journal of Applied Corporate Finance, 22*(2), 36–47.
Katz, J. A. (1991). Endowed positions: Entrepreneurship and related fields. *Entrepreneurship Theory Practice, 15*(3), 53–67.
Katz, J. A. (1992). Secondary analysis in entrepreneurship: An introduction to databases and data management. *Journal of Small Business Management, 30*(2), 74–86.
Katz, J. A. (2003). The chronology and intellectual trajectory of American entrepreneurship education 1876–1999. *Journal of Business Venturing, 18*(2), 283–300.
Katz, J. A. (2004). *2004 Survey of endowed positions in entrepreneurship and related fields in the United States*. Kaufmann Foundation, Kansas City. www.kauffman.org/uploadedfiles/survey_endowed_chairs_04.pdf. Accessed 02 March 2013.
Kauffman Foundation (2007). *Entrepreneurship in American Higher Education*. Kaufmann Foundation, Kansas City. www.kauffman.org/uploadedfiles/entrep_high_ed_report.pdf. Accessed 02 March 2013.
Kawasaki, G. (2004). *The art of the start: The time-tested, battle-hardened guide for anyone starting anything*. London: Penguin Books.
Kerr, C., Dunlop, J. T., Harbison, F. H., & Myers, C. A. (1960). *Industrialism and industrial man. The problems of labor and management in economic growth*. Cambridge: Harvard University Press.
Knight, F. H. (1921). *Risk, uncertainty, and profit*. Boston: Hart, Schaffner & Marx.
Koellinger, P., Minniti, M., & Schade, C. (2007). "I think I can, I think I can" – Overconfidence and entrepreneurial behavior. *Journal of Economic Psychology, 28*(4), 502–527.
Krueger, N. F., & Brazeal, D. V. (1994). Entrepreneurial potential and potential entrepreneurs. *Entrepreneurship Theory and Practice, 18*(3), 91–104.
Kuratko, D. F. (2005). The emergence of entrepreneurship education: Development, trends, and challenges. *Entrepreneurship Theory and Practice, 29*(5), 577–597.
Landes, D. L., Mokyr, J., & Baumol, W. J. (2010). *The invention of enterprise: Entrepreneurship from ancient mesopotamia to modern times*. Princeton: Princeton University Press.
Langlois, R. N. (2004). Chandler in a larger frame: Markets, transaction costs, and organizational form in history. *Enterprise and Society, 5*(3), 355–375.
Low, M. (2001). The adolescence of entrepreneurship research: Specification of purpose. *Entrepreneurship Theory and Practice, 25*(4), 17–25.
Mariotti, S., DeSalvo, D., & Twole, T. (2000). *The young entrepreneur's guide to starting and running a business* (2nd ed.). New York: Random House.
McDaniel, B. A. (2000). A survey on entrepreneurship and innovation. *The Social Science Journal, 37*(2), 277–284.
McGee, J. E., Peterson, M., Mueller, S. L., & Sequeira, J. M. (2009). Entrepreneurial self-efficacy: Refining the measure. *Entrepreneurship Theory and Practice, 33*(4), 965–988.
Mills, C. W. (1951). *White collar: The American middle classes*. New York: Oxford University Press.
Mowery, D. C. (1983). Industrial research and firm size, survival, and growth in American Manufacturing, 1921–1946: An assessment. *The Journal of Economic History, 43*(4), 953–980.
Murphy, P. J., Liao, J., & Welsch, H. P. (2006). A conceptual history of entrepreneurial thought. *Journal of Management History, 12*(1), 12–35.
National Venture Capital Association. (2009). Venture capital investments Q4 and full year 2008. Press release, http://www.nvca.org/index.php?option=com_docman&task=doc_download&gid=404&Itemid=93, www.nvca.org/index.php?option=com_docman&task=doc_download&gid=404&Itemid=93. Accessed 2 Mar 2013.
OECD. (1998). *Fostering entreprneurship*. Paris: OECD. doi:10.1787/9789264163713-en.

OECD (2004). *OECD Compendium II on SME and entrepreneurship related activities carried out by International and Regional Bodies*. www.oecd.org/dataoecd/27/20/36402632.pdf. Accessed 02 March 2013.

OECD (2009a). *Measuring entrepreneurship: A collection of indicators*. 2009 edition, OECD-Eurostat Entrepreneurship Indicators Programme (EIP). www.oecd.org/industry/business-stats/44068449.pdf. Accessed 02 March 2013.

OECD. (2010a). *SMEs, entrepreneurship and innovation*. Paris: OECD. doi:10.1787/9789264080355-en.

OECD. (2010b). *The OECD innovation strategy – getting a head start on tomorrow*. Paris: OECD. doi:10.1787/9789264083479-en.

Parker, S. C. (2008). Statistical issues in applied entrepreneurship research: Data, methods and challenges. In E. Congregado (Ed.), *Measuring entrepreneurship: Building a statistical system* (pp. 9–20). New York: Springer.

Peneder, M. (2009). The meaning of entrepreneurship: Towards a modular concept. *Journal of Industry Competition and Trade, 9*(2), 77–99.

Prahalad, C. K., & Bettis, R. (1986). The dominant logic: A new linkage between diversity and performance. *Strategic Management Journal, 7*(6), 485–501.

Prahalad, C. K., & Bettis, R. (1995). The dominant logic: Retrospective and extension. *Strategic Management Journal, 16*(1), 5–14.

Purrington, C., & Bettcher, K. E. (2001). *From the garage to the boardroom: The entrepreneurial roots of America's Largest Corporations*. Washington, DC: National Commission on Entrepreneurship. doi:10.2139/ssrn.1260383.

Reich, R. (1987). Entrepreneurship reconsidered: The team as hero. *Harvard Business Review, 65*(3), 77–83.

Reynolds, P., Bosma, N., Autio, E., Hunt, S., De Bono, N., Servais, I., Lopez-Garcia, P., & Chin, N. (2005). Global entrepreneurship monitor: Data collection design and implementation 1998–2003. *Small Business Economics, 24*(3), 205–231.

Riesman, D., Glazer, N., & Denney, R. (1953). *The lonely crowd: A study of the changing American character*. Garden City: Doubleday.

Schmookler, J. (1957). Inventors past and present. *The Review of Economics and Statistics, 39*(3), 321–333.

Schramm, C. J. (2004). Building entrepreneurial economies. *Foreign Affairs, 83*(4), 104–115.

Schramm, C. J. (2006). *The entrepreneurial imperative – how America's economic miracle will reshape the world (and change your life)*. New York: Collins.

Schramm, C. J. (2010) *2010 State of entrepreneurship address*. National Press Club, Washington, DC, 19 Jan, 2010. www.kauffman.org/uploadedfiles/state_of_entrepreneurship_2010.pdf. Accessed 02 March 2013.

Schumpeter, J. A. (1942). *Capitalism, socialism and democracy* (6th ed.). London: Routledge. Reprint 2006.

Schumpeter, J. A. (1947). The creative response in economic history. *The Journal of Economic History, 7*(2), 149–159.

Servan-Schreiber, J. J. (1968). *The American challenge*. (With a foreword by Arthur Schlesinger, Jr., Trans. from the French by Ronald Steel). New York: Atheneum.

Shane, S. (2008). *The illusion of entrepreneurship: The costly myth that entrepreneurs, investors, and policy makers live by*. New Haven: Yale University Press.

Shane, S., & Venkataraman, S. (2000). The promise of entrepreneurship as a field of research. *Academy of Management Review, 25*(1), 217–226.

Smith, D. K., & Alexander, R. C. (1999). *Fumbling the future: How xerox invented, then ignored, the first personal computer*. New York: toExcel.

Sohl, J. (2011). *The angel investor market in 2010: A market on the rebound*. Center for venture research, 12 April, 2011. www.unh.edu/news/docs/2010angelanalysis.pdf. Accessed 02 March 2013.

Solomon, G. T., & Fernald, L. W. (1991). Trends in small business and entrepreneurship education in the United States. *Entrepreneurship Theory Practice, 15*(3), 25–40.

Solomon, G. T., Weaver, K. M., & Fernald, L. W. (1994). A historical examination of small business management and entrepreneurship pedagogy. *Simulation Gaming, 25*(3), 338–352.

Stangler, D., & Kedrosky, P. (2010). Firm formation and economic growth exploring firm formation: Why is the number of new firms constant? Kauffman Foundation, Kansas City, http://www.kauffman.org/uploadedFiles/exploring_firm_formation_1-13-10.pdf, www.kauffman.org/uploadedFiles/exploring_firm_formation_1-13-10.pdf. Accessed 2 Mar 2013.

Summers, L. H. (2010). *Remarks of Lawrence H. Summers at the presidential summit on entrepreneurship*. Ronald Reagan Building, Washington, DC 27 April, The White House, Office of the Press Secretary. www.whitehouse.gov/the-press-office/remarks-lawrence-h-summers-presidential-summit-entrepreneurship. Accessed 02 March 2013.

Sutton, J. (1997). Gibrat's legacy. *Journal of Economic Literature, 35*(1), 40–59.

The Economist (2001) The future of the company: A matter of choice. 20 Dec 2001.

The Economist (2009a). *Special report on entrepreneurship*. An idea whose time has come: Entrepreneurialism has become cool. 12 March 2009.

The Economist (2011c). *Brain scan. Seer of the mirror world: David Gelernter, a pioneering computer scientist, foresaw the modern internet but thinks computers are still too hard to use. Technology Quarterly*, 3 Dec 2011.

The Economist (2011d). *Big and clever: Why large firms are often more inventive than small ones.* 17 Dec 2011.

The New Yorker (2011). *Creation myth: Xerox PARC, Apple, and the truth about innovation.* 16 May, by Gladwell, M. www.newyorker.com/reporting/2011/05/16/110516fa_fact_gladwell. Accessed 02 March 2013.

Thurik, A. R. (2008). *Entrepreneurship, economic growth and policy in emerging economies. ERIM report series research in management*, No ERS-2008-060-ORG. http://repub.eur.nl/oai oai:repub.eur.nl/res/pub/13318/. Accessed 02 March 2013.

Thurik, A. R. (2009). Entreprenomics: Entrepreneurship, economic growth, and policy. In Z. Acs, D. Audretsch, & R. Strom (Eds.), *Entrepreneurship, growth, and public policy* (pp. 219–249). Cambridge: Cambridge University Press.

UK Department of Trade and Industry (1998). *DTI white paper: Our competitive future: Building the knowledge driven economy*. Report presented to Parliament by the United Kingdom Secretary of State for Trade and Industry, December 1998. http://webarchive.nationalarchives.gov.uk/+/ http://www.dti.gov.uk/comp/competitive/. Accessed 02 March 2013.

van Praag, C. M., & Versloot, P. H. (2008). The economic benefits and costs of entrepreneurship: A review of the research. *Foundations and Trends in Entrepreneurship, 4*(2), 65–154.

van Stel, A. (2005). COMPENDIA: Harmonizing business ownership data across countries and over time. *International Entrepreneurship and Management Journal, 1*(1), 105–123.

van Stel, A. (2006). *Empirical analysis of entrepreneurship and economic growth*. New York: Springer.

Vesper, K. H., & Gartner, W. B. (1997). Measuring progress in entrepreneurship education. *Journal of Business Venturing, 12*(5), 403–421.

Wennekers, A. R. M., & Thurik, A. R. (1999). Linking entrepreneurship and economic growth. *Small Business Economics, 13*(1), 27–55.

Wennekers, S. (2006). *Entrepreneurship at country level: Economic and non-economic determinants*. Erasmus Research Institute of Management Ph.D. Series Research in Management 81. http://repub.eur.nl/res/pub/7982/EPS20060810RG9058921158Wennekers.pdf. Accessed 02 March 2013.

Whyte, W. H. (1956). *The organization man*. New York: Simon and Schuster.

Wilson, S. (1955). *The man in the gray flannel suit*. New York: Simon and Schuster.

Wingham, D. W. (2004). Entrepreneurship through the ages. In H. P. Welsch (Ed.), *Entrepreneurship: The way ahead* (pp. 27–42). New York: Routledge.

Wong, P. K., Ho, Y. P., & Autio, E. (2005). Entrepreneurship, innovation, and economic growth: Evidence from GEM data. *Small Business Economics, 24*(3), 335–350.

World Economic Forum (2009). *Educating the next wave of entrepreneurs.* REF:150409. www3.weforum.org/docs/WEF_GEI_EducatingNextEntrepreneurs_ExecutiveSummary_2009.pdf. Accessed 02 March 2013.

Zahra, S. A. (2006). Entrepreneurship and disciplinary scholarship: Return to the fountainhead. In S. A. Alvarez, R. Agarwal, & O. Sorenson (Eds.), *Handbook of entrepreneurship research: Interdisciplinary perspectives* (pp. 253–268). New York: Springer.

Open Innovation 4

How is innovation done today? The emergence of the entrepreneurial society has not gone unnoticed. Firms are realizing the need to open up their doors to ideas from the outside to compete with innovative start—ups and survive in a competitive market driven more and more by innovation. They are reacting to the increasing pressure from entrepreneurs and the promise of profit from external innovation. They are more open to ideas from the outside, and make better use of their own ideas. This offers new possibilities, new paths for ideas to be realised, and an alternative to entrepreneurship. It is reshaping the innovation landscape.

Clearly the considerable socio-economic changes that brought about and come with the arrival of the entrepreneurial society do not go unnoticed. As innovation is becoming an increasingly important factor, and entrepreneurship is becoming an increasingly important path of realizing innovation, it would be myopic to think this a static process of moving from one ideal type of closed/managed innovation to the other extreme of the entrepreneurial society. Despite its prevalence and its contribution to economic progress, the entrepreneurial society is not an end in itself. Stopping the analysis here falls considerably short. The notion of the entrepreneurial society builds on several assumptions that are eroding quickly, often precisely as a reaction to its own success. Entrepreneurship may currently be the most pervasive approach to realizing innovation, but to some extent this is less due to the nature of the innovation process, than due to the lack of alternatives. The pressure to rely so heavily on entrepreneurship is based on the notion that it is the only seemingly realistic prospect to profit from one's idea. Any idea outside a corporation, or even within that does not fit or is rewarded badly can only be realized as an entrepreneur. This notion is changing, and it is changing fast – and with it challenges some of the fundamental arguments underlying the entrepreneurial society. The more successful the entrepreneurial approach becomes, the more it is able to challenge existing innovation cultures, namely vertically integrated corporate innovation, the more likely firms adapt to this new environment. Surely, a reaction of those having the most to lose by such development is to be expected. Any incumbents, even fairly new ones, are increasingly threatened by a steady stream of aspiring entrepreneurs. It is only natural to expect some form of adaptation to the new situation. As the old concept of closed innovation is failing, alternative means are sought to compete in a changing innovation environment. 'The old model is dead.'[1] Refining their moribund approach of managed innovation to better utilize and get hold of promising ideas is an imperative. And many are doing just that.

The emergence of the entrepreneurial society is challenging and disrupting the established innovation structure. Incumbents are more and more threatened by new entrants. In general, markets have become increasingly contestable. Not even the largest corporations can feel secure. Complacency with size and dominance is no longer an option. Clever ideas emerge increasingly outside the incumbents. Disruptive innovation is becoming commonplace (dismal examples include US Steel and the spread of mini mills; IBM and the emergence of the personal computer; Microsoft and open source, Google, and Web 2.0).[2] But also, competition between firms is now often based primarily on innovation. Innovate or perish – to put is harshly. This 'new' mantra is all the more applicable the shorter the innovation cycle, the more democratized innovation, the larger the markets and more promising first mover advantages, and the more appealing entrepreneurship becomes. The more important and profitable innovation is, and the easier it gets to become an entrepreneur, the more pressure there is for firms to keep up and stay ahead. This places severe pressure on companies to innovate and partake in the fast-paced innovation markets. Businesses are placing increasing emphasis on innovation and the need to access and realize innovations to survive and prosper. More and more resources are devoted to corporate research and development (R&D). More is expected of these funds, more pressure is exerted to deliver results, to generate ideas. Increasingly firms are starting to look for opportunities to access also the pool of ideas outside their own confined research activities.[3]

To increase innovation the most obvious approach is indeed to scale up efforts and devote more funds to research and development. Given the very nature of innovation, this faces some serious constraints. Ideas cannot simply be prompted. Surely paying for research facilities and providing time and means for smart and proven researchers and inventors to generate bright ideas is a promising approach and likely to yield some results. Much effort has been put into trying to optimise this process, to increase creativity, to foster idea generation. But even then, given that many others are working on it, given that ideas are often generated outside the specific area of focus, there is a huge risk to this. It is often not the most effective and certainly not

---

[1] For a dismal commentary on corporate R&D efforts, see Anderson (2004, p. 59).

[2] See Drucker (1986), Foster (1986), Christensen (1997), and Crandall 1999. Also see Langlois (2003, 2004) for the notion of a post-Chandlerian economy. Also see Mowery (1990).

[3] See Andrew et al. (2010). In their global survey among senior executives in 2010, 72 % reported innovation among their top three priorities, coupled with a higher willingness to spend more on R&D. Data from the US support this observation over time. It shows a surge from $17 billion in 1968 to almost $270 billion in 2007 (See Data on Industrial R&D Expenditure by Size of Company provided by the National Science Foundation's (NSF) Industrial Research and Development Information System (IRIS) (nsf.gov/statistics/iris/ 02.03.2013)) and Survey of Industrial Research and Development (nsf.gov/statistics/industry/ 02.03.2013).

the most efficient process. Ideas are more and more linked and based on diverse sources. They increasingly seem more likely to originate outside the research focus and even outside the corporation.

Worse still, even if the researches do come up with ideas, especially bold, visionary, and unorthodox approaches and ideas often fall victim to internal knowledge filters, to path dependency and utter ignorance, and to inadequate compensation schemes. Xerox "PARC researchers invented the Ethernet, windowed computer applications, screen icons, and laser printers. Of the 10 most important developments in computing, Xerox PARC birthed at least half of them. And how did Xerox management handle this windfall? They blew it. Choked. Perhaps the biggest screw-up in technology history. Almost every other company in Silicon Valley benefited from PARC's innovations, but the only one Xerox managed to cash in on is the laser printer."[4]

To address these issues many promising proposals have been put forward to improve the corporate invention and innovation process to make the invested funds more effective, better leveraging the innovation potential of companies. This includes improved management of innovation, of personnel and processes. Research is becoming more and more applied, more outcome oriented, and at the same time more incremental. New business models are emerging to best utilize this potential, and new managerial approaches are developed to best harness innovation though dedicated corporate innovation management. Chief Innovation Officers are mushrooming to steer this process. New organizational structures are introduced to better manage and encourage innovation. The knowledge filters are increasingly addressed and reformed to allow for a better absorption of promising ideas. To expand the base of potential sources of ideas increasingly not just research and development departments are required to come up with new ideas, but more broadly any employees throughout the firms are strongly encouraged to come up with new ideas. Additional absorption structures are put into place to enable and promote the generation and sharing of ideas. This includes initiatives like workgroup feedback systems, internal idea/suggestion programmes, exchange and cross-pollination initiatives, and many more, often highly original approaches.[5]

Underlying these aspects again is the question of incentives. As alternative realisation paths are becoming more and more lucrative the easier and more readily available and acceptable entrepreneurship becomes, corporations need to come up with improved compensation schemes to compete and capture these ideas for their own use. Many highly valuable ideas come from employees. Often they are not shared with the employer. Adequate incentives to do so are missing. In many cases,

---

[4] Anderson (2004, p. 58).

[5] See, for example, Aiken, Bacharach and French (1980), Burgelman (1983), Cohen and Levinthal (1990), Kim and Mauborgne (1997), Troy et al. (2001), Burgelman (2002), Chesbrough and Rosenbloom (2002), Mumford and Hunter (2005), Chesbrough (2006a), Sakkab (2007), Prahalad and Krishnan (2008). Also see Nonaka (1994), Woodman et al. (1993).

they leave the company to form successful ventures, often rivals, some with, but many without the blessing and benefit participation of the former employer. Indeed, many if not most high growth companies replicated or modified an idea encountered through previous employment.[6] Think of Intel – its founders left first Shockley then Fairchild to set up their own computer chip company. While Shockley failed and Fairchild stagnated, Intel grew to de facto dominate the computer chip market today. Intel "is now one of the most profitable firms while Fairchild Semiconductor is virtually a footnote in business history"[7] New remuneration approaches to spur and incentivize the generation and sharing of ideas are needed to capture these ideas for the company. Approaches to achieve this include an increasing use of rewards and prizes, profit participation through innovative accounting measures, and outcome based remuneration.[8]

As the perception of inherent value in ideas spreads, and the economic potential and benefit becomes more and more apparent and appreciated, companies pay more attention and focus more on the possible value ideas can create, also outside their company. New opportunities and strategies have emerged, previously neglected by many companies. Increasingly use is made of intellectual property management: intellectual property trading, licensing, etc.[9] Previously dormant claims are now used more effectively and viewed as financial assets and its' return, in the context of the corporate strategy, should be maximised. Such 'Rembrandts in the Attic' are being put to better use to appropriate more of its value.[10] Similar holds for the

---

[6] Compare Bhide (1994), as well as (2000, p. 54).

[7] Rajan and Zingales (2001, p. 806). Based on Jackson (1997).

[8] See, for example, Hamel (1999), Rajan and Zingales (2001), Hellman (2007), Bernardo et al. (2009), and Hellman and Thiele (2011). In many countries, legal provisions have been put in place that governs compensation rights between employer and employee (For a brief overview, see Peberdy and Strowel 2009 or Harhoff and Hoisl 2006, with reference to Rebel 1993; also Merges 1999a and Cherensky 1993). Results, however, are mixed and highly dependent on the given context in which innovation can take place.

[9] "International licensing, for example, appears to be on the rise. International receipts for intellectual property (including patents, copyrights, and trademarks) increased from USD 10 billion in 1985 to approximately USD 110 billion in 2004." (Yanagisawa and Guellec 2009, p. 7. Similarly Kamiyama et al. 2006, pp. 17–18). Also, IP trading is gaining ground. IP has even been discovered as an investment opportunity. "What was once viewed as a stodgy legal asset is fast becoming a sought-after financial one. The market is still small but it is growing quickly – by perhaps 20–30 % a year, [...] $4 billion-worth were bought and sold last year overall" (The Economist 2009c). Services and marketplaces are being established (e.g. IP auctioning, brokering, trading, commercialization services: Yet2.com; ipotential.com; oceantomo.com; ipauctions.com; ip-auction.eu; tynax.com; thinkfire.com; ipvalue.com; redchalkgroup.com; qed-ip.com (also see Yanagisawa and Guellec 2009, pp. 10–11)). Academic interest is growing: e.g. Arora 1995; Arora and Gambardella 1994, 2010; Troy and Werle 2008; Gambardella et al. 2007; Park et al. 2007; Palomeras 2007; also see comments by Lemley and Myhrvold (2008).

[10] This cannot be viewed as independent of the corporate strategy as often such intellectual property has some potential connection with the core business and profit maximization of the asset may conflict with the overall corporate maximization. See especially Rivette and Kline (2000) and Lichtenthaler and Ernst (2007). Also see Palomeras (2007).

increasing acceptance and even encouragement of spin-offs and joint ventures to utilise ideas originating from the company. Instead of shelving seemingly non-core, not immediately applicable ideas and developments, corporations now encourage active uptake of ideas even outside the organization. Increasingly companies encourage employees to take up their idea – again, within the overall business framework – to start new ventures and realize their ideas. Corporations even offer assistance, initial funding and venture capital. This often offers promising opportunities for the company to participate in and profit from.[11] Think of SAP whose founders, inspired during their work at IBM, left IBM to start the software company – frustrated not to get any support for their idea internally.[12] SAP now is a company with some $2 billion annual operating profits and almost 50,000 employees. Though IBM reportedly participated somewhat in its growth, it lost out on huge profit opportunities. Today, likely, more encouragement, more support and better alignment would enable a higher appropriation of value for the originating firm. A better example is Amazon. "The company had developed ways to allocate computing capacity flexibly in order to deal with the mountains of data being generated by its retail operations. This led to the idea that the same know-how could be used to solve similar problems at other companies, too, and Amazon Web Services (AWS) was born. It is now used by hundreds of thousands of firms, ranging from start-ups such as Spotify, a music streaming service, to established companies like Ericsson, a Swedish telecoms giant. The firm does not break out AWS's revenues, but Gartner, a consulting and research outfit, has estimated that they exceeded $1 billion in 2011."[13] A great idea, even one outside of the core business, fully appropriated by the inventing firm. This of course depended highly on the still energetic risk taking culture of its boss and co-founder with a still entrepreneurial spirit himself, Jeff Bezos. Others, likely, would not have dared (think of Yahoo). But many are pursuing efforts to better appropriate such ideas, be it through diversification or spin-offs.

Significant efforts and much research have been put into an even more promising aspect of corporate innovation: accessing ideas outside the corporation. The possibilities to gain access to the vast spectrum of ideas outside the firm are being explored to take advantage of such broad innovation potential. As the generation of ideas becomes more and more distributed and democratized, they become increasingly likely to originate outside the corporation, with independent inventors, customers, suppliers, lead users, anyone really. Firms have become to understand this fact very well, aware of the threat, but also its potential. The different stages in the production process are more closely aligned. Suppliers and downstream partners (end producers, distributors, even consumers) often participate actively in the development process. Substantial efforts have been directed to accessing customer experience and ideas, tapping into customer feedback to improve products and generate

---

[11] For an interesting case study, see Chesbrough (2003c).
[12] See Handelsblatt (2006).
[13] The Economist (2012b).

inspiration for future developments. Many engage in active scouting. Markets and research activities in a broad range of possibly relevant areas are monitored. Often universities, research facilities, start-ups, and rivals are closely observed even encouraged and supported with the aim to get early and preferred access to the developments through hiring practices, research grants, IP trading, start-up acquisition, seed funding, and other forms of engagement. Special attention is increasingly paid to lead users to observe new trends, adaptations and modifications that may inspire or point the ways for new developments. Specialised services have been established to envision future developments and assess emerging trends. Others include cross-pollination and analogical learning efforts to spur creativity and unusual adaptations of technologies and innovation in originally unintended areas.[14] Corporations want to jump ahead of the trend, drive it and participate early in it to reap first mover advantages and an innovation head start. For example, Intel uses mentoring systems and operates university linked small research facilities to gain early access, and scouts start-ups to invest or acquire them early. Procter and Gamble has scouting teams all over the world to find exciting new products. Further examples of a range of such often explicit strategies include IBM, Lucent, Genzyme, and many more.[15]

In addition, new marketing and product concepts are devised to leverage external innovativeness that is bound to occur, and utilise it to increase their own product's value. In many cases, this creativity may thus be considered a compliment rather than a threat. The emergence of open development platforms (e.g. Apple's AppStore, Facebook Platform, or early concepts such as Cisco Hosted Applications Initiative (CHAI)) and other forms of crowdsourcing/external participation attest to this.[16]

In 2003 Henry Chesbrough from then Harvard, now Berkeley, published a seminal business book conceptualising this development and giving it a catchy label: 'Open Innovation' – gaining significant prominence and attention ever since.[17] As a self-styled definition, he proposes: "Open innovation is the use of purposive inflows and outflows of knowledge to accelerate internal innovation, and expand the markets for external use of innovation, respectively. Open Innovation is a paradigm that assumes that firms can and should use external ideas as well as internal ideas, and internal and

---

[14] On co-creation, see von Hippel (1982, 1988, 2001), Thomke and von Hippel (2002), and Prahalad and Ramaswamy (2004), as well as Day and Schoemaker (2005) or Enkel et al. (2005). On Ideation methods and approaches, see, for example, discussion in Cooper and Edgettt (2008), Björk et al. (2010).

[15] Also compare Chesbrough (2003a, 2006a) and Chesbrough and Crowther (2006). Also see von Zedtwitz and Gassmann (2002).

[16] Also see Howe (2009). Many public initiatives have now sprung up (Data.gov, data.worldbank.org, data.dc.gov, data.london.gov.uk, gov.opendata.at, dataTO.org, etc.) not only to become more transparent, but to utilize such platform effects.

[17] Chesbrough (2003a). Also see Chesbrough et al. (2006) and Chesbrough (2006a), 2009. Also see Christensen et al. (2005), Gassmann and Enkel (2006), Enkel et al. (2009), and Fredberg, Elmquist and Ollila-Chalmers (2009), as well as Dahlander and Gann (2010) including a literature overview.

external path to market, as they look to advance their technology. Open innovation processes combine internal and external ideas into architectures and systems. They utilize business models to define the requirements of these architectures and systems. The business model utilizes both external and internal ideas to create value. Open innovation assumes internal ideas can also be taken to market through external channels, outside the current businesses of the firm, to generate additional value. The Open Innovation paradigm treats R&D as an open system. Open Innovation suggests that valuable ideas can come from outside or inside the company and go to market from inside or outside the company as well. This approach places external ideas and external path to market on the same level of importance as that reserved for internal ideas and path to markets in the earlier era."[18] Now ideas can, and should come from outside. The source of ideas has expanded beyond firms confined R&D capacity. The firm takes them up and realizes them. Also, ideas from within the company that it does not deem fitting within its profile and competency need no longer be shelved or discarded. Its value is now harnessed through, for example, spin-off ventures or licensing (see Fig. 4.1[19]). "Firms can create and capture value from their new technology in three basic ways: through incorporating the technology in their current business, through licensing the technology to other firms, or through launching new ventures that exploit the technology in new business arenas."[20]

Open Innovation styles the company as a more fluid facilitator – open to influences from inside as well as outside, and possible effects inside and outside the firm. Firms must open up and be receptive, or even actively seek out ideas from outside. At the same time, they can no longer discard the immense value of ideas they generate that may lay dormant instead of reaping its benefits also if this means realizing it outside the company. This acknowledges the changing environment the firm today operates in. "At its root, Open Innovation assumes that useful knowledge is widely distributed, and that even most capable R&D organizations must identify, connect to, and leverage external knowledge sources as a core process in innovation. Ideas that once germinated only in large companies now may be growing in a variety of settings – from the individual inventor or high-tech start-up in Silicon Valley, to the research facilities of academic institutions, to spin-offs from large, established firms."[21]

It is an on-going process, featuring prominent examples of firms opening up the innovation process. Practical implementation is not easy. Policies and processes are

---

[18] Chesbrough et al. (2006, p. 1).
[19] Based on Chesbrough and Garman (2009, pp. 68–76, p. 71).
[20] Chesbrough (2003a, pp. 63–64).
[21] Chesbrough (2006a, p. 2). As MIT economist Quinn phrased the issues and the challenges now already a decade ago: "Innovate or die. That's a theme many executives support. How to keep ahead is the issue. [...] no one company acting alone can hope to out-innovate every competitor, potential competitor, supplier or external knowledge source around the world. But there is hope. Strategically outsourcing innovation", "develop sophisticated outsourcing partnerships with those billions of new minds who can provide innovative support. Developing the necessary management practices is not easy. But using only traditional internal innovation practices can be fatal." (Quinn 2000, p. 13, and p. 27).

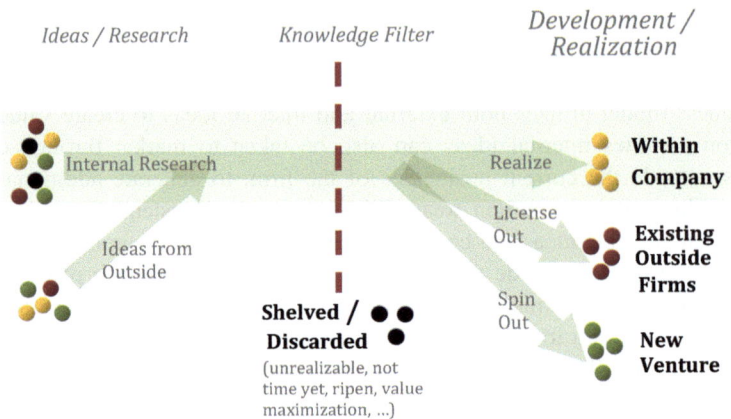

Fig. 4.1 Open innovation: going outside the frim to realize ideas

fashioned to support this transition. Many obstacles remain that need to be addressed. Solution and operation manuals are surfacing to enable such processes. Management is increasingly embracing such new mind-set, and open innovation seems increasingly adopted by existing business cultures.[22]

The notion of Open Innovation has come a long way.[23] It now offers many more ideas access to a more efficient, sometimes even the only possible realisation process available by offering an access to firms. It may also better leverage many ideas a company produces and more efficiently appropriate its value. Indeed, firms are often becoming facilitators. Innovation is opening up. Innovation becomes easier and more accessible. Firms adapt to the entrepreneurial threat and offer alternative innovation paths, somewhat reducing the outside innovation pressure.

Alas, it does not seem to get very far. Implementation is gradual, and mostly confined to but a special subset of ideas. 'Open' still typically refers to intellectual property, not ideas in general.[24] It is technology focussed, excluding a vast range of ideas. More work is needed on the granularity of the process, on how ideas flow. That they flow is good. But the intricacies of in- and outflow of ideas are demanding.[25]

---

[22] See Chesbrough and Crowther (2006), Chesbrough (2006a). Also Christensen et al. (2005), von Zedtwitz and Gassmann (2002), Lichtenthaler and Ernst (2007), Poot et al. (2009), Neyer et al. (2009), Rohrbeck et al. (2009), and Chiaroni, Chiesa, and Frattini (2010).

[23] Clearly, to some degree, companies have always been open (see Dahlander and Gann 2010). For an excellent early, IP focused discussion of what later became known as open innovation and its implications see Arora et al. (2001). Also somewhat Quinn (2000). Still, as Chesbrough argues, firms have been increasingly forced to open up, and the concept of open innovation has spread throughout the business world, embraced by ever more corporations (for a recent overview see Fredberg, Elmquist and Ollila-Chalmers (2009), Dahlander and Gann (2010), and Gassmann et al. (2010)).

[24] Especially see Chesbrough (2006a).

[25] See Zeckhauser (1996, p. 12746).

Anecdotes only go so far. Many obstacles exist that need to be identified and addressed. For example: what is fair? Trivial as it may seem, notions of fairness, as will be argued further down, can make or break cooperation, especially in such amorphous and intangible an environment as when it comes to ideas. How much is an idea worth? Who can make decisions? Who decides? Who 'owns' ideas? But this is not due to a lack of imagination, but a current absence still of proper processes that would make such true openness reasonable or realistic. To put it brusquely: Open innovation, as it stands today, is a maximisation exercise of a changing competitive environment. Maximise profit in circumstances where shelving ideas has become inefficient, where disruptive threats from outside have become commonplace, where all markets are contestable, where innovation dominates over size, market power, and economic and political prowess. Firms are forced to open up. And they have. Such openness is reactive, not proactive. It maximises profits in a new competitive setting. The processes and means have not much changed still. Open innovation captures and provides a solid portrayal of such current developments. 'Open Innovation' offers a label for on-going developments. It conceptualises it, phrases it into a coherent analytical framework, conveyable, replicable. Yet, unlike its claim, it is by far not an end, but rather a means in a broader argument. Openness is an essential building block of economic evolution, but the impact of this openness goes much beyond its own claim of transforming the firm. It is but a reaction to the entrepreneurial charge, and but an element of things to come. Openness itself will evolve; evolve beyond the confined contours of intellectual property, enabling ever more, and more fruitful cooperation.[26] The concept of open innovation as it is now, it just a first step. It set a motion a more ambitious, a more transformative process.

Open Innovation is an important step forward. It augments and expands the innovation process and challenges some of the fundamental assumption of the entrepreneurial society. A more open innovation culture in the firms is starting to rival the notion of the indispensable entrepreneurial realisation of ideas. While it was not possible to cooperate with firms, entrepreneurship seemed the only option. This is changing. Corporations are again gaining ground by offering alternative realisation paths and opening up to the outside. Within the spectrum of the two stereotypes of the lone-genius-gone-entrepreneur who best exemplifies the economic hero of the entrepreneurial society, and the typically but now enhanced internal/managed innovation process, new hybrid solutions increasingly emerge. For example: the possibilities of trading patents and selling start-ups to a big corporation to realise the idea in exchange for a fee or stake in the profits; or spinning off internal research via separate firms even with the support of the corporation. The spectrum of possible realisation paths is expanding as the closed/managed path is eroding and the entrepreneurial path is transformed and rivalled by an increasingly open innovation path (see Fig. 4.2).

---

[26] For now, Open Innovation remains IP focused (see Chesbrough 2006a; Dahlander and Gann 2010).

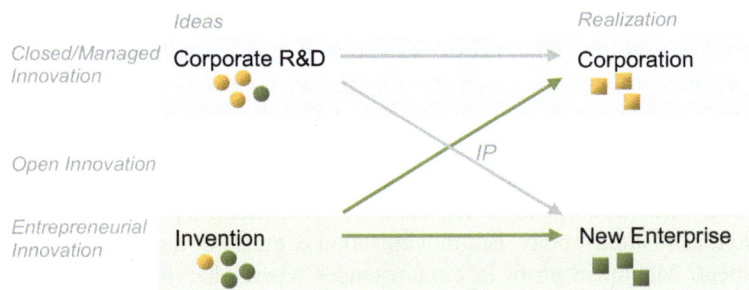

Fig. 4.2 More open innovation paths: moving IP in and out of the firm

It still has to come a long way, but the transformation has begun. For now it somewhat fends off and defers the entrepreneurial threat. Some of the pressure to realise innovations through entrepreneurship is relieved as corporate alternatives emerge. The possibilities for inventors are expanding. The entrepreneurial route is no longer the only one. Fewer inventors need to be entrepreneurs to see their ideas realised. Alternatives are increasingly available, alternatives that in many instances may be much more lucrative and promising than trying their entrepreneur luck on their own. More and more differentiated approaches emerge, more and more possibilities of approaching and cooperating with incumbents become available. Such alternative paths undermine the notion of the entrepreneurial society. The lone hero is being challenged. Firms are adapting to the entrepreneurial threat, dampening the impact of the entrepreneurial society, and with it again, changing the socio-economic characteristics away from the economic heroism of entrepreneurship. A new characteristic emerges.[27]

References

Anderson, H. (2004). Why big companies can't invent. *Technology Review, 107*(4), 56–59.
Andrew, J. P., Manget, J., Michael, D. C., Taylor, A., & Zablit, H. (2010). Innovation 2010: A return to prominence and the emergence of a new world order. *Boston Consulting Group Report*. www.bcg.com/documents/file42620.pdf. Accessed 02 Mar 2013.
Aiken, M., Bacharach, S. B., & French, J. J. (1980). Organizational structure, work process, and proposal making in administrative bureaucracies. *Academy of Management Journal, 23*(4), 631–652.
Arora, A. (1995). Licensing tacit knowledge: Intellectual property rights and the market for know-how. *Economics of Innovation & New Technology, 4*(1), 41–60.
Arora, A., & Gambardella, A. (1994). The changing technology of technological change: General and abstract knowledge and the division of innovative labour. *Research Policy, 23*(5), 523–532.

---

[27] Even an 'Era of Open Innovation' has been announced (see Chesbrough 2003b). Though its effect may indeed be just that, in terms of firms, and surely is a vital input, it does not reflect the larger socio-economic picture. It is a firm centric perspective.

References

Arora, A., & Gambardella, A. (2010). Ideas for rent: An overview of markets for technology. *Industrial and Corporate Change, 19*(3), 775–803.

Arora, A., Fosfuri, A., & Gambardella, A. (2001). Markets for technology: The economics of innovation and corporate strategy. *Industrial and Corporate Change, 10*(2), 419–450.

Bernardo, A. E., Cai, H., & Luo, J. (2009). Motivating entrepreneurial activity in a firm. *The Review of Financial Studies, 22*(3), 1089–1118.

Bhide, A. V. (1994). How entrepreneurs craft strategies that work. *Harvard Business Review, 72*(2), 150–161.

Bhide, A. V. (2000). *The origin and evolution of new businesses*. New York: Oxford University Press.

Björk, J., Boccardelli, P., & Magnusson, M. (2010). Ideation capabilities for continuous innovation. *Creativity and Innovation Management, 19*(4), 385–396.

Burgelman, R. A. (1983). Corporate entrepreneurship and strategic management: Insights from a process study. *Management Science, 29*(12), 1349–1364.

Burgelman, R. A. (2002). Strategy as vector and the inertia of coevolutionary lock-in. *Administrative Science Quarterly, 47*(2), 325–357.

Cherensky, S. (1993). A penny for their thoughts: Employee-inventors, preinvention assignment agreements, property, and personhood. *California Law Review, 81*(2), 595–669.

Chesbrough, H., & Rosenbloom, R. S. (2002). The role of the business model in capturing value from innovation: evidence from Xerox Corporation's technology spinoff companies. *Industrial and Corporate Change, 11*(3), 529–555.

Chesbrough, H. W. (2003a). *Open innovation: The new imperative for creating and profiting from technology*. Cambridge: Harvard Business School Press.

Chesbrough, H. W. (2003b). The era of open innovation. *Sloan Management Review, 44*(3), 35–41.

Chesbrough, H. W. (2003c). The governance and performance of xerox's technology spin-off companies. *Research Policy, 32*(3), 403–421.

Chesbrough, H. W. (2006a). *Open business models: How to thrive in the new innovation landscape*. Cambridge: Harvard Business School Press.

Chesbrough, H. W., & Crowther, A. K. (2006). Beyond high tech: Early adopters of open innovation in other industries. *R&D Management, 36*(3), 229–236.

Chesbrough, H. W., & Garman, A. R. (2009). How open innovation can help you cope in lean times. *Harvard Business Review, 87*(12), 68–76.

Chesbrough, H. W., Vanhaverbeke, W., & West, J. (2006). *Open innovation: Researching a new paradigm*. New York: Oxford University Press.

Chiaroni, D., Chiesa, V., & Frattini, F. (2010). Unravelling the process from closed to open innovation – evidence from mature, asset-intensive industries. *R&D Management, 40*(3), 222–245.

Christensen, C. M. (1997). *The innovator's dilemma: When new technologies cause great firms to fail*. Cambridge: Harvard Business School Press.

Christensen, J. F., Olesen, M. H., & Kjær, J. S. (2005). The industrial dynamics of open innovation: Evidence from the transformation of consumer electronics. *Research Policy, 34*(10), 1533–1549.

Cohen, W. M., & Levinthal, D. A. (1990). Absorptive capacity: A new perspective on learning and innovation. *Administrative Science Quarterly, 35*(1), 128–153.

Cooper, R., & Edgett, S. (2008). Ideation for product innovation: What are the best methods? *PDMA visions magazine*, March 2008. www.stage-gate.net/downloads/working_papers/wp_29.pdf. Accessed 02 March 2013.

Crandall, R. W. (1999). From competitiveness to competition: The threat of minimills to large national steel companies. *Resources Policy, 22*(1–2), 107–118.

Dahlander, L., & Gann, D. (2010). How open is innovation. *Research Policy, 39*(6), 699–709.

Day, G. S., & Schoemaker, P. J. H. (2005). Scanning the periphery. *Harvard Business Review, 83*(11), 135–148.

Drucker, P. F. (1986). *Innovation and entrepreneurship*. New York: HarperCollins. Reprint 2006.

Enkel, E., Grassmann, O., & Chesbrough, H. W. (2009). Open R&D and open innovation: Exploring the phenomenon. *R&D Management, 39*(4), 311–316.

Enkel, E., Kausch, C., & Gassmann, O. (2005). Managing the risk of customer integration. *European Management Journal, 23*(2), 203–213.

Fredberg, T., Elmquist, M.,& Ollila-Chalmers, S. (2009). Managing open innovation: Present findings and future directions. VINNOVA Report Nr VR 2008:02, http://www.vinnova.se/upload/EPiStorePDF/vr-08-02.pdf, www.vinnova.se/upload/EPiStorePDF/vr-08-02.pdf. Accessed 2 Mar 2013.

Foster, R. N. (1986). *Innovation: The attacker's advantage*. New York: Summit Books.

Gambardella, A., Giuri, P., & Luzzi, A. (2007). The market for patents in Europe. *Research Policy, 36*(8), 1163–1183.

Gassmann, O., & Enkel, E. (2006). Open innovation: Externe Hebeleffekte in der Innovation erzielen. *Zeitschrift Führung + Organisation, 3*, 132–138.

Gassmann, O., Enkel, E., & Chesbrough, H. (2010). The future of open innovation. *R&D Management, 40*(3), 313–321.

Hamel, G. (1999). Bringing silicon valley inside. *Harvard Business Review, 77*(5), 70–84.

Handelsblatt (2006). *60 Jahre deutsche Wirtschaftsgeschichte: SAP: Langsam, aber gewaltig.* Handelsblatt, 07.04.2006. by Nonnast, T.

Harhoff, D., & Hoisl, K. (2006). Institutionalized incentives for ingenuity – patent value and the German Employees' Inventions Act. http://epub.ub.uni-muenchen.de/1262/1/German_Inventor_Compensation_230106_DP_LMU.pdf. Accessed 02 March 2013.

Hellman, T. (2007). When do employees become entrepreneurs? *Management Science, 53*(6), 919–933.

Hellman, T., & Thiele, V. (2011). Incentives and innovation: A multitasking approach. *American Economic Journal: Microeconomics, 3*, 78–128.

Howe, J. (2009). *Crowdsourcing: Why the power of the crowd is driving the future of business*. New York: Random House.

Jackson, T. (1997). *Inside intel: Andrew grove and the rise of the world's most powerful chip company*. New York: Dutton Books.

Kamiyama, S., Sheehan, J., & Martinez, C. (2006). Valuation and exploitation of intellectual property. *OECD science, technology and industry working papers 2006/5*. Paris: OECD. doi: 10.1787/307034817055.

Kim, W. C., & Mauborgne, R. (1997). *Harvard business review on innovation*. Boston: Harvard Business School Publishing.

Langlois, R. N. (2003). The vanishing hand: The changing dynamics of industrial capitalism. *Industrial and Corporate Change, 12*(2), 351–385.

Langlois, R. N. (2004). Chandler in a larger frame: Markets, transaction costs, and organizational form in history. *Enterprise and Society, 5*(3), 355–375.

Lemley, M. A., & Myhrvold, N. (2008). How to make a patent market. *Hofstra Law Review, 36*(2), 257–259.

Lichtenthaler, U., & Ernst, H. (2007). External technology commercialization in large firms: Results of a quantitative benchmarking study. *R&D Management, 37*(5), 383–397.

Merges, R. P. (1999a). The law and economics of employee inventions. *Harvard Journal of Law & Technology, 13*(1), 2–53.

Mowery, D. C. (1990). The development of industrial research in U.S. manufacturing. *The American Economic Review, 80*(2), 345–349.

Mumford, M. D., & Hunter, A. T. (2005). Innovation in organizations: A multi-level perspective on creativity. *Research in Multi Level Issues, 4*, 9–73.

Nonaka, I. (1994). A dynamic theory of organizational knowledge creation. *Organization Science, 5*(1), 14–37.

Neyer, A. K., Bullinger, A. C., & Moeslein, K. M. (2009). Integrating inside and outside innovators: A sociotechnical systems perspective. *R&D Management, 39*(4), 410–419.

Palomeras, N. (2007). An analysis of pure-revenue technology licensing. *Journal of Economics and Management Strategy, 16*(4), 971–994.

References

Park, J., Shin, S. K., & Lawrence, G. L. (2007). Impact of international information technology transfer on national productivity. *Information Systems Research, 18*(1), 86–102.

Peberdy, M., & Strowel, A. (2009). *Employee's rights to compensation for inventions – A European Perspective. PLC Life sciences handbook 2009*/10. www.cov.com/files/Publication/4ffe8880-deba-493a-8994-2a69f0da78dd/Presentation/PublicationAttachment/3b0e8983-fe2a-41b5-9e96-2fb6c8a3a8c1/Employee%E2%80%99s%20Rights%20to%20Compensation%20for%20Inventions%20-%20A%20European%20Perspective.pdf. Accessed 02 March 2013.

Poot, T., Faems, D., & Vanhaverbeke, W. (2009). Toward a dynamic perspective on open innovation: A longitudinal assessment of the adoption of internal and external innovation strategies in the Netherlands. *International Journal of Innovation Management, 13*(2), 177–200.

Prahalad, C. K., & Ramaswamy, V. (2004). Co-creation experiences: The new practice in value creation. *Journal of Interactive Marketing, 18*(3), 5–14.

Prahalad, C. K., & Krishnan, M. S. (2008). *The new age of innovation: Driving co-created value through global networks.* New York: McGraw-Hill.

Quinn, J. B. (2000). Outsourcing innovation: The new engine of growth. *Sloan Management Review, 41*(4), 13–28.

Rajan, R., & Zingales, L. (2001). The firm as a dedicated hierarchy. *Quarterly Journal of Economics, 116*(3), 805–851.

Rebel, D. (1993). *Handbuch Gewerbliche Schutzrechte – Übersichten und Strategien: Europa, USA, Japan.* Wiesbaden: Gabler.

Rivette, K. G., & Kline, D. (2000). *Rembrandts in the attic: Unlocking the hidden value of patents.* Boston: Harvard Business School Press.

Rohrbeck, R., Hoelzle, K., & Gemuenden, H. G. (2009). Opening up for competitive advantage – how Deutsche Telekom creates an open innovation ecosystem. *R&D Management, 39*(4), 420–430.

Sakkab, N. Y. (2007). Growing through innovation. *Research Technology Management, 50*(6), 59–64.

The Economist (2009c). *Trolls demanding tolls: Intellectual property comes of age as an alternative investment.* 10 Sept 2009.

The Economist (2012b). *Brain scan. Taking the long view: Jeff Bezos, the founder and chief executive of Amazon, owes much of his success to his ability to look beyond the short-term view of things. Technology Quarterly.* 3 Mar 2012.

Thomke, S., & von Hippel, E. (2002). Customers as innovators: A new way to create value. *Harvard Business Review, 80*(4), 74–81.

Troy, I., & Werle, R. (2008). *Uncertainty and the market for patents.* Max-Planck-Institut für Gesellschaftsforschung working paper 08/2. www.mpifg.de/pu/workpap/wp08-2.pdf. Accessed 02 March 2013.

Troy, C., Szymanski, M., & Varadarajan, P. (2001). Generating new product ideas: An initial investigation of the role of market information and organizational characteristics. *Journal of the Academy of Marketing Science, 29*(1), 89–101.

von Hippel, E. (1982). Get new products from customers. *Harvard Business Review, 60*(2), 117–122.

von Hippel, E. (1988). *The sources of innovation.* New York: Oxford University Press.

von Hippel, E. (2001). Perspective: User toolkits for innovation. *Journal of Product Innovation Management, 18*, 247–257.

von Zedtwitz, M., & Gassmann, O. (2002). Market versus technology driven in R&D internationalisation: Four different patterns of managing research and development. *Research Policy, 31*(4), 569–588.

Woodman, R. W., Sawyer, J. E., & Griffin, R. W. (1993). Toward a theory of organizational creativity. *The Academy of Management Review, 18*(2), 293–321.

Yanagisawa, T., & Guellec, D. (2009). *The emerging patent marketplace* (Science, Technology and Industry Working Papers 2009/9). Paris: OECD. doi:10.1787/218413152254.

Zeckhauser, R. (1996). The challenge of contracting for technological information. *Proceedings of the National Academy of Sciences of the United States of America (PNAS), 93*, 12743–12748.

The Idea Economy 5

What will change in the future? Innovation is evolving. As firms are opening up to allow in ideas from inventors, an increasing division of labour will take place. This opens up an entirely new perspective on innovation. An Idea Economy emerges where ideas are traded. Anyone with an idea can approach a firm a or gifted entrepreneur to realize the innovation together. Anyone can profit from their ideas even without the skills and resources to be an entrepreneur themselves. This new division of labour will lead to a new kind of innovation: cooperative innovation. This development has profound implication for the innovation process, it will reshape the nature of the firm, and will influence the way we think about innovation.

Barely unfolding, the entrepreneurial society is being undermined and threatened. It is already changing. Entrepreneurship has changed and is changing the economy – but not merely in the predicted way. It is not an end in itself, but rather a transition phase. It is, to some extent self-defeating. Entrepreneurship is not the only path for ideas to be realized. Corporations are adapting, offering alternative innovation paths for inventors. Such alternatives are becoming more readily available, more adapt, and more open to innovations from different sources including from outside the firms. The single outlet pressure for ideas via entrepreneurship is diminishing fast. This does not replace entrepreneurship – it remains an important aspect – but it seizes to be the defining characteristic. A fundamental shift is happening that will change the underlying characteristics of the economy, moving further and further away from the entrepreneurial society. These developments are on-going and progressing fast. The heralded entrepreneurial society, just emerging, is in demise.

The nature of innovation is changing. As innovation seizes to be equated with entrepreneurship, and alternative means open up for the inventors,[1] the focus will shift towards the invention and idea itself, and away from the realisation process.

---

[1] Note, for a lack of a better word – 'Ideaist', though it has significant appeal, is unsuited – inventor is used in a fairly broad manner to represent anyone with an idea, and any idea, not merely a narrow technological or otherwise product oriented connotation it may often have. In many cases 'inventor' implies some degree of development, beyond the initial hunch or formulation of the

The idea has value. The way it is realized, the entrepreneurial sweat is secondary. The choice of how to realise ideas will become a task of profit maximisation rather than a question of whether or not the inventor has the required attributes and/or means to succeed as an entrepreneur or just happens to be part of a the ideally suited company. The inventor might be entrepreneurial, or engage in cooperation with others who are entrepreneurs, or collaborate with existing firms to realise their idea. It certainly is no longer a binary question of everyone with an idea being an entrepreneur or not. The path from ideas to market can now be manifold. In can be entrepreneurial innovation, but just as well cooperation with a firm or expert.

As the different paths become more and more interchangeable, as they increasingly compete with each other, and innovation becomes more fluid, any such differentiation may remain an interesting aspect to ponder, may remain an intriguing feature of research, and indeed, it may even still matter for the different innovations, but it loses its overall significance. The possibilities of an idea to be realised have become relatively insignificant. Wherever they may come from, ideas now have a realistic chance of being realised. As the corporations are opening up to new ideas, are redefining their knowledge filters, are allowing for alternative realisation path, as entrepreneurship is becoming widely accessible and itself an increasingly open innovation process, more and more ideas have a good chance of being realised. Its origin no longer constrains its possibilities. Its origin no longer predefines its realization path. Be it internal ideas, previously shelved or filtered, be it the necessity and coincidence of idea and entrepreneurial skills, they all now come within reach of realization. As an increasing range of realisation paths open up, neither one – entrepreneurship or open corporate innovation – can be considered a defining characteristic. They are circumstance specific; they become a matter of convenience and profit maximisation.

What does gain significance is the notion of ideas themselves. Ideas compete with each other. They are no longer confined by the possibilities of realising them. Ideas themselves become the defining characteristic. What once was a constrained and narrow managed realisation path soon became augmented by the entrepreneurial one. This in turn transformed and opened up a vast array of different paths. Now there are multiple, an entire spectrum of possible realisation paths. In such a setting segmentation between the different innovation paths (horizontal segmentation of entrepreneurial/collaborative/corporate/etc. innovation) becomes less and less meaningful. Instead distinguishing between the generation of idea and realisation process per se will become more defining (vertical segmentation of ideas/realization).[2] The emphasis will be on the division of labour between the idea and its realization (see Fig. 5.1). Ideas will be seen as the original source of prosperity,

---

idea. Here this may be the case, but need not. It has a broader meaning focusing on the idea, not including any further developments.

[2] Shane and Eckhardt (2003) hint at this development, proposing a vertical disintegrated process. Arora and Gambardella (1994 and 2010, also see Arora and Merges 2004) explicitly argue for such division of labour.

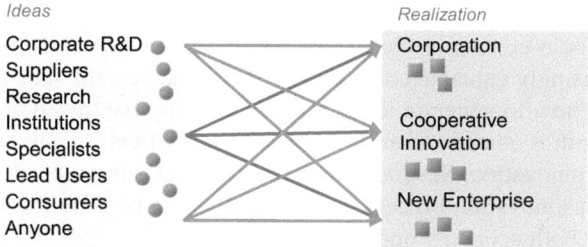

Fig. 5.1 Cooperative innovation paths in the Idea Economy

while the realisation process will be more a 'means', a 'process' and 'tool' to succeed. Anyone can have a brilliant idea, and anyone can realise it – through a variety of processes. Not everyone needs to, though they could be, an entrepreneur, not everyone needs to have access to all the necessary resources to realise it, or need to battle incumbents. Ideas can be realised in cooperation with experts, can be traded to existing firms. Ideas are now independent, no longer path dependent. Ideas themselves are now valuable. Not only its realisation.[3]

This notion gives rise to a different perspective on the economy, a different vision of a socio-economic environment of the future. It goes a step further, reconsidering the concept of the entrepreneurial society, its founding claim, and its development and prospects. The discussed economic, technological, and societal developments are leading to yet another defining characteristic: the *Idea Economy*. In the *Idea Economy*, the realisation path is secondary. The focus is on ideas. The manner in which they are realised is no longer the defining characteristic. Ideas are. Ideas are the seeds of prosperity – realisation merely the fertile soil they are planted on. Ideas are the focal point; realization becomes a question of optimization, not of feasibility. Anyone with an idea can get it realized – in a myriad of ways. Neither does the corporation need to realise only its own inventions itself, nor does the independent inventor need to become an entrepreneur. Alternatives are viable. Ideas are no longer bound to specific paths. They are not inherent to the ideas. Realization is flexible, idea and circumstance specific – whichever one appears most promising and most lucrative. A division of labour become the norm. The more evolved open innovation in firms and the more readily available the entrepreneurial path become, the more the focus shifts to ideas. Ideas are the scarce resource, not the implementation. Those able to realize them compete for ideas. Given an appropriate framework, ideas are being commoditized. They are becoming an increasingly valuable commodity. Now anyone can have ideas. Anyone can realise them or get them realised. Anyone can profit from their ideas.

---

[3] This contradicts Schumpeter;s assertive claim of stressing the realization much more than the invention (Schumpeter 1934, pp. 88–89). With good reason, ideas in themselves have little value so long as they are not realized. But as realization becomes more and more available and realistic to any viable and promising idea, the emphasis shifts towards invention. Ideas become relatively more scarce. With it, the focus shifts towards ideas.

Extrapolating this concept, and envisioning resulting changes yields a glimpse of the possible socio-economic characteristics of such an *Idea Economy*. As ideas become increasingly valuable commodities in themselves, the scope of innovation, the process of how to generate ideas, the manner they are traded and realised will change, and with it other economic, institutional, political, social, and psychological aspects of innovation. Several conceptual implications arise substantially altering the current innovation process, enhancing the prospects but also creating new challenges for both economy and society:

Scope (more ideas) Anyone now has the chance to see his or her ideas realised and realistically profit from them. This will vastly expand the scope of ideas to be realised. It will expand the source of ideas by including a tremendous range of untapped ideas – to everyone, as opposed to a confined number of fortunate few. The source of ideas should no longer matter, but the ideas themselves. Thus, anyone can participate, any ideas can be considered. At the same time, also the efforts of generating ideas will increase. With the realistic prospect of profiting from ones ingenuity developing ideas will become a more active endeavour. Increased attentiveness to spot and generate ideas will lead to even more ideas. Some may specialise in such search and idea generation. Different techniques and approach will be developed to foster such search. A whole new perspective opens up. Though possibly still somewhat serendipitous ideas are more actively sought and longed for, no longer a lucky happenstance and by-product but a dedicated and attentive search. More opportunities will be recognized, more ideas will be generated.

Nature (different ideas) With the increasing focus on and value of ideas, the nature of the generation and the kind of ideas sought after will change. Now that research has an ever-greater opportunity to result in substantial economic gain, it is becoming more commercialized and more outcome oriented. To a large extent, commoditization implies commercialisation. The link of ideas and value is strengthened. As ideas gain in value, the focus will be on generating value. Ideas no longer exist for their intellectual stimuli, but for their potential economic gain. This does surely not preclude the generation of such ideas, but shifts the focus of inquiry to generating gain, not per se ideas. The lure of gain from research increases. The drive to commercialize and cash in on research is strengthened. The choice what ideas to pursue will be determined by economic potential more than by curiosity, intellectual repute, or greater public value. Faster ways of applying basic research will be sought. Technology transfer services are and will be progressively more a key element at universities and research facilities. Ideas may be more applied and focussed, even segmented to explore the different applications, and thus diffuse faster throughout the different sectors and application possibilities. With the accelerated turnover of ideas and the prospects and value, the nature of ideas may equally shift towards more incremental ideas. The opportunity costs and the competitive risk of holding on and/or focussing on the big idea rather than trading in several more modest approaches may influence this. Together with a changing more modular production structure, this will further intensify.[4]

---

[4] Yet, at the same time, enable the realistic realization of bolder and more comprehensive ideas, if the different components become readily available.

Value As the focus and emphasis on ideas shifts, so does its value. Ideas are seen as the true sources of innovation. Accordingly, they should be rewarded. An increasing portion of the profits will remain with the originating source of the idea. Less is left with the implementers and risk-bearers. Ideas will receive increasing credit and be rewarded for the innovation as the competition among implementation options, realisation paths, and the range of actors increases.[5] Implementation will be a service, competing for ideas. Valuable ideas are the scarcer factor as implementation professionalises, shifting the value allocation towards ideas. The implementation mark-up that currently absorbs most of the value is competed away. Ideas are attributed true value. In a broader sense, this also means shifting the notion of fairness from labour to ideas. The intellect will be rewarded more, relative to 'work' and 'capital'. This also means the incentive structure in firms for internal innovation will have to adapt.

Implementation Efficiency At the other end, the realisation path will further expand and become more equipped to handling and realising ideas. Companies and partners will become more agnostic about the origin of ideas, accepting and trading them, overcoming an often widespread not-invented-here attitude. The mechanism of handling and trading ideas will become more sophisticated further increasing the possibilities of realising ideas, further commoditizing ideas, and in turn increasing efforts to generate and process ideas. Competition for ideas will increase and new innovative approaches will emerge to attract, handle, and to remunerate ideas. This will make the process of realisation even more efficient and more attractive for those with ideas. The different approaches will compete among each other and competition within each path will push for more and more innovation and efficiency. Especially cooperative innovation will increase, matching up corporations, small firms, entrepreneurs, experts, anyone able and willing to assist in the realisation with those with ideas – rather than the inventors doing it all on their own. The division of labour is strengthened further. Now that ideas can be separated from the realisation process, many more different forms of innovation paths may open up, hybrids, complementary services, intermediaries, etc. New services and professions will emerge: e.g. to match partners, to mediate and trade ideas, to assess, develop, arbitrage, leverage, diffuse, apply, handle ideas, etc. (see Fig. 5.2[6]). This will also transform the conceptual framework.

Such a division, almost fragmentation of labour may also change some of the more fundamental characteristics of the participants' tasks. Firms will become more implementers than innovators to best utilize their resources, skills, production

---

[5] Though, it is not clear to what extent this will be. It will be context dependent on the market structure, the nature of the idea, etc. From a simple economic modelling approach it could be argued, the full amount of residual profit (profit less risk premiums) would be attributed to the idea. At the same time, generating ideas may become increasingly complicated (but need not be), thus also increasing the costs possibly in many cases to an extent of the expected value of the idea less a risk factor.

[6] This is indicative at best. Innumerous services and activities could be listed. It merely shows a sample of a few services.

Fig. 5.2 Help along the innovation path

potential – the use of existing expertise in design, supply chains, marketing channels, production facilities, etc. benefiting from scale and scope to compete with peers and market entrants.[7] They lose the stigma of lethargic behemoth, becoming more agile, responsive and proactive. The focus will be much more on the firms' core competences and competitive advantage as alternative options become available and competition intensifies. Venture capitalists will become more service providers that compete with one another. They will have to seek out ideas rather than act as passive Grand Seigneurs. Also increasingly new forms of

---

[7] Paraphrasing Chesbrough's distinction between integrators and innovators (see Chesbrough 2003a). 'Integrators', however, does not go far enough, and is more likely a transitory state, given the large contextual changes. Not only the entry point will change, but the function and position of the firm.

venture finance and alternative innovative financial sources will emerge (e.g. threshold pledging, micro venture capital, crowd funding, incentive prizes, leveraged and hybrid banking-VC, etc.), strengthening competition and requiring more engagement and itself innovative approaches.[8] With such increased division of labour and the availability of new resources, implementation will become more efficient and effective.[9] It becomes faster – and easier than ever before. Ideas are realised more efficiently. Skills are better utilized. Ideas will be realised cheaper, and with a better chance of success. Using existing structures, they can be applied more quickly and on a broader scale, better utilizing first mover advantages, and thus higher appropriation. In effect, ideas will be more attractive, more valuable. With competition comes better services, new and itself innovative approaches to handle and realise ideas. It is a self-perpetuating dynamic.

Economy The changing innovation process affects not only firms. It alters economic processes in general. Competition in general based on innovation will intensify. First mover advantages for goods, more efficient production structures, etc. will dominate over more traditional competitive factors of size and market access. As multiple realisation paths open up incumbent firms become more cooperative facilitators than top dogs. Any market is contestable, may be disrupted. Shifts and shocks can come from anywhere, within the sector but also increasingly from outside. The boundaries of confined industries become more permeable as cross-pollination increases and competition becomes a question of ideas, less a matter of given implementation and production structures. Competition intensifies and becomes far less predictable. Leaps and jolts are more frequent; unimpeded steady progression of products and innovations is a thing of the past. This goes hand in hand with a reorganization of production structures. With the division of labour deepening, new production and value chains arise. Broader not deeper integration takes place. Horizontal bundling is more likely than vertical one. Vertical integration dissolves. A more composite path of specialists replaces fully integrated structures to best leverage value creation. The core business is important. Peripheral activity will decouple. Internalising externalities by range of ideas becomes part of profit maximisation, beyond boundaries of own firm expertise. More transitory linkages between firms and sectors arise. Co-production, shorter life spans of organisations will occur. More frequent reorientation becomes necessary. Production is likely to become ever more modular. Trade in tasks will intensify. More standardised, less integrated components become the norm that can be used in a wider range of products.[10] This allows for a better balance of scale and the uptake of

---

[8] See, for example, Ueda 2004; Sørensen 2007; or Plehn-Dujowich et al. 2010.

[9] See, for example, also discussion in Silvera and Wright 2010.

[10] This may have significant impact. To get a hunch of what such increasing modularity means and the impact it could have on a macroeconomic, but also on a firm strategic level, see the work by Cesar Hidalgo et al. (2007; Hidalgo and Hausmann 2009; Hidalgo 2009). Also see the classic article on modularity in production by Leonard Read (1958) on both the complexity of modular production and the potential.

innovation. This enables innovation to diffuse faster. Though it may appear less effective in a static setting, it will prove superior and thus dominant in a dynamic one. Much value still lies in new production methods, in ways of combining and assembling inputs that is more flexible to adapt to frequently changing production to absorb new innovations. This may also lead to a more far-reaching global division of labour further 'flattening the world'.[11] The vertical integration from idea to market dissolves, at least to the extent of decoupling ideas from its realisation. More fluid idea realization chains mean better optimisation of production, unbound by local externalities and proximities to invention. Production costs matter, not the integration of idea and producer. Production is a service not a vertically integrated implementation element.[12] As externalities can be internalised by more differentiated and encompassing realisation processes involving a larger range of partners, and higher standards of handling ideas become common, the services compete by costs, and may thus still be often location specific. Especially downstream and more basic modular production processes become 'outsourced' and global trade intensifies further. Local specialisation in terms of value creation path, of idea economies and production clusters may be conceivable. Research and idea generation will become a highly profitable sector in its own right, independent of the implementation structure. As more value will remain with such idea generators as externalities can be internalised, as implementation risk is reduced and an efficient idea market is established, production and employment structures will shift further towards the knowledge and idea workers. Possibly thick market externalities may still hold in some areas, while in others proximity to implementation remains essential. Certain types of clusters may still hold value. Given rising attentiveness, the better the enabling environment, the more exposure to processes, to inspiring basic research, other disciplines, etc., may provide a competitive edge. Thus, in certain areas, location may matter still – an aspect economic policy will focus on. Creative clusters may emerge, idea clusters as opposed to production locations. Local segmentation may go hand in hand with a more cooperative production process and segmented value chain. Creativity itself becomes valuable, no longer just in connection with local production.[13] With the increased

---

[11] For this compelling description, see Friedman 2005.

[12] Note, though, this cannot be equated with unit production costs in terms of labour conditions and cost of capital around the world. Other factors, such as distance to markets, time, conditions of doing business, political, environmental, regulatory etc., risk and other factors still need to be taken into account. None the less, the basic argument of comparing costs of production instead of being producer bound by in-house processes still holds. A counterargument may hold in terms of proximity to local markets that may be stronger than scale effects and geographic location of production in terms of the ability to cater and adapt to local needs.

[13] Though, surely ideas have become more amorphous, certain rigidities, certain externalities may still be locally bound to a certain degree. Especially, the generation of ideas may exhibit co-location externalities, both with other inventors, but also, in many cases with producers (more exposure to the process, deeper insights, etc.). For example, Glaeser identifies this as a distinct advantage of cities. "As the world has become flatter, the ability to export new ideas has grown and

importance, associated value and political will generated, more focus will be placed on ideas and knowledge in general. Faster diffusion and application of ideas will be fostered, demanded and sought after. Higher growth will be expected. While it may appear all will want to focus on ideas, it seems likely some economies will specialize more on idea generation while others may focus more on implementation and/or production.

Socioeconomic Impact The *Idea Economy* goes beyond confined economic impact. It undermines and shifts the socioeconomic status quo. With the widened and more active source of ideas, the velocity of innovation is likely to increase, and with it, provided ideas originate from diverse sources, increase the velocity of wealth turnover. Innovation is no longer confined to the privileged few who combine an idea with implementation capacity and/or access, but realistic for all. It offers a more meritocratic element. It may not be egalitarian as it still places additional premium on exposure, on education, understanding of principles, or expertise, but it is relatively more permeable than before. Provided that anyone can now reasonably profit from their idea, this will generate hope and prospects for anyone to advance and benefit from their ingenuity, fostering a spirit of possibility and endeavour. Inventors are the new heroes.[14] Ideas will be seen as means of economic and social mobility and thus much more focus will be placed on generating them. This will be especially pronounced in political debate on fostering access and equal opportunity to participate in such a changing environment. Creativity is still little explored and means of fostering it and will be given more weight. Hopefully, with more attention, better interventions become available beyond a simplistic focus on throwing money at pre-identified innovation potentials.

This sounds alluring. It promises many attractive features and achievements. But such developments may not be without drawbacks and risks that may threaten and undermine its development and success. This includes procedural aspects facing the ones with ideas and the ones realising them, as well as conceptual challenges that threaten the entire approach:

Process As ideas are increasingly assigned value, they may become more guarded. Less openness, less sharing, certainly less creative altruism in submitting ideas will be the consequence. Customers will provide less feedback, share less freely with producers. The same holds for suppliers. More secluded development of ideas ensues, possibly also diminishing the potential of cross-pollination and open

---

the returns to innovation have increased as well. As the returns to innovation increase, the returns to locating in urban places that specialize in innovation also increase. As the world has become flatter, the returns to becoming the smartest person in the world increase, and you can only become the smartest person by being close to other smart people". (Glaeser 2009, p. 27; also see Glaeser 2011 and Landry 2000).

[14] Also see Florida 2002 on the rise of the 'creative class'. Those realizing the ideas, firms and entrepreneurs are more at risk of becoming the new villains, robber barons, impeding the realization of great ideas, of lacking the vision and willingness to implement anyone's ideas. Also see Horx 2011.

discussion of ideas. Ideas will be pressured to be realised faster, without maturing and refinement. Attempts will be made to internalize externalities, limiting exposure to others. Almost worse, as the search for ideas becomes all the more rewarded and glorified, some sort of cognitive dissonance may occur, suddenly spotting ideas everywhere, also where there are none. This may flood and overwhelm the recipient and/or make cooperative process more costly. Though better screening and more advanced services may emerge to cope with such likely initial surge, the effects it has on the individual are not to be underestimated. Clinging to ideas, angling from one idea to the next can be highly problematic.[15] Hope can be exhilarating, but it can be daunting just as well. With fully dispersed access to ideas and remuneration, a risk of a lotto-mentality arises. A prevailing faith it fate, karma, providence, and the hope for chance and serendipity may foster an undue confidence in 'striking it rich' from coming up with a 'game changing' idea, at the costs of neglecting other things, neglecting more conventional ways to prosperity, neglecting the often laborious task of coming up, developing and exploring ideas – the risk of ideas as manna from heaven. This may be particularly pronounced especially with those still relatively most disadvantaged.[16] Immense conflict potential arises in connection with the generation and processing of ideas. Many questions of fairness arise, especially in terms of complementarities and causal attribution of idea generation. Building on someone else's ideas, applying it to other circumstances, etc. all call for a clear understanding of fairness. Who should be the beneficiary? To what extent does this warrant participation? Cause and effect may not always be clear, and even where they are, ambiguity remains. New questions of fairness arise requiring new frameworks. And even when these become more established, a considerable conflict residual lingers. As the focus shifts more on internalising externalities to maximise profit, precisely ideas with still high externalities, particularly more fundamental ideas, are neglected. This may risk broader progress, and may warrant more policy attention to spur the growth of ideas. Incrementalism is a risk as the immediate value is more obvious. Basic research becomes more and more of a global public good as its impact becomes more valuable, but less location bound. New approaches, new policies will be needed.

Concept As the velocity of innovation increases the lifecycle of products will shorten. This may have adverse effects as it may have limits. The turnover of new ideas is contingent on the willingness of consumers to be venturesome. This may be naturally limited by the opportunity costs of keeping existing products. Constantly

---

[15] See, for example, Lowe and Ziedonis 2006 or Asterbro et al. 2007 on inventors overconfidence – especially when they have invested time and money (also see Asterbro 2003). The data of course is biased as it is based on a low success rate and on a self-selection bias. With better means, more cooperative innovation and more optimism may be warranted, and also less overoptimistic people may pursue their ideas. Still, the risk of overconfidence (especially when ideas become more valuable) can create new challenges.

[16] See research on the regressive nature of lotteries and gambling behaviour (i.e. demand for lottery tickets exhibit low income elasticities). See Kitchen and Powells 1991; Farrell and Walker 1999. For an interesting extension, see Oster 2004.

upgrading to new products may also have learning and adaptation costs overburdening the user. This may cap or slow down the turnover of certain ideas. The same may hold for production. Though its nature may change and become more adaptable and upgradable, it often incurs sunk costs. Still turnover is likely to increase. This also incurs environmental costs of waste and use of scarce resources. A likely trade-off between the two will further constrain the full advancement of ideas (though may itself offer a vast emerging field and demand for smart ideas). More challenging than such inherent constraints is the risk of severe pushback, questioning and attacking the very fundamentals of the *Idea Economy*. The sheer notion of ideas as commodities may be denounced as "commoditization of intellect" and be seen as going too far. First commoditizing leisure, etc. now ideas.[17] With the exuberance of the prospects of ideas, also comes the risk of a backlash. This may be in form of a conflict between idea-generation and implementation about primacy, but also between idea haves and have-nots and the adverse effect an increasing pace of innovation may have on the economy, the environment, and peoples' lives. Ideas may not be perceived as hard work or deserving, seen as neglecting and under-appreciating the laborious task of implementation as a key ingredient. A potential conflict for remuneration, recognition and social status looms. Idea generation may appear, and often indeed may be decoupled from input intensity. Some stumble across ideas with ease and little effort while some labour tirelessly without success. Especially the disruptive nature of general access sand entry into previously closed professional streams will, and already has generated considerable pushback. Already today, via crowdsourcing anyone can come up with a design, a slogan, or sell their photos, without being a professional designer, advertising specialist or photographer.[18] This is often perceived as a threat by the respective community. This will intensify severely as ubiquitous competition not only within a professional stream but from anywhere will increase. Envy ensues as no clear factor of deserving can be identified than seemingly luck or predisposition. This may also include that those fortunate to gain from ideas may be seen as a threat to the status quo, as imposters, as 'newly rich'. On the other hand, if precursors exist, the effect may still be the same. Though the generation of ideas may be more meritocratic, to a certain degree, it may still – and possibly more factors will be identified – be influenced by certain facilitating factors. Such facilitating factors are advantages that will be exploited by those with means, undermining the meritocratic nature, sustaining, rather than undermining the status quo, thereby self-sustaining an idea plutocracy. Thus if the generation of ideas is dependent on specific attributes such as education, access to knowledge, facilities,

---

[17] This can be a substantial threat. As many examples show, social attitudes matter greatly. The best example, that demonstrates such, at least from an economists point of view, startling and counterintuitive effect is that of blood donation (e.g. see Titmuss 1971). Also see discussion in Gans and Stern 2010.

[18] For a more details on these and more examples, see Howe 2009 (on istockphoto.com, crowdspring.com, threadless.com, etc.).

etc., the resulting idea-gap may breed conflict. All such developments are valid concerns and may breed powerful opposition to the evolution towards and deepening of the *Idea Economy*. This also has another facet. The rising importance, prominence and value of ideas occur at the costs of other elements, predominantly labour inputs. Not only does it loose relative importance and thus its share of the overall output, but innovation may, and typically indeed often does entail also negative economic effects. Unemployment, especially skill biased technological change affects those with the least skills, often also the ones least flexible to adapt to increasingly changing economic environments. Companies are no safe havens, but far less even than before. Job turnover will increase and flexibility will be the norm. But it can also affect others. Sectoral shift will become more frequent. Capital owners will have to bear more risks for relatively less reward. Wealth turnover will accelerate. The pressure to perform in such an environment increases. Complacency is punished. People are expected to perform despite the genuinely unpredictable nature of ideas. The risk of inventing will have to be borne more and more by the inventors, rather than the firms. Employment even for the best and the brightest will become more outcome-based, less steady and less reliable. Performance becomes relative and less absolute. Having an idea before someone else creates value. Having it second is not good enough. The 'economics of superstars' will dominate where the winner takes almost all.[19] This may lead to macroeconomic inefficiencies, with too many focussing on similar high profile inventions, 'squandering' economic resources.[20] The most severe threat to the *Idea Economy* however is a different one: Its own success. As corporations become more and more open and the entrepreneurial approach is less and less used, it may lose its viability, and thus undermine the credibility of a threat of disruptive innovation and alternatives to work around corporate knowledge filters. This would be a threat to the *Idea Economy* as a whole and a risk of gradually returning to a more closed innovation model. Though alternative paths open up that are likely more efficient, the entrepreneurial one is utmost vital to maintain a credible threat to any not fully open corporate innovation paths. Only if entrepreneurship remains an option, the easier the better, does it keep cooperative innovation honest and efficient. Should entrepreneurship lose its appeal, become more demanding, less used, etc. this would risk shifting power back to internal innovation, closed and more path dependent innovation, gradually undermining cooperative and open innovation. It would again confine innovation, reduce its effectiveness, and threaten the *Idea Economy*.

Despite these risks, the *Idea Economy* is alluring – anyone having a fair chance of realising ideas, more and better innovations, growth, progress and prosperity. It better utilizes the skills of all through an increasing division of labour, making also the realisation process and production more efficient. The potential of such developments is vast, and its prospects appealing. And yet, though meandering

---

[19] With small differences in talent at the top of the scale, or in this case small advantages of timing and strategy of the idea, will reap disproportionately large returns (see Rosen 1981).

[20] See for example discussion in Scotchmer 2004.

towards it, it has not yet materialized. Much progress has been made towards it, but several issues remain unresolved. The generation of ideas is being increasingly distributed, broad based, and democratized. Ideas can come from anywhere and anyone. At the other end, a competitive market for different implementation path is increasingly opening up, ranging from the lone entrepreneurial path, to cooperative innovation, to trading the idea to a corporation to realise it. For now, this is but a faint trend. Admittedly, it is an increasingly colourful one, with fascinating developments and new possibilities, but a trend, and/or an extrapolation of such a trend, nevertheless.[21] The *Idea Economy* has not fully unfolded yet. It has not yet arrived. There is still a shortfall: how best to connect the ones with idea to the ones able to realize them? True competition for implementation needs an open market for ideas. Anyone needs to be able to pursue any path. This is not the case yet. The missing link: connecting those with ideas to those able to realize them – being able to actually trade ideas. This is still not fully possible. Early beginnings exist and some approaches are making progress. But the innovation process is still not open to most – and for good reasons. Some highly challenging issues exist in bridging this gap. Without robust solutions, it may postpone and hinder a full unfolding of the *Idea Economy*. Identifying and addressing these issues is subject of Part II.

References

Arora, A., & Gambardella, A. (1994). The changing technology of technological change: General and abstract knowledge and the division of innovative labour. *Research Policy, 23*(5), 523–532.
Arora, A., & Gambardella, A. (2010). Ideas for rent: An overview of markets for technology. *Industrial and Corporate Change, 19*(3), 775–803.
Arora, A., & Merges, R. P. (2004). Specialized supply firms, property rights and firm boundaries. *Industrial and Corporate Change, 13*(3), 451–475.
Asterbro, T. (2003). The return to independent invention: Evidence of unrealistic optimism, risk seeking or skewness loving? *The Economic Journal, 113*(484), 226–239.
Asterbro, T., Jeffrey, S. A., & Adomdza, G. K. (2007). Inventor perseverance after being told to quit: The role of cognitive biases. *Journal of Behavioral Decision Making, 20*(3), 253–272.
Chesbrough, H. W. (2003a). *Open innovation: The new imperative for creating and profiting from technology*. Cambridge: Harvard Business School Press.
Dyer, J. H., Gregersen, H. B., & Christensen, C. M. (2009). The innovator's DNA. *Harvard Business Review, 87*(12), 61–67.
Farrell, L., & Walker, I. (1999). The welfare effects of lotto: Evidence from the U.K. *Journal of Public Economics, 72*(1), 99–120.

---

[21] This may initially vary in the different sectors, depending on the nature of production, the market competition and resource intensity of research, but as experience has shown, no market may be fully excluded – it can profit still from a range of ideas, and may mostly be contestable and disrupted. (e.g. IBM – see Dyer et al. 2009). As long as a range of viable paths exists, a pressure to compete for ideas exists. (To be precise: the pressure for incumbents exists to compete with each other and to compete with the alternative of the entrepreneurial (be it alone or cooperative) innovation path.)

Florida, R. (2002). *The rise of the creative class: And how it's transforming work, leisure community and everyday life*. New York: Basic Books.

Friedman, T. L. (2005). *The world is flat. A brief history of the twenty-first century*. New York: Farrar, Straus and Giroux.

Gans, J. S., & Stern, S. (2010). Is there a market for ideas? *Industrial and Corporate Change, 19*(3), 805–837.

Glaeser, E. L. (2009). The death and life of cities. In R. P. Inman (Ed.), *Making cities work – prospects and policies for Urban America* (pp. 22–62). Princeton: Princeton University Press.

Glaeser, E. L. (2011). *Triumph of the city. How our greatest invention makes us richer, smarter, greener, healthier, and happier*. New York: Penguin Press.

Hidalgo, C. A. (2009). *The dynamics of economic complexity and the product space over a 42 year period. CID working paper no 189*. www.hks.harvard.edu/centers/cid/publications/faculty-working-papers/cid-working-paper-no.-189. Accessed 02 March 2013.

Hidalgo, C. A., & Hausmann, R. (2009). The building blocks of economic complexity. *Proceedings of the National Academy of Sciences of the United States of America, 106*(26), 10570–10575.

Hidalgo, C. A., Klinger, B., Barabasi, A. L., & Hausmann, R. (2007). The product space conditions the development of nations. *Science, 317*(5837), 482–487.

Horx, M. (2011). *Das Megatrend-Prinzip: Wie die welt von morgen entsteht*. Muenchen: Deutsche Verlags-Anstalt.

Howe, J. (2009). *Crowdsourcing: Why the power of the crowd is driving the future of business*. New York: Random House.

Kitchen, H., & Powells, S. (1991). Lottery expenditures in Canada: A regional analysis of determinants and incidence. *Applied Economics, 23*(12), 1845–1852.

Landry, C. (2000). *The creative city: A toolkit for urban innovators*. London: Earthscan Publications.

Lowe, R. A., & Ziedonis, A. A. (2006). Overoptimism and the performance of entrepreneurial firms. *Management Science, 52*(2), 173–186.

Oster, E. (2004). Are all lotteries regressive? Evidence from the powerball. *National Tax Journal, 57*(2), 179–187.

Plehn-Dujowich, J. M., Serfes, K., & Thiele, V. (2010). *Competing for entrepreneurial ideas: Matching and contracting in the venture capital market. Searle center on law, regulation, and economic growth working paper no 2010–031*, Northwestern University, Chicago. www.economie.uqam.ca/pages/docs/Thiele_Veikko.pdf. Accessed 02 March 2013.

Read, L. E. (1958). *I, pencil. My family tree as told to Leonard E. Read*. www.fee.org/the_freeman/detail/i-pencil/#axzz2HPCcc6O2. Accessed 02 March 2013.

Rosen, S. (1981). The economics of superstars. *The American Economic Review, 71*(5), 845–858.

Schumpeter, J. A. (1934). *Theory of economic development*. New Brunswick: Transaction Publishers. Reprint 2004.

Scotchmer, S. (2004). *Innovation and incentives*. Cambridge: MIT Press.

Shane, S., & Eckhardt, J. (2003a). Opportunities and entrepreneurship. *Journal of Management, 29*(3), 333–349.

Silvera, R., & Wright, R. (2010). Search and the market for ideas. *Journal of Economic Theory, 145*(4), 1550–1573.

Sørensen, M. (2007). How smart is smart money? A two-sided matching model of venture capital. *Journal of Finance, 62*(6), 2725–2762.

Titmuss, R. (1971). The gift of blood. *Society, 8*(3), 18–26.

Ueda, M. (2004). Banks versus venture capital: Project evaluation, screening, and expropriation. *Journal of Finance, 59*(2), 601–621.

More and Better Innovation 6

What will be the impact of more cooperative innovation? The implications are far-reaching. Innovation will change. There will be more and better innovation. Ideas will be used better as anyone can now see their ideas realised. Ideas will be realized by professionals with a better chance to succeed. Ideas will be traded, as ideas themselves will be valuable. Innovation becomes faster.

The nature of innovation has changed. Innovation has, and continues to evolve. What was once a mere beginning is becoming the essence of innovation, the true value: the idea. The process, whether it gets realized by a firm or an aspiring entrepreneur, no longer characterizes innovation, but is becoming an afterthought of how best profit from the idea. Ideas now define innovation. The realisation path optimizes its value. This, in short, is what the *Idea Economy* represents. Innovation through ideas. Not innovation as a process. The path is secondary. Ideas are key.

This is nothing surprising. This is nothing new – no epiphany, no sudden occurrence. Innovation never has been a static process. It has evolved from a predominantly corporate, to one characterized by entrepreneurship, to lately an increasingly open innovation process. This has partly been driven by a changing environment but also as a natural evolution of the process itself. The corporate innovation process was too narrow. Entrepreneurship was the alternative. In nascent sectors, even a necessity. It opened up new prospects, new chances, and indeed has contributed to prosperity by enabling more ideas to be realized. Entrepreneurship became glorified when corporate behemoths were struggling to keep up. Now firms are being forced to adapt – and still are struggling to find a good model to keep up and innovate as quickly and openly as entrepreneurship promises to do. Innovate or perish. Firms can be as good, if not often better a partner in realizing ideas. Companies are becoming more and more open to ideas from the outside, seeking to cooperate with inventors rather than be threatened by new ventures. Firms adapt. The economy adapts. Innovation adapts. Innovation evolves. It changes. A new age of innovation emerges (see Fig. 6.1).[1] Ideas are

---

[1] With a kind reference to Prahalad and Krishnan (2008).

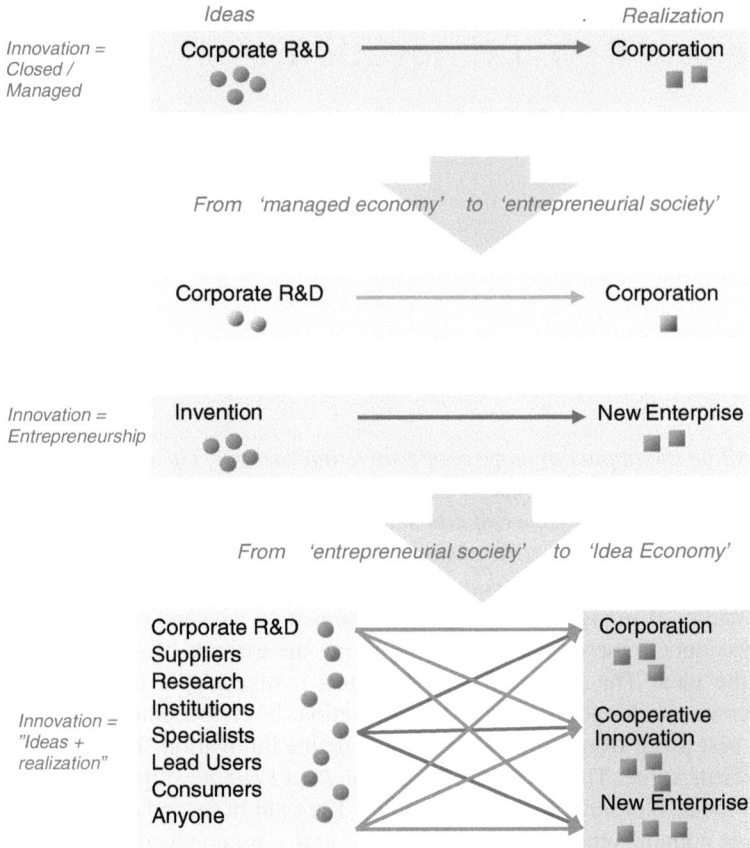

Fig. 6.1 The evolution of innovation: from the managed to the Idea Economy

increasingly disentangled from a seemingly predetermined realisation path. Not all ideas in a firm are realised there or shelved, not all ideas outside need to be realised via entrepreneurship. If it becomes possible to freely bring ideas together with such existing firms or eager established entrepreneurs, this opens up an array of new realization paths. It leads to a new dynamic, and enables something new: cooperative innovation. Ideas are valuable now in their own right. Anyone can realise them: alone, or more likely in cooperation with entrepreneurs, experts, or firms. Ideas may even be traded to those best suited to realise them. A division of labour now characterizes innovation. Some invent. Others realize. All involved benefit from it. This is the *Idea Economy*. It is on its way. It is becoming reality. To enable its arrival, to foster it, cooperation needs to be strengthened. Then ideas can be traded freely, cooperation can be fair; innovation can be promoted, and the *Idea Economy* can unfold.

This matters not for the sake of analysis, to draw a picture in broad strokes, to tell a story of how innovation evolves. It is more than a story of dialectics.[2] The entrepreneurial society as the famous thesis, soon to be challenged, challenged as a reaction to it, by the increasingly open innovation process of those it challenged, the corporations; its culmination in a synthesis, a setting where its former essential feature, the realisation path, becomes secondary, if not negligible: the *Idea Economy*. As intriguing as such dialectic is it is not essential. The implications are. We use stereotypes to conceptualize. We use concepts to engage with the world. We stereotype in how we approach innovation, what prospects we see, what we think we can achieve. We use such concepts to construct policy, and gear our attention towards it. All too often, we still equate innovation to entrepreneurship – for good reason. We infer from success, but fail to notice wasted potential. We observe the road more travelled. We derive policies from our preconceptions; we set up support structures and dedicate vast sums of money to encourage entrepreneurship. We fall short, neglecting what could be. Entrepreneurship is just one aspect of innovation, one of many paths to realize ideas. New concepts are emerging, offering unprecedented opportunities, changing the very nature of innovation and how each and every one can and will participate.

The emerging *Idea Economy* will change the way we innovate, and change the very process of innovation per se. Its impact will be significant. In many ways, it represents a major expansion of innovation processes, better capturing and better utilising the innovation potential. In a theoretical jargon, this could be phrased as: (i) improving the effectiveness and return of ideas, and (ii) a vast expansion of access and utilization of the potential idea basis. Put differently: the existing ideas are realised better, and more ideas are realised – better and more innovation.

Better innovation implies two things: The first is the impact of innovation. The prevalent innovation landscape is largely path dependent. Its origin determines its prospects. In a firm, the realisation path is clear. The firm realises the idea, or shelves it. For ideas outside a firm, if feasible, it means entrepreneurship. These need not be the most efficient paths. Though successful, someone may not be the most effective entrepreneur or best suited firm to realise the idea. Alternative paths could have been far superior. An incumbent could have brought a product to market much faster, more cost effective leveraging capacity, expertise, and market penetration. An able serial entrepreneur and market veteran could be a much better partner than trying to try ones luck in teaching oneself business 101. Now these are available. Ideas will be realised by those best suited. Ideas are realised more

---

[2] The open innovation approach rather styles itself as the antithesis to closed innovation (See Chesbrough 2006b, p. 1). This to some extent is true, and seems to hold from a very narrow firm perspective, but lacks the dynamic effect and the bigger picture of innovation (that is not in the sense of corporate innovation, but rather in terms of innovation per se). Only the changes in the socioeconomic fabric, and subsequent threat of the entrepreneurial path, have necessitated such a move. It is rather a reaction to the emergence of the entrepreneurial society than to the closed innovation. Entrepreneurship can rather be seen as the antithesis to the managed innovation process, and in turn, open innovation the reaction to the entrepreneurship.

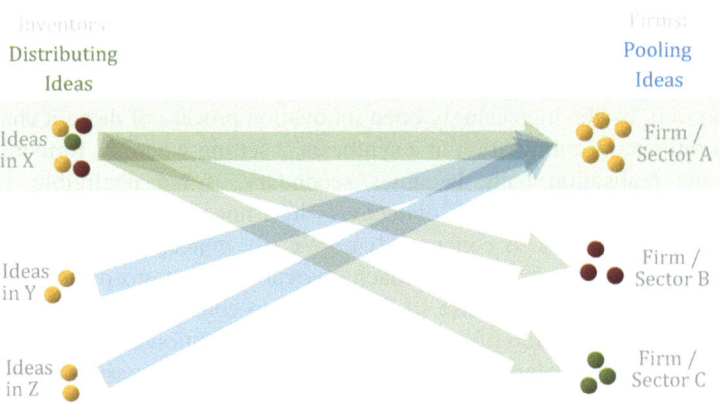

Fig. 6.2 Making the most of an idea: applying ideas across firms and industries

effective and efficiently, increasing the impact of innovation. The second aspect is the scope of realisation. It is more complicated. The previous, and to a large extent still the existing realisation path is linear. It is mostly narrowly defined along existing organization and/or industrial demarcations. Especially in still vertically integrated industrial structures, ideas are underexploited. Low degrees of cross-pollination, little exploitation of ideas in other segments, etc. remains the norm. The same holds of course the other way around. Ideas are taken up gradually by firms, confined often to internal idea generation. Path dependency is the norm. By disentangling the two – ideas and the realisation path – ideas can be used in different firms and sectors. An idea X that would previously have been realized by its corresponding firm in sector A, can now also be applied to other sectors, sectors the firm may not be active in (see Fig. 6.2: Distributing Ideas to not only Sector A, but also B and C). The value of the idea increases with a broader use and higher uptake. For the one with the idea, this means internalising what often used to be externalities, what often was only absorbed with a lag, or sometimes not at all. Ideas are now utilized more effectively by exploring the spectrum of options, the different areas where it can generate value – and appropriate some of it. In other words, a firm realizing an idea can now also sell it to other companies in different sectors, sectors it could never enter itself and now make a profit from applying the idea also there. This maximises profit for the ones with ideas, and better utilizes ideas in an economy.

Similarly holds in effect for those implementing ideas. A firm could buy ideas from other sectors and combine several ideas, not just ideas that come up in its own sector. The firm active in sector A would not have had access to ideas from other sectors. Now that it has, it can realize more ideas in its sector (see Fig. 6.2: Pooling ideas, drawing on not only idea X, but also Y and Z). This means absorbing relevant ideas, pooling ideas beyond internal confinement, seeming sectoral knowledge path dependence, and so on, thus making better use of their expertise and resources.

Consider Play-Doh, the colourful modelling compound used by kids. It was developed by a soap manufacturing company, the Kutol Products Company, as wallpaper cleaner. Its use as a toy is just as, if not far more lucrative. Expanding the soap manufacturer into toys is an unlikely move. Rather a spinoff (as indeed was the case with the Rainbow Crafts Company), or by cooperating with an established toymaker could much better leverage the value of such invention.[3] Thus by using the invention in several sectors, its value can be much better appropriated. The more cooperation is available, the more effective such inventions will be used, the more effective innovation will become. Play-Doh was indeed later bought by Hasbro, a giant toymaker. Hasbro may never have come up with the idea, nor most likely would it have thought of collaborating with a soap maker to discover new toys. Having open access to ideas from the outside and using them in their field of expertise makes better use of the idea for the ones inventing and better use of the resources and expertise of the ones implementing them.

Similarly, the story of Mike Henderson, who started a firm called SmartTruck to develop and install wind-deflecting devices on trucks to make them more aerodynamically efficient. "The UnderTray System can easily be added to an existing trailer or included with new or refurbished trailers"[4] It can help realize "16 % reduction in drag and 10 % improvement in fuel milage".[5] With estimated 1.3 m large trucks in America alone, "the Department of Energy estimates that if all the semis in America had such devices installed it would produce fuel savings of 1.5 billion gallons of diesel a year, worth about $6 billion at the current diesel price of $4 a gallon."[6] This is all the more relevant as the idea comes from a seemingly unrelated industry. "Before he started the firm, Mr Henderson ran a Boeing research unit that investigated aircraft aerodynamics using elaborate computer models"[7] He applied his ideas to trucks. While the aerospace industry might not have any use for or competence in establishing truck parts, Mr Henderson saw an opportunity and set up a SmartTruck – distributing ideas across industries. Boing seems an unlikely manufacturer of car parts. Fortunately, Mr Henderson not only had a great idea, but also the means and skill to become an entrepreneur. Were he able to trade ideas he could just as well have sold it to or cooperated with large truck manufacturers – pooling ideas from across different sectors and firms.

Linear innovation neglects all linkages from the idea to the outside, and from the outside that could contribute internally. The opening up of innovation through entrepreneurship and more open firms allows capturing some of this potential. Further opening up can leverage it even more. Truly disentangling ideas from the realization

---

[3] See ohiohistorycentral.org/entry.php?rec=2623 (02.03.2013).
[4] The Economist (2011b).
[5] smarttrucksystems.com/brochure 02.03.2013.
[6] The Economist (2011b).
[7] The Economist (2011b).

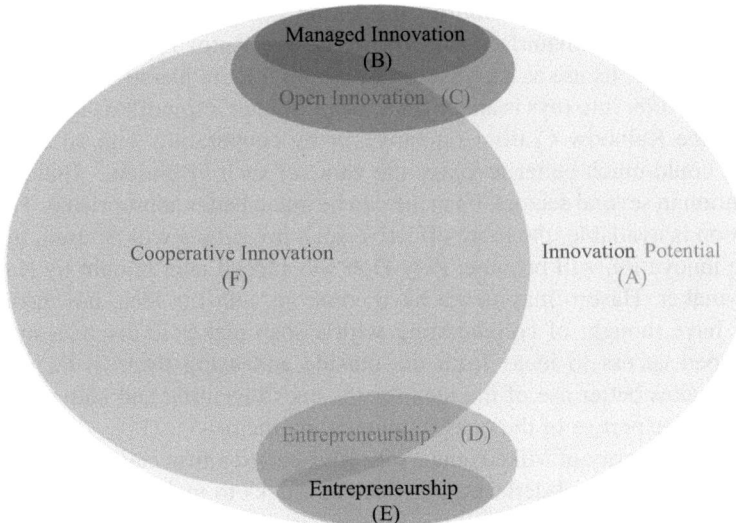

Fig. 6.3 Making better use of the innovation potential: realizing anyone's ideas

process can fully leverage this potential through far-reaching internalization.[8] Especially in the context of an increasingly vertically disintegrated value chain and a more modular production, both the firm and the inventor can better internalize such externalities. The two developments are mutually reinforcing. Disintegration of production chains enables disentangling of ideas and its realization; disentangling fosters disintegration.

More important, possibly, is the increase in the utilization of ideas – more ideas are realized (see Fig. 6.3). Many ideas, previously neglected ideas, now can be realized. No longer do existing realisation paths determine innovation possibilities. The realisation of ideas becomes open to all, in all kinds of ways. Inventions no longer have to be shelved if not by chance thought of in just the right firm, confining researches and firms alike; independent inventions do not necessarily have to be realized through entrepreneurship as the only outlet – irrespective of the skills and means of the inventor to become an entrepreneur. Firms are starting to open up more and more. They start allowing ideas to be spun off, and somewhat allow for spin-in of ideas. The Open Innovation mantra has succeeded in putting it on the corporate map, allowing for a better use of ideas (C). Many efforts have gone into creating a more conducive enabling environment and support system for

---

[8] Appropriability increases – overall, through a better use of ideas, and for the inventor, through better bargaining position (The best available estimate, though to be viewed critically, which may at least hint at the under-appropriation of ideas see Nordhaus 2004). This, of course, assumes these opportunities are exploited. This is not trivial. Many opportunities may exist, but might not easily be discovered (see Shane 2000). Likely, this will improve with more available support services and more transparent market opportunities.

entrepreneurs. The infatuation with entrepreneurship has borne fruit – though falling way short of its presumptuous claim. This has allowed for more and more people to pursue their luck as an entrepreneur and realize their idea (E). Both approaches complement rather than really replace existing innovation mechanisms. They expand and refine existing approaches. This allowed for more innovation. Both developments have expanded the use of the potential innovation pool.

Still, both fall short of a huge proportion of potential innovations. Ideas are still seriously under-realized (A). Even easier access to entrepreneurship still role-forces often ill-equipped inventors to pursue entrepreneurial endeavours, and Open Innovation remains mostly IP focussed.[9] Cooperative innovation can vastly expand this pool, realizing more ideas, and realizing them more effective and efficiently (F). This will alter existing structures and possibly absorb some parts of both managed innovation and entrepreneurship. Managed innovation will have to open up further, and in many parts, it will be replaced with a vertically more disentangled innovation structure. Ideas increasingly come from outside. There will be less internal R&D spending in the traditional sense. More cooperation will be the norm. Firms will have to adapt. Similar holds for entrepreneurship. To many, though ill equipped, it is currently the only alternative. It is inefficient. Cooperative approaches will replace many such futile or inefficient endeavours. Entrepreneurship will be an option, but likely alternative paths will often be preferred. It was often a choice out of desperation, out of a lack of alternatives. Now, more specialized and experienced entrepreneurs will emerge cooperating in the realization of ideas instead of all with ideas attempting their luck as entrepreneurs; firms will compete for ideas offering much better terms of realization than ill-fated attempts to make entrepreneurs out of everyone.

Together both effects – more ideas are realized, and ideas are realized more effectively – promise enormous economic gain. In the managed economy, good ideas outside the relevant corporation had nowhere to go. Entrepreneurship relieved this pressure to some extent opening up a presumed floodgate of ideas to be realized. And true enough, more can now be realized, but many barriers still exist. Many are unable or unwilling to become entrepreneurs; many ideas cannot be realized via the entrepreneurial path. Open innovation reduces some of these barriers. Some. For a certain range of ideas, it becomes possible to cooperate with firms, and for some internal ideas, it becomes possible to pursue the realization outside the corporation or license out such inventions. Cooperative Innovation promises to reduce these berries still further, enabling the realization of an immensely underutilized innovation potential. And these ideas are realized more successfully through a better division of labour.[10] Together the impact can be significant: truly promoting innovation.

---

[9] See Chesbrough (2006a), as well as Dahlander and Gann (2010). Also Zeckhauser (1996).

[10] To be fair, Chesbrough did hint at this development: "Hidden among these worrisome trends are other developments that are perhaps more hopeful for the future of innovation. One such development is the growing division of innovation labor. By a "division of innovation labor." I mean a system where one party develops a novel idea but does not carry this idea to market itself. Instead, that party partners with or sells the idea to another party, and this latter party carries the idea to market. This new division of labor is driving a new organizational model of innovation, one that may offer more hopeful

To reflect these significant changes, the dominant characterization of innovation in terms of the entrepreneurial society seems to fall considerably short. It captures but one aspect – an aspect that is losing ground. Innovation can no longer, if ever if could, be simply equated with entrepreneurship. It was convenient a formula while it lasted. But this simple mental abode is crumbling. A broader picture is needed to capture innovation; a broader picture is needed to conceptualize it in order to address it adequately. Open innovation is an important extension to this. But Open Innovation is portrayed as the corporate response and a particular business mantra rather than a socioeconomic setting. Analytically, the concept of the *Idea Economy* provides for a more encompassing characterization of the socio-economic situation than either one approach may offer, still including both corporate as well as entrepreneurial innovation. It better captures the essential features of the innovation process. No longer is innovation segmented by the different realisation paths. These are increasingly indistinct and blurred into each other. The defining features are the steps within the process itself. What seemed an intangible unity of idea and realization path (corporate innovation, entrepreneurship) can now be broken up. Innovation is increasingly differentiated horizontally (origin of ideas and the means of realization) instead of vertically (by different innovation paths). Ideas are the key. Not the path to its realisation. This matters not for its own sake, not for dialectic and etymological vanities. It matters for the way we approach innovation, we conceive it, we discuss and politicise it; the way we innovate; the way we devise policies; the manner in which we bring it about; the manner we utilize and cope with it; the chances we get, and the opportunities we can seize. We act on such concepts. Too narrow a concept means too narrow our approach, too narrow our policies, to narrow our perceived choices. This misses out on a large innovation potential. It misses out on prosperity.

This should not diminish the merits or belittle the significance of entrepreneurship. Entrepreneurship is and remains an integral component. It may lose its glamour of a seemingly necessary entanglement between ideas and its realisation. But without a strong such alternative path, progress towards and the deepening of the *Idea Economy* would be at risk. Entrepreneurship still plays a critical role to challenge rigidities, and offers an effective alternative to other innovation processes. While it may no longer be the only general choice, it needs to remain a viable threat by offering an alternative realization path. The more available, the more efficient and effective the entrepreneurial approach is, the higher pressure it exerts on corporate or cooperative innovation paths. Any path remains contestable. And yet, despite its vital importance, the concept itself loses prominence. It no longer characterizes the economy. Ideas do. The Entrepreneurial Society is evolving towards an *Idea Economy*. Ideas gain importance. They, rather than the manner of realisation are credited for the innovation. Ideas are the new mantra; generating ideas the new objective. Realisation is a service. Entrepreneurship is but one of them.

---

prospects for innovation in the future." Chesbrough (2006a, p. 2). Unfortunately, he viewed this only from a corporate perspective, and only in connection with IP. Similarly, Silvera and Wright (2010).

References

Chesbrough, H. W. (2006a). *Open business models: How to thrive in the new innovation landscape*. Cambridge: Harvard Business School Press.

Chesbrough, H. W. (2006b). Open innovation: A new paradigm for understanding industrial innovation. In H. W. Chesbrough, W. Vanhaverbeke, & J. West (Eds.), *Open innovation: Researching a new paradigm* (pp. 1–12). New York: Oxford University Press.

Dahlander, L., & Gann, D. (2010). How open is innovation. *Research Policy, 39*(6), 699–709.

Nordhaus, W. D. (2004). *Schumpeterian profits in the American economy: Theory and measurement*. NBER working paper no 10433. www.nber.org/papers/w10433. Accessed 02 March 2013.

Prahalad, C. K., & Krishnan, M. S. (2008). *The new age of innovation: Driving co-created value through global networks*. New York: McGraw-Hill.

Shane, S. (2000). Prior knowledge and the discovery of entrepreneurial opportunities. *Organization Science, 11*(4), 448–469.

Silvera, R., & Wright, R. (2010). Search and the market for ideas. *Journal of Economic Theory, 145*(4), 1550–1573.

The Economist (2011b). *Rig on a roll. Transport: Computer modelling is being used to improve the airflow around big trucks and reduce their fuel consumption*. 2 Jun 2011.

Zeckhauser, R. (1996). The challenge of contracting for technological information. *Proceedings of the National Academy of Sciences of the United States of America (PNAS), 93*, 12743–12748.

Part III

Cooperative Innovation

Setting the Stage 7

Why is cooperation necessary? Many people have ideas. Mostly, unfortunately, they are not businessmen or those best able to realize ideas in the market. Bringing the two together – ideas and companies – and getting them to cooperate could seriously foster innovation and offer many more people the chance to benefit from their ideas. A division of labour seems the most promising way to give everyone a better chance to succeed and innovation to flourish. Everyone could do what they do best, invent or realize – all too rarely do they coincide.

Ugly! Embarrassing! Such and more, reportedly, were many men's reactions. Strollers had unfortunate colours with little teddy bear prints, or other cute and cuddly creatures, to hallmark the adorability of the toddler inside. "The image didn't fit: it was like seeing a guy in an apron beating a doormat."[1] Men felt awkwardly wrong to push their infant along in public, just as, for sure, did many fashion conscious women. Those were the strollers of old. It all changed in 1999. Max Barenbrug had neither a child then nor any affinity to pushchairs. He was a design student at the Academy of Industrial Design in Eindhoven, The Netherlands. Given the appalling state of stroller design, he made it his graduation project, redesigning a stroller parents would be proud – and grateful for its convenience – to push around the city. He graduated with honours for designing a pushchair that was both more convenient, and definitely more stylish than what was around. He approached many of the established stroller manufacturers. No one wanted it. They ignored him or rejected the design as 'too radical'. Fortunate for him he knew Eduard Zanen, his then brother in law, a physician with a knack for business, offering him financial backing and emboldening perseverance. Lacking other options, they assembled the first stroller in an attic room in Amsterdam. In 1999, at a fair in Germany, they launched their product. They set up a company, curiously named 'Bugaboo', and gradually expanded from the Netherlands to neighbouring markets, assisted by word of mouth advertising. Their clever American distributer

---

[1] See interview with Max Barenbrug in the Boo (2012).

managed to feature the stroller on *Sex and the City* – enough said. By 2009, Bugaboo had sold more than 500,000 strollers with a presence in over 50 countries. It now is one of the leading manufacturers for pushchairs for infants and toddlers. The still privately owned company of Mr. Barenbrug and Mr. Zanen runs its own factory in Taiwan, with an annual turnover of over $100 mio. and more than 800 employees. A marked success. More is surely to come.[2]

Two aspects here are striking. No one would have pegged the student Max Barenbrug to redefine a business segment. No one would have asked him to come up with a new stroller (indeed, none of the incumbents listened to him after he did). This unlikely candidate with no exposure to the stroller industry took the initiative. The second aspect: he could not have done it without his partner Eduard Zanen and later people such as Kari Boiler, their US representative – a serendipitous and utmost successful match. Neither one without the other would likely have been so successful. Mr Barenbrug himself claims, "I am the worst manager you can imagine."[3] A professional management team today runs the business side of Bugaboo.

Max Barenbrug is not alone. More often than not, someone has an idea, yet neither the talent nor means to realize it themselves. Others often have the skills and means to realize it most effective and efficiently. This raises serious questions. This means enormous potential is wasted. Too many ideas never stand a chance. Who are the people that come up with ideas and who are the best able to realize them? Why are they not the same? Who comes up with ideas? Why? Most of these questions are still unanswered or the debates surrounding them remain utterly unresolved. Can anyone come up with ideas or can certain features, traits, circumstances or even processes be identified that determine or encourage the generation of ideas? Can anyone have a flash of genius? Is it possible to train for it? Such questions certainly depend on the nature of ideas, the original character, the depth of insight, and so forth. It likely depends on very specific personal and environmental aspects. Much effort has gone in to trying to identify if, and if, what kind of underlying characteristic and circumstances may serve as antecedent for the generation of ideas.[4] Not to get lost in these on-going, more and more specialised debates, consider maybe two somewhat extreme ideal types that both seem quite reasonable propositions. One case is where really anyone can have a great idea, but only some are businessmen. It would be good for both to work together. The other case is where only some, the most dedicated and specialised,

---

[2] See Bugaboo Magazine (Boo 2012) and Financial Times (2008).

[3] See Financial Times (2008)

[4] For some overview and insightful discussions, see, for example, Root-Bernstein (1989) or Koellinger (2008). On the elaborate adjacent discussion on creativity, see Amabile (1983, 1996, 1998), also Sternberg (1988), Sternberg and Lubart (1995), Kaufman and Sternberg (2006), Csíkszentmihályi (1996), Pink (2005), Robertson (1984), Arthur (2007), Yusuf (2009). For an interesting discussion, also see Johnson (2010). Also Drucker (1986) and von Hippel (1988). For an Economics perspective, also see Galenson (2010), or Magee (2005).

come up with certain ideas. They may not, nor should they be the ones realizing them – both they and businessmen would benefit if only they could work together.[5]

In the first case, it could be hypothesised that ideas are distributed rather randomly throughout the population. Anyone can come up with a brilliant idea. This may include deliberate problem solving, but equally a 'flash of genius'. Anyone, without special training or focus can come up with the most brilliant ideas, anyone can discover, anyone can invent. Distinctive predicators of ideas are indiscernible.[6] This does not mean that he/she needs to be able to realize the idea, but 'merely' come up with the originating idea. For example: any avid shopper could have conceived something like Groupon without having to be a marketing expert or microeconomist – it was started by Andrew Mason, a web designer and policy major, who reportedly got inspired when having problem with his cell phone contract.[7]; any movie fan could come up with a concept for a video rental-by-mail such as Netflix without being a movie executive or video rental or logistics expert – this is indeed what Reed Hastings did, a software engineer, who came up with the idea after being charged a huge fee for returning a video, "Apollo 13", late.[8]; any coffee drinker could come up with the idea for a coffee cup sleeve (the paper based sleeve you wrap around your coffee to handle the hot paper cup) without being an expert cup engineer, a producer of cups or a coffee chain manager – as for example Jay Sorensen, one of the many patent holders of a lucrative coffee cup sleeve. At the time he was in real estate, and claims the idea came to him when he "dropped a cup of hot coffee in his lap because the paper cup was too hot"[9]; any student could envision the convenience and appeal of 'Facebook' without being a programmer, or Silicon Valley serial start-up specialist – as indeed was the case if you think of the original idea for HarvardConnection; any practical person could come up with the idea of a multi tool, without being a tool manufacturer – while on travel Timothy Leatherman needed better tools to repair the car and hotel plumbing and came up with the now commonplace Leatherman multi-tools[10]; and so on and so forth. Glance over illustrious publications such as Entrepreneur or FastCompany Magazines and you can find innumerous such examples of ordinary people coming up with highly innovative and lucrative ideas.[11] This is a highly democratic view on ideas, especially when thinking of ideas, less of technological gadgetry and frontier

---

[5] The use of ideal types is somewhat particular, and not without its challenges (see Weber 1949, pp. 80–81).

[6] For the generation of ideas in general this may hold. However, some research has been done into identifying predicators of alertness, and circumstances in which opportunities may arise (see, for example, Tang 2008; Foss and Klein 2010; Shane and Eckhardt 2003; Yu 2001; Gaglio and Katz 2001; Gaglio and Winter 2009).

[7] See Chicago Magazine (2010)

[8] See New York Times (2006) and USA Today (2006).

[9] See javajacket.com/about/ 02.03.2013

[10] See leatherman.com/about/history 02.03.2013

[11] See entrepreneur.com and fastcompany.com

science. Instead, think of a broader range of potentially lucrative ideas. But even high-tech ideas are becoming within reach to anyone with ubiquitous availability of cheaper computing and affordable small-scale research equipment. And indeed, what used to be the view that innovation comes mainly from highly specialised research facilities and corporate research and development is now increasingly shifting outside to the general public. The generation of ideas is becoming more democratic. A particular focus has been placed on users, which are argued to be "perhaps the *most* important developers of innovations".[12] Users invent: open source software, cooking recipes, car upgrades, and much, much more. Thus, it is not unreasonable to think of ideas as coming from anywhere. This also means that it becomes less and less likely that a suitable idea is generated within a particular firm able and willing to realise it. It is more likely it will arise somewhere else. Ideas may coincide with the firm looking for it, they may coincide with the gifted entrepreneur, but that is actually does, seems highly unlikely.

With such a broad basis of innovation, hardly any characteristics apply which indicate or even predetermine the ability to come up with a brilliant and most lucrative idea. But, if this does hold true, then apart from the appeal and hope that it could be anyone, including you, it has another more striking implication for the innovation process. If indeed ideas come from anywhere, it is unlikely to correspond with the special sub-segment that at the same time is also able to realise this idea, be it the specific researcher, developer or entrepreneur. Anyone can come up with an idea, but surely not everyone has the ability to realise it, not everyone has the required skills and competencies. Not everyone is an entrepreneur, not everyone works in just that firm that can realise the idea. Just from the basic argument assume the following: a certain group of people, by far not all, have a certain ability to realise ideas, say, have the required abilities and willingness to be successful entrepreneurs (and this may hold only in a special market segment where they are experienced and skilled), or are existing firms in the desired market segment. If ideas are distributed randomly over the entire population, it is unlikely that both, the idea and the ability to realise it coincide. If they do, all the better, but for many if not most ideas this is not the case (see Fig. 7.1).

---

[12] von Hippel and Jin 2009, p. 2 (emphasis in original). Also see von Hippel (2005). Also see findings in Poetz and Schreier (2012) and Parker and Udell (1996). This view is sometimes ridiculed from a corporate perspective. "The independent inventor often has been portrayed as something of a mad scientist-type individual or an uneducated dreamer in search of the Holy Grail. The result of these perceptions is that the independent inventor no longer is viewed as a serious source of product innovation. . . . On occasion, the "nut inventor" working in a basement comes up with an innovation that goes on to great success. However, these days, . . . most innovations come from serious research and development efforts undertaken with the support of formal organizations. The amateur inventor may still enjoy considerable success, but this has become an increasingly rare occurrence (Zikmund and D'Amico 1989, p. 268)." Parker and Udell 1996, p. 7. It may still hold true for various industries where invention requires large funding volumes. On the other hand, given the increased potential of appropriation also outside of corporations, independent inventors are likely to become even more prevalent (also see Wagner Weick and Eakin 2005; Udell 1990).

Fig. 7.1 Anyone can come up with an idea, but only few have the skills and means to realize it

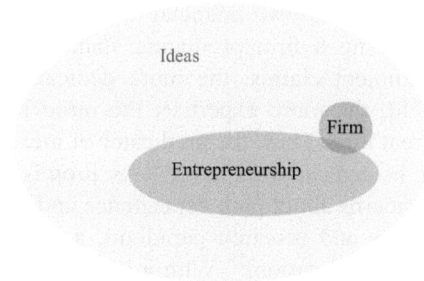

Only by chance is the person with the idea also the one best able to realise it. But then a great many ideas, all that do not coincide, never are realised, or realised far from effective and efficiently. Not everyone comes up with ideas. Not all are firms, not all have entrepreneurial ambitions. Yet many that are not may just as well have good ideas- without being able to realize them. Ideas are left without a chance to be realized, and entrepreneurs and firms left without sufficient ideas. If on the other hand the one with the idea, and those best able to realise it were to cooperate, many more ideas could be realised and realised more effectively. The ones with ideas would benefit from the abilities of those able to realize it, and those able, those in need of ideas, would benefit from realizing them. It constitutes a division of skills, and requires a 'division' of labour, or more correctly a 'cooperation' of labour.

Alternatively, as a second case, consider the hypothesis of idea clusters. Assume it is most likely for ideas to be generated by experts and dedicated practitioners. The more people focus on a particular aspect, or the more involved they are, the more likely they are to come up with a brilliant and informed idea. This proposition has considerable merit. The more exposure someone has to a particular issue, the more involved in the problem solving process, the more knowledge of the underlying mechanisms and process someone has, the more focussed they are on a particular outcome, the more likely it seems they are to find solutions, to envision new uses, improve processes or products, etc. Uninformed have often no overview of what already may exist, and often lack the expertise to spot the feasible, and understand the constraints, the practical impediments and restricting market realities. The category of such informed and predicated sources of new ideas may include researchers, dedicated experts, lead users, and all those with a high exposure to the field or subject. Many areas clearly require such research expertise, areas for example such as chemistry and drugs – it is unlikely that just anyone could come up with a new anti-cancer drug; many benefit from very high degrees of exposure as in the famous surfer innovation example of specialised lead users coming up with footstraps for surfboards or leading surgeons design new surgical equipment[13]; or market experts reconfiguring products to service the observed needs of clients (e.g. Michael Bloomberg after leaving the investment bank Salomon Brothers

---

[13] von Hippel (2005, pp. 1–2).

founded his own financial-data service firm; or the vacuum cleaner testing expert founding a firm producing standardised dirt (artificial test soil)). In short, the argument claims: the more dedicated and exposed someone is to a particular field, the more expertise, the more likely to have a hunch, to come up with a great idea. Thus, the predicator of ideas is expertise and exposure. It is not anyone. It is dedicated experts.[14] As promising as it sounds, there are however sever concerns about path dependence and assimilation with the dominant way of doing things and research paradigm, a lack of exposure to enable recombination and cross-pollination.[15] With a limited exposure to ideas in other areas, with limited diversity of user expertise, with limited heterogeneity of application and accidental exposure, dedicated search for ideas is often limited, and especially visionary and unorthodox ideas, and as the name suggests, out-of-the-box ideas are less likely to originate from it.[16] None the less, undoubtedly expertise, and no doubt also access to resources, often do foster the generation, sometimes it may even seem a precondition for new ideas, especially in highly specialised areas. This sounds promising to all those dedicating time and effort to generate ideas, and holds promise for funding directed research and development to maintain a competitive edge or to generate a new idea for entrepreneurial endeavour. And yet, wherever it does indeed succeed it faces a similar concern as in the case of randomly distributed ideas. Especially those most specialised, most apt to generate new ideas in this perspective of dedicated idea generation, are typically differently qualified than those best suited to realise them (see Fig. 7.2).

Two arguments apply here: First, if entrepreneurship is considered a special ability, meaning the desired abilities differ mostly from those of researchers and experts, than the two skills not only will, but more importantly *should* not coincide in one person.[17] The more focussed and dedicated the search for ideas, the more likely it is to succeed. Hence, their time would best be spent developing ideas, rather than becoming entrepreneurs. Think of a brilliant researcher, a gifted professor. Surely, their expertise and time would be better spent in discovering new ideas

---

[14] To provide it with a little more context, it could be limited to only certain kinds of experts, perhaps the top performing of those with the best track record of previous creativity. The argument may still hold as long as this pool is still wide enough and the variance of discovery broad enough (see, for example, Ernst et al. 2000, but also Audia and Goncalo 2007). On the other hand, also see research indicating a strong clustering on inventions (see Lotka 1926, also Pao 2007; Narin and Breitzman 1995; Allison and Stewart 1974). However, such results are typically applicable to envelop invention and highly scientific fields only (see, for example, Coile 1977).

[15] Also see Cohen and Levinthal (1989, 1990), von Hippel et al. (1999), Page (2007), or Burgelman 2002.

[16] See Martin and Mitchell (1998), von Hippel et al. (1999), Audia and Goncalo (2007), Carbonell et al. (2009), and Magnusson (2009). Also see, for example, Page (2007).

[17] Such normative claim derives from the objective of maximizing innovation, and via this fostering prosperity (see Katz and Shapiro 1986; Silvera and Wright 2010. From a theoretical perspective also see Michelacci 2003 (If it requires a choice of focussing on one or the other, it requires an appropriate balance of research and entrepreneurs)).

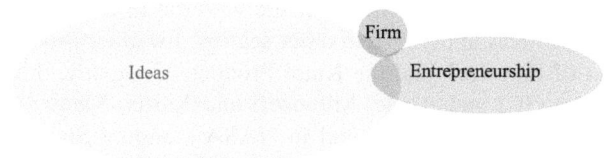

Fig. 7.2 Some people come up with ideas, others are best able to realize them

than spending time in boardrooms, with management and business administration.[18] The same holds the other way around for the entrepreneurs. If they have this skill or gift, it seems it would best be spent realising ideas rather than trying to become an expert and generate ideas. Think of bright and ambitious MBA graduates; think of successful serial entrepreneurs, able managers, and practical doers. They are good at what they do. They would be best suited pursuing this, than trying to do research and desperately trying to come up with ideas. In effect, also here a division of labour would be most effective to maximise the innovation potential. The second argument is more complicated. For firms you may think this is a good thing. Hire experts to generate ideas. If indeed expertise is the dominant predicator for ideas, this seems to hold true. It is already a division of labour. However, this applies only to a certain extent. For one thing, it is neither possible to absorb all the smartest people in a specific field, nor for that matter distinguish between those intelligent, and those creative.[19] As a famous business quote puts it: "No matter who you are, most of the smartest people work for someone else."[20] This becomes all the more difficult as the pool of researchers grows, and solutions become increasingly interwoven. As the experience from crowdsourcing shows: "Solutions come from places you'd never have imagined."[21] Solutions come from everywhere. Solutions to many of the issues faced now come from related fields outside of dedicated subgroups occupied with their own inquiry.[22] Then also, the reverse applies: solutions emerge within fields of expertise that have applications far beyond its narrow realm. Innovation potential becomes more and more interwoven. Ideas may come from dedicated experts, but their application is unlikely to coincide with the mandate of the firm alone. A researcher may solve the problem, may discover new products and may come up with a brilliant idea. In a vertically integrated firm, where the researchers interests and outputs is confined to the sector and production of the firm he/she works for, the

---

[18] This is particularly pressing if the inventors do not even value success as highly as inventing itself. Often, it seems inventing is the priority (see Taussig 1915; Rossman 1931. For a more critical view, also see Nelson 1959).

[19] It may be true that such distinction may be determinable through observed traits and behaviours (e.g. Root-Bernstein 1989), but the theory and process so far seem inapt.

[20] Lakhani and Panetta (2007, p. 97). This is also referred to as "Joy's Law", attributed to Sun Microsystems cofounder Bill Joy. This holds for most industries, and almost any size of company.

[21] Paul Stiros, CEO NineSigma in Business Week (2007). Also see Lakhani, and Jeppesen (2007).

[22] See especially Jeppesen and Lakhani (2010)

likelihood however that these do not coincide with the firm's line of business, the likelihood of these ideas apply also in other sectors, for other products, is substantial. Think also of Play-Doh and the Kutol Products Company, think of SAP and IBM, think of the GUI and Apple, Microsoft and Xerox. Many entertaining and illustrious examples can also be found in NASA's annual Spinoff report where since 1976 it publishes the spinoffs it creates from its research for public commercialization many outside of space technology, including things like sunglasses, swimsuits, ski boots, gardening devices, hairstyling tools, and many more.[23] For these discoveries, for these ideas to be utilized efficiently, partners are needed. Similar holds to the entrepreneur-inventor argument: The two are unlikely to coincide, and perhaps should not in order to best utilize the idea potential.[24]

The truth possibly lies somewhere in-between, or more likely is a mix of both. For some ideas, expertise is required, for some it makes it more likely, and for some it may even be better to have no predisposition.[25] But in either case, it is unlikely that the ones generating the idea necessarily or significantly coincide with those best suited and able to realise them. As Schumpeter stressed: "entrepreneurs of course may be inventors just as they may be capitalists, they are inventors not by nature of their function but by coincidence and vice versa."[26] Many ingenious ideas have failed or have been neglected because of this. "Most of Silicon Valley – but most of the new biological high-tech companies as well – are still inventors rather than innovators, still speculators rather than entrepreneurs."[27] It is unlikely that the two skills coincide. In rare cases, the required skills do coincide, holding for magnificent success (e.g. Henry Ford, Walt Disney, Carl von Linde, or Elon Musk). In other cases often, success is dependent on a good combination, partnership, and cooperation between complementary skills (e.g. Google: Larry Page/Sergey Brin, together with

---

[23] See sti.nasa.gov/tto (02.03.2013)

[24] In any case, even working in a firm is a form of cooperative innovation especially when it comes to the unmonitorable nature of idea generation.

[25] This may still depend on access to information, cognitive ability, alertness, etc. but may not be confined to specific exposure, dedication, etc. It seems highly unlikely a clear ex ante identification is possible. Certain 'likelihoods' may be discovered, but the variance is likely to be high. The extent may however differ depending on the nature of the idea, the area of inquiry and the objective. Some innovations may require high degree of expertise, some may require none (see, for example, Marvel and Lumpkin 2007). Some innovation processes (most likely to produce incremental innovations) may be 'mechanized', many others may not be. Indeed, more and more people have access to highly specialized knowledge, expertise, and even the facilities to conduct research and development, independently allowing also for more expertise driven innovations to come from a wider range of sources (see von Hippel 1988). Galbraith discouraging assessment is unlikely to hold: "There is no more pleasant fiction than that technical change is the product of the matchless ingenuity of the small man forced by competition to employ his wits to better his neighbor. Unhappily, it is a fiction. Technical development has long since become the reserve of the scientist and engineer." (Galbraith 1952, p. 86)

[26] Schumpeter (1934, pp. 88–89). More forcefully still in Schumpeter (1947, p. 152).

[27] Drucker (1986, p. 12–13). For another tragic example see the fate of Shockley Semiconductor Laboratory (though it could be argued that special circumstances applied here). See Shurkin (2006).

Eric Schmidt; or Siemens: Werner von Siemens and Johann Georg Halske)). Indeed, more often than not, successful ventures are the result of cooperation. This holds beyond an implementer/investor divide. There is a strong need for complementarily, often exactly between those with ideas and expertise, and those willing and able to realise it, entrepreneurs and firms.[28] Alas, such cooperation is mostly more by chance than design. It is typically a friend, a relative, a chance encounter. It surely is not a systematic approach.[29] A similar need to cooperate exists when it comes to ideas in firms that can only, or would better be realized by other firms, or outside the existing firm, and ideas outside of firms that could best be realized by that firm. Many great ideas are discovered in firms, but not used, though they could be valuable to others. Many great ideas are discovered outside the firm that would be ideally suited to realise them, but they have no access to the particular firm.[30]

Many people have neither the skills nor access to the partners to realise their idea. Mostly they do not even try to realise their idea, in view of the seemingly insurmountable challenges that they appear to lack the skills to tackle. In effect, many innovations are only realised with a considerable lag, and to the detriment of the original inventor.[31] This is often a problem and can lead to significant waste and inefficiency when it comes to ideas and innovation processes. It would be better for the amount of ideas and the effectiveness of their implementation to acknowledge and target this division of labour than focus on the few instances where it does coincide. It would be more efficient and effective. More ideas would be realised, and possibly more successfully.

---

[28] This is not merely a question of logic. Much debate surrounds this topic (see for examples also Holmes and Schmitz 1990 as well as Silvera and Wright 2010). But given the absence of sound cooperation mechanism it is hard to assess. Several findings indicate positive effects of cooperation, showing higher performance (Cooper and Bruno 1977; Feeser and Willard 1990; Eisenhardt and Schoonhoven 1990; Schutjens and Wever 2000; Ruef et al. 2003; also Cooper and Saral 2010). This rests on various underlying effects from cooperation and interaction. Diversity of teams covers and combines more needed skills; discussion can refine the approach; teamwork can motivate, can sustain 'collective passion'; can help make decisions; etc. (see for example Vesper 1990; Kerr and Tindale 2004; Lechler 2001; Aldrich et al. 2004; Kim et al. 2005; Biais and Perotti 2008; Drnovsek et al. 2009; also somewhat Hagel and Brown 2005; Labianca 2004).

[29] In fact "less than 10 % of all new businesses are founded by teams of nonrelatives" Shane (2008, p. 75). Also Cooper and Saral (2010). Despite the notion that teams have a higher success rate (see Feeser and Willard 1990, also Cooper and Bruno 1977). Several attempts to systematize such matching exist (e.g. Entrepreneurship Center Vienna, or Aalto University School of Science and Technology). But, as will be argued below, simple matching does not suffice, and there are good reasons why such cooperation is typically confined.

[30] See Chesbrough (2003a, c, 2006a) and Leonard-Barton (1992). Also see Henderson (1993), Henderson and Clark (1990), also Christensen (1997) and Chesbrough (2001) on the discussion of the 'incumbent's curse'. For more ambiguous results, also see Chandy and Tellis (2000). Also Henderson (2006).

[31] Such a lag can be substantial. It is sometimes estimated at between 5 and 30 years before the idea is realised. See Rice et al. (1998), Utterback and Brown (1972).

But this also requires the availability and willingness of those with the desired skills to cooperate. It requires able entrepreneurs and willing incumbents to effectively realise ideas – ideas of others. Realisation becomes less and less intertwined with the generation of ideas, and more and more a service. This calls for an even further disentangling of entrepreneurship from the content, promoting a new kind of serial entrepreneurship, and an opening up to external source of ideas, for both entrepreneurs and firms, willing to trade ideas, to realise ideas with and for others. As entrepreneurship is becoming more and more accessible with a wide range of support mechanisms it could be argued the necessary skills become also less important. On the other hand, as ideas are becoming increasingly commoditised, the innovation cycle shortens, the spectrum of ideas widens and the division of labour increases, the competition between entrepreneurs, between firms, between entrepreneurs and firms, increases and possibly more than ever the special skills become important, as does the appeal and offer of corporations to realise ideas. It is no longer a question of mere access to the realisation path. It is a competition between the different path, and between the different actors. Entrepreneurs and corporations compete among each other, and between them for access to ideas. This requires positioning, appeal, and considerable comparative advantage in realising ideas.

For entrepreneurship the question is: Can the necessary skills be fostered, or does it require inherent traits and aptitudes. As training increases and research on the underlying mechanism and process increases it might become more structured, better captured, codified and conveyable. Some skills seem very general, some more specific; some seem possible to be learned, others less so.[32] In either case the claim remains: Entrepreneurship remains a special skill set, rare and precious. Trained or by natural ability, it is best used in pursuing just that – realising ideas in the market. Chances that it coincides with the generation of ideas are slim.[33]

---

[32] The debate whether entrepreneurs are born or made is still on-going (for an overview see for example Henry et al. 2005; also Baumol 1993. Also see Ernst and Young 2011). The extreme positions, often proclaimed in the past, are vanishing fast. Yet, a certain level of reservation prevails, a perception that not all can be nurtured, and may be limited to a considerable extent by inherent traits. Especially Schumpeter proposed that possibly, each generation or cohort only has a certain amount of true entrepreneurial potential. (Schumpeter 1931, p. 119. Also see Haberler 1937, p. 78). Others have tried to identify such 'traits', or better 'enduring characteristics', commonly deemed relevant for entrepreneurial success, many of which indeed seem not easily attained through training (e.g. such as overconfidence, risk loving, etc.) For a more extensive list of 'motivations', many of which seem more like traits than abilities easily learned, also see Shane et al. 2003; Such characteristics include psychological and sociological ones that seem hard to address through learning (see for example Frese 2009; Thornton 1999; Ruef and Lounsbury 2007; Shaver and Scott 1991; Carsrud and Brännback 2009). As a contrasting view, today to many a consensus seems to be emerging that many needed abilities, at least the needed techniques, can indeed be learned (see especially Drucker 1986).

[33] This may also add to the explanation why often it is so challenging in entrepreneurial research to clearly identify entrepreneurial aptitudes, and identification of skill. This may to a large extent be due to circumstance specific requirement, but also that it is not entrepreneurial ability alone but the situational cues and opportunity, the coincidence of access to ideas that rather determines the occurrence of entrepreneurial activity in a random set of potential entrepreneurs rather than a general uptake of activity due to potential ability.

Ideas come increasingly from outside. Entrepreneurs need to compete for them, appeal to the idea inventors to entrust them with the realisation process. The pressure to perform is increasing. Participants need to specialise, build, retain, and increase skills and competencies to be successful, provide much more targeted offers to differentiate and find a competitive advantage. With an increasing division of labour also in the realisation process, the entrepreneurial skills will be more focussed, focussed more on the essential entrepreneurial skills. As do other service providers to compete in the realisation market, compete for ideas.

Similar could be argued for incumbent firms. As ideas are increasingly, being generated outside the traditional internal R&D departments and by increasingly diverse sources, they need to compete for realisation in order to gain access and profit from them. They too need to attract ideas – competing with other firms and other realisation paths such as entrepreneurs. They need to open up, specialise, attract, and deliver. In other words, also the implementers and 'realizers' increasingly focus and specialise, and with it offer increasingly complementary services to the generation of ideas. As it is unlikely that the two coincide they are increasingly dependent on the access to ideas from outside.

In short: opportunities to realize an idea need not coincide with the necessary skills, knowledge, or access. A division of labour applies. The increasing commoditization of ideas, the broadening scope and intensified search as well as the increasing competition of the realisation path is best served by, even demands for a better division of labour between ideas and its realisation. Not everyone with an idea is an entrepreneur or manager. Not every entrepreneur or company comes up with ideas.[34] In trying it alone either one can go terribly wrong. Individuals try to realise ideas without means or skills. Entrepreneurs fail to generate ideas and waste their time and skills on the realisation of inferior ideas or other pursuits. Firms fail to innovate and fail in the market. All, it seems, would be better served by collaborating, by cooperating in the innovation process. More ideas could be realised, and realised more effectively. This serves both inventor and 'realiser' – and society as a whole.

But how to achieve this? For all the promises cooperative innovation holds, the challenges are equally apparent. A division of labour seems superior to relying on chance where ideas and skills coincide. Without the likes of Mr. Zanen, all those Barenbrugs in the world do not stand much of a chance. Yet, only for a few chance encounters, such cooperation is still not the norm. Why? The quest is to enable such division of labour, to make cooperative innovation work. But how? How to match and more demanding yet how to enable cooperation between the two – those with ideas and those with the skills and willingness to realise them? How to ensure successful cooperation? Many obstacles still remain and hinder the full unfolding of such approach, hinder the unfolding of the *Idea Economy*. How this can indeed be perused is the discussion of part II.

---

[34] Indeed, it appears often the desire to start a business comes first, without having an idea (see Gartner and Carter 2003).

References

Aldrich, H. E., Carter, N. M., & Ruef, M. (2004). Teams. In W. B. Gartner, K. G. Shaver, N. M. Carter, & P. D. Reynolds (Eds.), *The handbook of entrepreneurial dynamics: The process of organization creation* (pp. 229–310). Thousand Oaks: Sage.

Allison, P. D., & Stewart, K. A. (1974). Productivity differences among scientists. Evidence for accumulative advantage. *American Sociological Review, 39*(4), 596–606.

Amabile, T. M. (1983). The social psychology of creativity: A componential conceptualization. *Journal of Personality and Social Psychology, 45*(2), 357–376.

Amabile, T. M. (1996). *Creativity in context: Update to the social psychology of creativity.* Boulder: Westwood Press.

Amabile, T. M. (1998). How to kill creativity. *Harvard Business Review, 76*(5), 76–87.

Arthur, W. B. (2007). The structure of invention. *Research Policy, 36*(2), 274–287.

Audia, P. G., & Goncalo, J. (2007). Past success and creativity over time: A study of inventors in the hard disk drive industry. *Management Science, 53*(1), 1–15.

Baumol, W. J. (1993). Formal entrepreneurship theory in economics: Existence and bounds. *Journal of Business Venturing, 8*, 197–210.

Biais, B., & Perotti, E. (2008). Entrepreneurs and new ideas. *The RAND Journal of Economics, 39*(4), 1105–1125.

Boo (2012). *The celebration issue.* Issue 1, March, by Bugaboo International. http://issuu.com/bugaboo/docs/boo-us?mode=window&backgroundColor=%23222222. Accessed 10 Dec 2012.

Burgelman, R. A. (2002). Strategy as vector and the inertia of coevolutionary lock-in. *Administrative Science Quarterly, 47*(2), 325–357.

Businessweek (2007). *NineSigma: Nurturing 'Open Innovation'.* Special report 12, June 2007, by Scanlon, J. www.businessweek.com/stories/2007-06-12/ninesigma-nurturing-open-innovation-businessweek-business-news-stock-market-and-financial-advice. Accessed 02 March 2013.

Carbonell, P., Rodríguez-Escudero, A. I., & Pujari, D. (2009). Customer involvement in new service development: An examination of antecedents and outcomes. *The Journal of Product Innovation Management, 26*(5), 536–550.

Carsrud, A. L., & Brännback, M. (2009). *Understanding the entrepreneurial mind: Opening the black box.* New York: Springer.

Chandy, R. K., & Tellis, G. J. (2000). The incumbent's curse? Incumbency, size, and radical product innovation. *The Journal of Marketing, 64*(3), 1–17.

Chesbrough, H. W. (2001). Assembling the elephant: A review of empirical studies on the impact of technical change upon incumbent firms. In R. A. Burgelman & H. W. Chesbrough (Eds.), *Comparative studies of technological evolution* (pp. 1–36). Amsterdam: Emerald Group Publishing Limited.

Chesbrough, H. W. (2003a). *Open innovation: The new imperative for creating and profiting from technology.* Cambridge: Harvard Business School Press.

Chesbrough, H. W. (2003c). The governance and performance of xerox's technology spin-off companies. *Research Policy, 32*(3), 403–421.

Chesbrough, H. W. (2006a). *Open business models: How to thrive in the new innovation landscape.* Cambridge: Harvard Business School Press.

Chicago Magazine (2010). *On groupon and its founder, Andrew Mason.* August 2010, by Coburn, M. F. www.chicagomag.com/Chicago-Magazine/August-2010/On-Groupon-and-its-founder-Andrew-Mason/. Accessed 02 March 2013.

Christensen, C. M. (1997). *The innovator's dilemma: When new technologies cause great firms to fail.* Cambridge: Harvard Business School Press.

Cohen, W. M., & Levinthal, D. A. (1989). Innovation and learning: The two faces of R & D. *The Economic Journal, 99*(397), 569–596.

Cohen, W. M., & Levinthal, D. A. (1990). Absorptive capacity: A new perspective on learning and innovation. *Administrative Science Quarterly, 35*(1), 128–153.

Coile, R. C. (1977). Lotka's frequency distribution of scientific productivity. *Journal of the American Society for Information Science, 28*(6), 366–370.
Cooper, A. C., & Bruno, A. V. (1977). Success among high-technology firms. *Business Horizons, 20*(2), 16–22.
Cooper, A. C., & Saral, K. J. (2010). Entrepreneurship and team participation: An experimental study. *Kauffman foundation small research projects research paper.* doi:10.2139/ssrn.1547186.
Csikszentmihalyi, M. (1996). *Creativity: Flow and the psychology of discovery and invention.* New York: Harper Perennial.
Drnovsek, M., Cardon, M. S., & Murnieks, C. Y. (2009). Collective passion in entrepreneurial teams. In A. L. Carsrud & M. Brännback (Eds.), *Understanding the entrepreneurial mind: Opening the black box* (pp. 191–215). New York: Springer.
Drucker, P. F. (1986). *Innovation and entrepreneurship.* New York: HarperCollins. Reprint 2006.
Eisenhardt, K., & Schoonhoven, C. B. (1990). Organizational growth: Linking founding team, strategy, environment, and growth among U.S. semiconductor ventures, 1978–1988. *Administrative Science Quarterly, 35*(3), 504–529.
Ernst & Young (2011). *Nature or nurture? Decoding the DNA of the entrepreneur.* www.ey.com/Publication/vwLUAssets/Nature-or-nurture/$FILE/Nature-or-nurture.pdf. Accessed 02 March 2013.
Ernst, H., Leptien, C., & Vitt, J. (2000). Inventors are not alike: The distribution of patenting output among industrial R&D personnel. *IEEE Transactions on Engineering Management, 47*(2), 184–199.
Feeser, H., & Willard, G. (1990a). Founding strategy and performance: A comparison of high and low growth high tech firms. *Strategic Management Journal, 11*(2), 87–98.
Financial Times (2008). *Buggy maker pushes ahead.* By Steen, M., 15 July. www.ft.com/intl/cms/s/0/aa44c280-5288-11dd-9ba7-000077b07658.html#axzz2EjJcHKKK. Accessed 02 March 2013.
Foss, N. J., & Klein, P. G. (2010). Entrepreneurial alertness and opportunity discovery: origins, attributes, critique. In H. Landström & F. Lohrke (Eds.), *The historical foundations of entrepreneurship research* (pp. 98–120). Cheltenham: Edward Elgar.
Frese, M. (2009). Towards a psychology of entrepreneurship – an action theory perspective. *Foundations and Trends in Entrepreneurship, 5*(6), 435–494.
Gaglio, C. M., & Katz, J. (2001). The psychological basis of opportunity identification: Entrepreneurial alertness. *Small Business Economics, 16*(2), 95–111.
Gaglio, C. M., & Winter, S. (2009). Entrepreneurial alertness and opportunity identification: Where are we now? In A. L. Carsrud & M. Brännback (Eds.), *Understanding the entrepreneurial mind, opening the black box* (pp. 305–325). New York: Springer.
Galbraith, J. K. (1952). *American capitalism: The concept of countervailing power.* Cambridge: The Riverside Press.
Galenson, D. W. (2010). Understanding creativity. *Journal of Applied Economics, 13*(2), 351–362.
Gartner, W. B., & Carter, N. M. (2003). Entrepreneurial behavior and firm organizing processes. In Z. J. Acs & D. B. Audretsch (Eds.), *Handbook of entrepreneurship research. An interdisciplinary survey and introduction* (pp. 195–221). Boston: Kluwer.
Hagel, J., & Brown, S. J. (2005). Productive friction: How difficult business partnerships can accelerate innovation. *Harvard Business Review, 83*(2), 82–91.
Henderson, R. M. (1993). Underinvestment and incompetence as responses to radical innovation: Evidence from the photolithographic alignment equipment industry. *The RAND Journal of Economics, 24*(2), 248–270.
Henderson, R. M. (2006). The innovator's dilemma as a problem of organizational competence. *Journal of Product Innovation Management, 23*(1), 5–11.
Henderson, R. M., & Clark, K. B. (1990). Architectural innovation: The reconfiguration of existing product technologies and the failure of established firms. *Administrative Science Quarterly, 35*(1), 9–30.

Henry, C., Hill, F., & Leitch, C. (2005). Entrepreneurship education and training: Can entrepreneurship be taught? Part I. *Education and Training, 47*(2), 98–11.

Holmes, T. J., & Schmitz, J. A. (1990). A theory of entrepreneurship and its application to the study of business transfers. *Journal of Political Economy, 98*(2), 265–294.

Jeppesen, L. B., & Lakhani, K. R. (2010). Marginality and problem solving effectiveness in broadcast search. *Organization Science, 21*(5), 1016–1033.

Johnson, S. (2010). *Where good ideas come from: The natural history of innovation*. New York: Penguin.

Katz, M. L., & Shapiro, C. (1986). Technology adoption in the presence of network externalities. *Journal of Political Economy, 94*(4), 822–841.

Kaufman, J. C., & Sternberg, R. J. (2006). *The international handbook of creativity*. Cambridge: Cambridge University Press.

Kerr, N. L., & Tindale, R. S. (2004). Group performance and decision making. *Annual Review of Psychology, 55*(1), 623–655.

Kim, P. H., Aldrich, H., & Ruef, M. (2005). *Fruits of co-laboring: Effects of entrepreneurial team stability on the organizational founding process*. Frontiers of entrepreneurship research 2005. Wellesley: Babson College. http://fusionmx.babson.edu/entrep/fer/2005fer/chapter_iii/paper_iii1.html. Accessed 2 Mar 2013.

Koellinger, P. (2008). Why are some entrepreneurs more innovative than others. *Small Business Economic, 31*(1), 21–37.

Labianca, J. (2004). The ties that blind. *Harvard Business Review, 82*(10), 19–19.

Lakhani, K. R., & Panetta, J. A. (2007). The principles of distributed innovation. *Innovations Technology Governance Globalization, 2*(3), 97–112.

Lakhani, K. R., & Jeppesen, L. B. (2007). Getting unusual suspects to solve R&D puzzles. *Harvard Business Review, 85*(5), 30–32.

Lechler, T. (2001). Social interaction: A determinant of entrepreneurial team venture success. *Small Business Economics, 16*(4), 263–278.

Leonard-Barton, D. (1992). Core capabilities and core rigidities: A paradox in managing new product development. *Strategic Management Journal, 13*(2), 111–126.

Lotka, A. J. (1926). The frequency distribution of scientific productivity. *Journal of the Washington Academy of Science, 16*(2), 317–323.

Magee, G. B. (2005). Rethinking invention: Cognition and the economics of technological creativity. *Journal of Economic Behavior and Organization, 57*(1), 29–48.

Magnusson, P. R. (2009). Exploring the contributions of involving ordinary users in ideation of technology-based services. *Journal of Product Innovation Management, 26*(5), 578–593.

Martin, X., & Mitchell, W. (1998). The influence of local search and performance heuristics on new design introduction in a new product market. *Research Policy, 26*(7–8), 753–771.

Marvel, M. R., & Lumpkin, G. T. (2007). Technology entrepreneurs' human capital and its effects on innovation radicalness. *Entrepreneurship Theory and Practice, 31*(6), 807–828.

Michelacci, C. (2003). Low returns in R&D due to the lack of entrepreneurial skills. *The Economic Journal, 113*(484), 207–225.

Narin, F., & Breitzman, A. (1995). Inventive productivity. *Research Policy, 24*(4), 507–519.

Nelson, R. R. (1959). The economics of invention: A survey of the literature. *Journal of Business, 32*(2), 101–127.

New York Times (2006). *The boss – out of Africa, onto the web*. 17 Dec, by Zipkin, A. www.nytimes.com/2006/12/17/jobs/17boss.html. Accessed 02 March 2013.

Page, S. E. (2007). *The difference: How the power of diversity creates better groups, firms, schools, and societies*. Princeton: Princeton University Press.

Pao, M. L. (2007). An empirical examination of Lotka's law. *Journal of the American Society for Information Science, 37*(1), 26–33.

Parker, R. S., & Udell, G. G. (1996) The new independent inventor: Implications for corporate policy. *Review of Business, 17*(3), 7–13.

Pink, D. H. (2005). *A whole new mind*. New York: Penguin.

Poetz, M. K., & Schreier, M. (2012). The value of crowdsourcing: Can users really compete with professionals in generating new product ideas? *Journal of Product Innovation Management, 29*(2), 245–256.
Rice, M., O'Connor, P., Colarelli, G., Peters, L. S., & Morone, J. G. (1998). Managing discontinuous innovation. *Research Technology Management, 41*(3), 52–58.
Robertson, A. (1984). Characteristics of the successful inventor: Some notes on the nature of creativity and the creative mind. *Technovation, 2*, 141–145.
Root-Bernstein, R. S. (1989). Who discovers and invents. *Research Technology Management, 32*(1), 43–50.
Rossman, J. (1931). The motives of inventors. *Quarterly Journal of Economics, 45*(3), 522–528.
Ruef, M., & Lounsbury, M. (2007). The sociology of entrepreneurship. *Research in the Sociology of Organizations, 25*, 1–29.
Ruef, M., Aldrich, H. E., & Carter, N. M. (2003). The structure of founding teams: Homophily, strong ties, and isolation among U.S. entrepreneurs. *American Sociological Review, 68*(2), 195–222.
Schumpeter, J. A. (1931). *Theorie der wirtschaftlichen Entwicklung* (3rd ed.). München: Duncker und Humbolt.
Schumpeter, J. A. (1934). *Theory of economic development*. New Brunswick: Transaction Publishers. Reprint 2004.
Schumpeter, J. A. (1947). The creative response in economic history. *The Journal of Economic History, 7*(2), 149–159.
Schutjens, V. A. J. M., & Wever, E. (2000). Determinants of new firm success. *Papers in Regionla Science, 79*(2), 135–159.
Shane, S. (2008). *The illusion of entrepreneurship: The costly myth that entrepreneurs, investors, and policy makers live by*. New Haven: Yale University Press.
Shane, S., & Eckhardt, J. (2003). Opportunities and entrepreneurship. *Journal of Management, 29*(3), 333–349.
Shane, S., Locke, E. L., & Collins, C. J. (2003). Entrepreneurial motivation. *Human Resource Management Review, 13*(2), 257–279.
Shaver, K. G., & Scott, L. R. (1991). Person, process, choice: The psychology of new venture creation. *Entrepreneurship Theory & Practice, 16*(2), 23–45.
Shurkin, J. N. (2006). *Broken genius: The rise and fall of William Shockley, creator of the electronic age*. New York: Palgrave Macmillan.
Silvera, R., & Wright, R. (2010). Search and the market for ideas. *Journal of Economic Theory, 145*(4), 1550–1573.
Sternberg, R. J. (1988). *The nature of creativity: Contemporary psychological perspectives*. Cambridge: Cambridge University Press.
Sternberg, R. J., & Lubart, T. (1995). *Defying the crowd: Cultivating creativity in a culture of conformity*. New York: The Free Press.
Tang, J. (2008). Environmental munificence for entrepreneurs: Entrepreneurial alertness and commitment. *International Journal of Entrepreneurial Behaviour & Research, 14*(3), 128–151.
Taussig, F. W. (1915). *Inventors and money-makers*. New York: Macmillan.
Thornton, P. (1999). The sociology of entrepreneurship. *Annual Review of Sociology, 25*, 19–46.
Udell, G. G. (1990). It's still caveat, inventor. *Journal of Product Innovation Management, 7*(3), 230–243.
USA Today (2006). *'Charismatic' founder keeps netflix adapting*. 23 April, by Hopkins, J. http://usatoday30.usatoday.com/tech/products/services/2006-04-23-netflix-ceo_x.htm. Accessed 02 March 2013.
Utterback, J. M., & Brown, J. W. (1972). Profiles of the future monitoring for technological opportunities. *Business Horizons, 15*(5), 5–15.
Vesper, K. H. (1990). *New venture strategies* (2nd ed.). Englewood Cliffs: Prentice Hall.
von Haberler, G. (1937). *Prosperity and depression*. Geneva: League of Nations.
von Hippel, E. (1988). *The sources of innovation*. New York: Oxford University Press.

von Hippel, E. (2005). *Democratizing innovation*. Cambridge: MIT Press.
von Hippel, E., Thomke, S., & Sonnack, M. (1999). Creating breakthroughs at 3M. *Harvard Business Review, 77*(5), 47–57.
von Hippel, E., & Jin, C. (2009). The major shift towards user-centered innovation: Implications for China's innovation policymaking. *Journal of Knowledge-based Innovation in China, 1*(1), 16–27.
Wagner Weick, C., & Eakin, C. F. (2005). Independent inventors and innovation: An empirical study. *International Journal of Entrepreneurship and Innovation, 6*(1), 5–15.
Weber, M. (1949). The methodology of social sciences. (trans: Shilz, E.S. & Finch, H.A.). Excerpts reprinted in Hausman, D.M. (1994) *The philosophy of economics – an anthology* (2nd ed., pp. 69–82). Cambridge: Cambridge University Press.
Yu, T. F. L. (2001). Entrepreneurial alertness and discovery. *The Review of Austrian Economics, 14*(1), 47–63.
Yusuf, S. (2009). From creativity to innovation. *Technology in Society, 31*(1), 1–8.
Zikmund, W., & D'Amico, M. (1989). *Marketing* (3rd ed.). New York: Wiley.

The Missing Link 8

Why are inventors and businessmen not already cooperating? When it comes to ideas cooperation is tricky. There is a strong risk of others stealing your idea. You cannot just offer it openly. Once shared there is no taking it back. But without sharing the idea the partner cannot decide if they want it or how much it is worth. A paradox arises, called Arrow's Fundamental Paradox of Information, which makes trading ideas very difficult. This is the reason a market for ideas has not yet evolved.

Innovation equals ideas plus implementation (innovation = ideas + implementation). Simple. Intuitive. Concise. Often claimed and more often quoted.[1] Yet, such simple math is faulty. It does not add up – literally. Increasing the number of ideas does not translate into an increase in innovation. Expanding the stock of entrepreneurs or eager firms does not either. Ideas need implementation. Implementation needs ideas. An idea in someone's drawer may be a great and valuable one, but if not realized it remains just that, an idea in someone's drawer. It does not add value. It does not increase innovation. A highly trained and capable would-be-entrepreneur without an idea remains a would-be-entrepreneur. It shows wasted potential, not an increase in innovation. Neither one alone, more ideas nor more implementation capacity, translates into more or better innovation. Some invent. Others realize. Only if by chance the two coincide – an entrepreneur with a bright idea – does this mean more innovation. In all other cases ideas and implementation potential is wasted. Thus, innovation is surely not additive. But the math is not the point. The implications are. If innovation requires both – and indeed ideas and realization are separate – how then to foster innovation? If some have ideas and others the means and skills to best realize them, then a gaping disconnect, a gap in execution becomes apparent. The question thus becomes: how to bring the two together, ideas and implementation, to truly foster innovation; to make ideas available to entrepreneurs and firms, and entrepreneurs and firms available to ideas.

---

[1] Roberts (1988, p. 36)

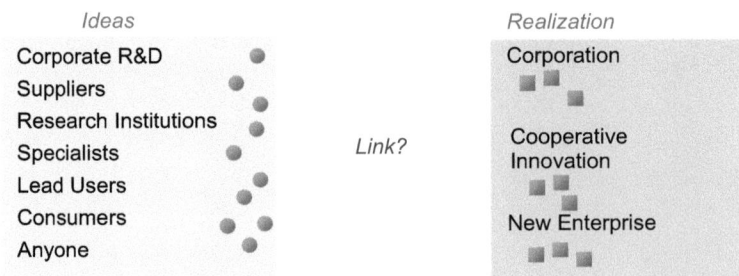

Fig. 8.1 The missing link between inventors and implementers

A practical gap exists between those with ideas, and those best suited to realise them. How can inventors cooperate with implementers? How can someone contact, negotiate, cooperate and trade ideas with others, others able to realize the idea? How can this gap be overcome? There is no clear linkage between the ones with ideas and those able to realise them (see Fig. 8.1). How to bring the two together? Engaging in cooperation in practice is complicated. It is even more complicated when it comes to ideas. Connecting inventors to implementers and fostering cooperation between them is problematic. As long as this gap remains, the only viable innovation path for most inventors outside a mandated corporate path remains the entrepreneurial one. And given its complexity and taxing constraints on the entrepreneurs, this will not get far. Even the most ambitious policies to foster entrepreneurship, the substantial efforts to create a broad based entrepreneurial society will still fall considerably short of the economic potential of a truly open and cooperative innovation process in the *Idea Economy*. Though surely promising and a significant improvement to models of a closed corporate innovation society, the true innovation potential of an economy cannot yet unfold. It remains inefficient and largely ineffective. Without a solution to enable cooperative innovation is will remain stuck in the entrepreneurial society, even if augmented by a more open corporate innovation process – underutilising resources, under-realising ideas.[2]

Addressing this gap is a challenge. For cooperative innovation to thrive and for ideas to become true commodities it has to become possible to connect and enable cooperation between those with ideas and those able to realize them. Here a fundamental disconnect exists. Even if possible to find suitable partners, the question remains: how to cooperate with them? How to trust them?

It seems a common notion among inventors that revealing ideas to others is a bad idea. They can easily take it up, use the idea for their own gains without rewarding the inventor for it. They can steal the idea. Nothing is there to keep them from doing so. No legal provisions apply. This notion that ideas cannot simply be traded seems, in one form or another, generally acceptable. It coincides with common perceptions

---

[2] Also see, for example, analysis by Astebro (1998), or Udell et al. (1993).

about ideas. Popular culture, media, and the relevant literature take the problem of stealing ideas for granted. It is a recurring and accepted premise: Ideas cannot be traded.[3]

This general sense of secrecy and the fear of stealing have been widely popularized via the internet, newspapers and movies. Popular films such as The Social Network and Flash of Genius stress and nourish this fear. In the Academy Awards winning movie The Social Network[4] Tyler and Cameron Winklevoss together with Divya Narendra, business students at Harvard, approach the gifted programmer and computer science student Mark Zuckerberg to help them set up a social network, HarvardConnection, that would better allow fellow students to connect to one another, provide a coherent social network where people can keep in touch and stay informed about their friends social life. Zuckerberg sees the potential in the idea, and instead of programming a site for HarvardConnection, he develops his own social network, The Facebook. The rest is history. The network spread virally across Harvard, soon after Stanford, Columbia, and Yale, other universities, globally, eventually opening up to anyone. Facebook is the by far dominant social network with more than one billion users, with a market value when it went public at some $100 billion. Marc Zuckerberg is now worth some $14 billion.[5] Though he settled with the Winklevoss Twins and Narendra whose own site withered and eventually failed due to the rise of Facebook, they 'merely' received a tiny fraction of several million dollars.[6] In short: they had the idea, shared it, and lost because of it. Their idea was stolen.

An article in The New Yorker magazine and the movie Flash of Genius portray the travails of Robert Kearns, the inventor of the intermittent windshield wiper.[7] In the film, he is portrayed as David fighting Goliath in form of the 'Big Three': Ford, Chrysler, and General Motors. Kearns developed the first fully functioning intermittent windshield wiper and together with the Tann Corporation he actually held a patent for it. He approached Ford and demonstrated his invention. They too were working on something similar, but so far were unable to solve the challenges it posed. Ford was happy to take up the design and integrate it into their cars, soon followed by General Motors, later Chrysler, and then the rest of the automobile industry, selling the invention with a huge profit mark-up. "It cost Ford about 10 dollars to make, and it sold for thirty-seven dollars. [...] By 1989, Ford alone had sold 20.6 million cars with the intermittent wiper, and made a profit that has been calculated at five hundred and fifty-seven million dollars. Altogether, about thirty

---

[3] For example, (Anton and Yao 1994, 2004, 2005; Biais and Perotti 2008; Hellmann and Perotti 2011; Rajan and Zingales 2001; Ueda 2004; Yosha 1995; Atanasov et al. 2008).

[4] See Fincher (2010). Based on Mezrich (2010)

[5] See Facebook Statistics at newsroom.fb.com/content/default.aspx?NewsAreaId = 22 (02.03.2013). For the valuation, see Bloomberg (bloomberg.com/news/2012-05-17/facebook-raises-16-billion-in-biggest-technology-ipo-on-record.html 02.03.2013). Form Marc Zuckerberg's profile see http://www.forbes.com/profile/mark-zuckerberg (02.03.2013). Also see Byrne (2011).

[6] New York Times (2009).

[7] The New Yorker (1993) and Abraham (2008)

million intermittent wipers are sold around the world each year."[8] The car producers thought they could get away with it, due to a technicality in the patent law stressing obviousness and incremental nature of the invention. They tried stalling Kearns' lawsuit until he ran out of money. Eventually they lost in court, Chrysler and Ford settled for 'a few' million Dollars, but only after many years of expensive legal struggle, over which Kearns' risked his career, his savings, his wife left him, and he suffered a nervous breakdown.

Both cases received wide attention, and have reinforced a seemingly common notion and fear of stealing ideas. Robert Kearns legal battle "made him one of the most famous inventors in the country, a hero to thousands of inventors with their own patent-infringement horror stories to tell."[9] Both stories still seem to have a 'happy-ending' as they were fortunate enough to appropriate some of the profit generated by their idea. Most others fare even worse. They get nothing. Any avid inventor that does any kind of search on the internet will read such stories, and is warned again and again by the respective community: never share your ideas! On the website of Dyson Ltd., the firm the inventor of the bagless vacuum cleaner James Dyson (now worth an estimated \$4.2 billion[10]) founded, it is put thus: "During the five years it took to develop his first vacuum, James was also battling. First to convince other manufacturers to embrace his new technology. Then to protect his invention when they copied it. It's enough to give you a complex. And it did. James' experience informs the way Dyson works today. Keeping our inventions secret. Protecting our ideas."[11]

This fear is warranted. It is inherent to the nature of ideas. Ideas are intangible and amorphous. Ideas mostly have no clear ownership rights. This poses special challenges to any attempt at cooperation in the innovation process. To cooperate, people need to share their idea with experts. But this is problematic. The expert could simply steal the idea and make away with a nice profit from it. This is the fear of all too many inventors. No wonder than, the one with the idea would be reluctant to reveal it. Only if the other commits to share the spoils or pay for the idea will the inventor want to share it. But this does not work either. The 'buyer' would be reluctant to engage in such arrangement without knowing the content and having a chance to evaluate the idea. How could they commit to pay for something that they could not possibly know the value of? The idea could be absolutely worthless or incredibly profitable. How could anyone guess the value up front without knowing the idea? A paradox arises – a paradox formulated in 1962 by Kenneth Arrow as the

---

[8] The New Yorker (1993).

[9] See The New Yorker (1993). For other accounts also see Rajan and Zingales (2001), p. 806: "'Intel was founded to steal the silicon gate process from Fairchild' [Jackson (1997), pp. 26–27]. [...] Many during the Industrial Revolution, including Arkwright who "invented" the water frame, appropriated rather than discovered the technological advances their names are associated with".

[10] See Forbes (forbes.com/profile/james-dyson 02.03.2013)

[11] See content.dyson.co.uk/insidedyson/article.asp?aID = jamesdyson&hf = 0&js = (02.03.2013)

'Fundamental Paradox of Information' (see box 8.1). Neither the one with the idea can share it with the partner in fear of the idea being stolen, nor can the partner commit to anything without knowing the idea. Either scenario is problematic. Not sharing would leave considerable abuse potential with the one with the idea, sharing would tip all negotiation power in favour of the expert, entrepreneur or firm. In either case, an asymmetry arises with considerable potential for free riding, abuse, and conflict.

> **Box 8.1 Arrow's Fundamental Paradox of Information**
> In 1962 Kenneth Arrow, a famous economist and Nobel Laureate, identified a fundamental paradox pertaining to ideas, or information in general: *"there is a fundamental paradox in the determination of demand for information; its value for the purchaser is not known until he has the information, but then he has in effect acquired it without cost. Of course, if the seller can retain property rights in the use of the information, this would be no problem, but given incomplete appropriability, the potential buyer will base his decision to purchase information on less than optimal criteria. He may act, for example, on the average value of information in that class as revealed by past experience. If any particular item of information has differing values for different economic agents, this procedure will lead both to a nonoptimal purchase of information at any given price and also to a nonoptimal allocation of the information purchased. It should be made clear that from the standpoint of efficiently distributing an existing stock of information, the difficulties of appropriation are an advantage, provided there are no costs of transmitting information, since then optimal allocation calls for free distribution. The chief point made here is the difficulty of creating a market for information if one should be desired for any reason. It follows from the preceding discussion that costs of transmitting information create allocative difficulties which would be absent otherwise. Information should be transmitted at marginal cost, but then the demand difficulties raised above will exist. From the viewpoint of optimal allocation, the purchasing industry will be faced with the problems created by indivisibilities; and we still leave unsolved the problem of the purchaser's inability to judge in advance the value of the information he buys."*[12]
>
> It pinpoints fairly accurately the dilemma that arises when trying to establish a market for ideas. Ideas have different properties to most other goods and services. You have an idea and want to sell it to someone. You can tell that person your idea, and then start to negotiate on the value of the idea.
>
> (continued)

---

[12] Arrow (1962, p. 615–616). Though typically known as Arrow's Fundamental Paradox of Information, some have used different expressions: e.g. Gans and Stern (2010) refer to it as a 'disclosure problem'

Box 8.1 (continued)
The problem here: unlike say a car or computer, you cannot take it back. The car you can keep, the computer you can keep, the idea, once fully revealed, is gone. The other person could offer you nothing and just take the idea, realize it on his/her own, try to sell it to others, etc. In most cases where no clear intellectual property exists – and even where it does it is not often clear or prohibitively expensive to enforce – this seems a rational reaction. And why not? It is free in a sense as it cannot be claimed. As a ruling on exactly this issue by the California Supreme Court puts it bluntly: "The idea man who blurts out his idea without having first made his bargain has no one but himself to blame for the loss of his bargaining power"[13] Clearly then, you are not stupid either. Knowing this will be the case, knowing people will 'steal' your idea, you, nor anyone else, would share it in such a manner. No market of ideas would exist. The alternative would be to demand a certain price up front for the idea of what you think it is worth. A car the buyer could inspect, kick the tires, test drive, etc. or the computer could be turned on, and some diagnostics run on. Even when you order services, say repair the plumbing or fix your car, you have a good idea of what to expect, and possibly more important, a general idea of what it will cost. A typical market for it exists. But why would anyone purchase an idea that he/she does not know the content of and thus also does not know what it is worth? There is no show and tell, no kicking the tires. You could be selling nonsense to them, or even if meaningful, place a value on the idea that is considerably different from what the recipient would be willing to pay if only he/she knew the content. You could and probably would overcharge. You could deliberately offer nonsense to milk the market. You could, even in best conscience, expect returns that, in practice, are unrealistic. In either case the market would be highly imbalanced – either in favour of the seller or the buyer. If the information were revealed, the buyer would have an advantage, offering you nothing or anything between nothing and the true value of the idea. If the idea is not revealed the seller would have an advantage asking any amount, likely much higher than the actual value of the idea. Adverse selection becomes the norm. Sellers would not want to reveal it without a set price, and buyers will not want to set a price without knowing the idea first. No market would come about, and no trading of ideas, no commoditization of ideas would take place. This challenge can be considered the underlying fundamental paradox that inhibits a broad based unfolding of the *Idea Economy*. It confines the possibilities of cooperative innovation to very narrow and special circumstances; it prevents the emergence of idea markets, and thus an unfolding of a broad based participation in innovation and lively trading of ideas.

---

[13] Desny v. Wilder, 299 P. 2d 257 – Cal: Supreme Court 1956. Also see Gunther-Wahl Productions, Inc. v. Mattel, Inc., 128 Cal. Rptr. 2d 50 – Cal: Court of Appeals, 2nd Appellate Dist., 8th Div. 2002

Arrow's paradox is fundamental for good reason. It poses a major obstacle to trading ideas. Markets for ideas will not arise. Ideas will be horded, concealed; many will remain dormant in someone's drawer. Ideas will remain underutilized. Innovation will be confined to the narrow realm of intellectual property and traditional means of entrepreneurship or closed corporate innovation processes. The *Idea Economy* will not unfold. This makes the paradox fundamental to the *Idea Economy*, to trading ideas. Only if adequate solutions are found can any meaningful broad based form of a division of labour in the innovation process progress.

Clear property rights would be one solution. They provide legal ownership of the idea. Legal remedies become available. But intellectual property (IP) rights only cover a fraction of the potential range of ideas. Most lie outside. Ideas are typically still too unrefined or vague.[14] Given a wide spectrum of ideas and the development spectrum of ideas from a hunch to a fully workable prototype of some sorts, only a small part of ideas are covered by IP law. Mere ideas generally fall outside. Indeed, as for example the United States Patent and Trademark Office (USPTO) confirms: "abstract ideas are not patentable subject matter. A patent cannot be obtained upon a mere idea or suggestion."[15] Especially at an early stage in the innovation process it is mostly not possible to protect ideas through legal means. Ideas would not be specific enough (without necessarily a proven usefulness and attainability or while still in development) and it may also not be economically meaningful to patent in such early stage (due to risk and the fact that if the development process requires time, the legal monopoly is shortened). But this is often the critical stage where expertise is needed. Many other ideas, even if more refined, simply are not covered under IP, and cannot be realised without a partner. IP only covers a fraction of possible ideas. Given the often adverse effects of intellectual property legislation and enforcement, its prohibitive costs and economic implications, its current range is already heavily criticised. It is unlikely – and it also seems unadvisable – to expand.[16] This leaves a vast range of ideas unprotected or protected ineffectively.

---

[14] Also see Hellmann and Perotti 2011, p. 1813 who argue such ideas are still "too vague to be granted patent rights."

[15] USPTO (uspto.gov/patents/resources/general_info_concerning_patents.jsp#heading-4 02.03.2013). For a discussion, also see Kaplan (1958). Also see Copyright Law of the United States of America and Related Laws Contained in Title 17 of the United States Code (copyright.gov/title17/92chap1.html 02.03.2013), especially § 102 Subject matter of copyright: In general: "(b) In no case does copyright protection for an original work of authorship extend to any idea, procedure, process, system, method of operation, concept, principle, or discovery, regardless of the form in which it is described, explained, illustrated, or embodied in such work."

[16] See, for example, discussion in Shulman (1999), Boyle (2003, 2008), Boldrin and Levine (2008), Gallini (2002). Also see Jaffe and Lerner (2006), Scotchmer (2004), Merges (1999), Besen and Raskind (1991), Arora and Merges (2004), Merges and Reynolds (2000), Machlup and Penrose (1950).

Considering recent examples of questionable and often seemingly counterproductive patentability – for example Amazon's 'one-click' purchasing, Priceline's patent on reverse auctions, DNA sequencing, etc. – and abusive use of patent systems (e.g. portfolio wars, patent pooling, patent trolls), a thorough discussion seems called for.

Alternative means to address this challenge need to be found – alternatives that allow for an easier and earlier trade in ideas, not only in patents. This problem is nothing new, nothing surprising. Yet, despite its fundamental importance and a broad awareness of this problem, no ready solution has sprung up; no market for ideas has evolved naturally over time, through trial and error, through experimentation and learning from failed attempts. Economists acknowledge: "The lack of organized markets is not simply an historical accident or a reflection of the fact that a market would have little value; instead, there are significant limitations on the feasibility of the market for ideas given the inherent challenges in market design. In other words, in the absence of specific institutional mechanisms to overcome these challenges, the nature of ideas undermines the spontaneous and uncoordinated evolution of a corresponding market for ideas. [...] ideas possess particular characteristics that make the efficient design of markets challenging and impede the unplanned emergence of markets."[17] This is not to say it has not been tried. Many economists have been grappling with this problem. If it does not evolve on its own, despite its huge promises, maybe it can be designed and instigated. Several solutions have been put forward. Vastly complex models to overcome the information paradox have been proposed and some elegant practical approaches to work around the problem to a certain degree have emerged.[18] None has jet been able to overcome it. Theoretical approaches have contributed greatly to the understanding of the problem, but so far have been unable to present a workable solution that truly overcomes the paradox. The practical approaches have typically worked around the problem, but still offer many insights into important aspects of the problem. Indeed, prospective approaches are a many. Just one solution is needed. None is yet in place.

It is a tricky problem – a fundamental one. It would be grossly negligent to consider the problem straightforward. It surely is not. In practice, it is much more complex, much more demanding than first it may appear. It is more than simply matching two parties, and enabling an exchange. It cannot be solved that simply. Many adjacent problems have to be taken into account. Many detailed questions arise that have to be addressed. How to evaluate an idea? How to assess the risk?

---

[17] Gans and Stern (2010, p. 822/p. 831)

[18] The most promising discussion addressing this problem can be found in Anton and Yao (1994, 2002, 2004, 2005). The basic argument is intriguing. In the absence of enforceable intellectual property rights, there may still be a way of being rewarded for the innovation by a firm. To put it in simple terms (though it cannot do justice to the complexity of the analysis): The inventor would reveal the idea to a firm. The firm would reward the inventor because he/she could threaten to reveal it also to the competitors. This would lower the return. This is an elegant approach to a one shot game, and would at least set the framework of the bargaining situation. This is an elegant approach, but neither does it seem practical or possible to be generalized nor does it address several other factors and challenges that would still occur (e.g. relative distribution within the margin, fairness considerations, etc.). All of these approaches add considerable insights into the problem. They offer valuable contributions to a more general solution and offer complementary mechanisms to reinforce other solutions to ensure a more robust mechanism and enable a better bargaining position to allow for higher appropriability for inventors.

How to handle multiple ideas and its relation to other similar or complementary ideas? How to assess its potential? How to account for uncertainties and risk of other ideas? How to split the benefits between the partners? These and many more questions arise, and need to be addressed if a solution is to be sought of how to best bridge the gap between inventers and implementers.

Only if addressed in a diligent and comprehensive manner can the theoretical intricacies and practical challenges that arise with them be overcome. Details matter here. Even the best, the most elegant solution may fail due to tiny practical 'technicalities'. This is a demanding task. But prominent solutions are emerging and many useful insights point the way forward. Putting them together, and analysing the practical examples in more depth can help identify a more general solution.

The importance and the fundamental character of this paradox cannot be overstressed. If it cannot be overcome, or bridged to a sufficient extent, the evolution towards an *Idea Economy* will stall. Open innovation will fail, and the entrepreneurial path will remain the only meaningful alternative to internal corporate innovation. Ideas will continue to be lost. The innovation potential will remain massively underutilised and confined to specialists and fortunate few with ideas and the skills of realising them. The difficulty of connecting ideas to implementers, of trading ideas, is at the heart of this malaise. It is the missing link. Addressing it is essential. It is conceivable this paradox cannot be fully overcome. However, this does not mean the gap cannot be narrowed, and, through clever design, bridged to a large extent. The analysis of this and the discussion of possible approaches to narrow and bridge this gap will be presented in the following chapters.

References

Abraham, M. (2008). *The flash of genius* [Movie]. Directed by Abraham, M., screenplay by Railsaback, P., Universal Pictures/Spyglass Entertainment/Strike Entertainment.

Anton, J. J., & Yao, D. A. (1994). Expropriation and inventions: Appropriable rents in the absence of property rights. *The American Economic Review, 84*(1), 190–209.

Anton, J. J., & Yao, D. A. (2002). The sale of ideas: Strategic disclosure, property rights, and contracting. *The Review of Economic Studies, 69*(3), 513–531.

Anton, J. J., & Yao, D. A. (2004). Little patents and big secrets: Managing intellectual property. *The RAND Journal of Economics, 35*(1), 1–22.

Anton, J. J., & Yao, D. A. (2005). Markets for partially contractible knowledge: Bootstrapping versus bundling. *Journal of the European Economic Association, 3*(2–3), 745–754.

Arora, A., & Merges, R. P. (2004). Specialized supply firms, property rights and firm boundaries. *Industrial and Corporate Change, 13*(3), 451–475.

Arrow, K. J. (1962). Economic welfare and allocation of resources for invention. In: R. R. Nelson (Ed.) *The rate and direction of inventive activity: Economic and social factors* (pp. 609–626). New Jersey: Princeton University Press. www.nber.org/books/univ62-1. Accessed 02 Mar 2013.

Astebro, T. (1998). Basic statistics on the success rate and profits for independent inventors. *Entrepreneurship Theory and Practice, 23*(2), 41–48.

Atanasov, V. A., Ivanov, V. I., & Litvak, K. (2008). The effect of litigation on venture capitalist reputation. *EFA 2009 Bergen meetings paper.* www.efa2009.org/papers/SSRN-id1343981.pdf. Accessed 02 Mar 2013.

Besen, S. M., & Raskind, L. J. (1991). An introduction to the law and economics of intellectual property. *The Journal of Economic Perspectives, 5*(1), 3–27.

Biais, B., & Perotti, E. (2008). Entrepreneurs and new ideas. *RAND Journal of Economics, 39*(4), 1105–1125.

Boldrin, M., & Levine, D. K. (2008). Market size and intellectual property protection. *International Economic Review, 50*(3), 855–881.

Boyle, J. (2003). The second enclosure movement and the construction of the public domain. *Law and Contemporary Problems, 66*, 33–74.

Boyle, J. (2008). *The public domain: Enclosing the commons of the mind*. New Haven: Yale University Press.

Byrne, J. A. (2011). *World changers: 25 entrepreneurs who changed business as we knew it*. New York: Penguin.

Fincher, D. (2010). *The social network*. [Movie] Directed by D Fincher, screenplay by Sorkin, A. Columbia Pictures/Sony Pictures International.

Gallini, N. T. (2002). The economics of patents: Lessons from recent U.S. patent reform. *The Journal of Economic Perspectives, 16*(2), 131–154.

Gans, J. S., & Stern, S. (2010). Is there a market for ideas? *Industrial and Corporate Change, 19*(3), 805–837.

Hellmann, T. H., & Perotti, E. P. (2011). The circulation of ideas in firms and markets. *Management Science, 57*(10), 1813–1826.

Jackson, T. (1997). *Inside intel: Andrew grove and the rise of the world's most powerful chip company*. New York: Dutton Books.

Jaffe, A. B., & Lerner, J. (2006). Innovation and its discontents. *Innovation Policy and the Economy, 6*(1), 27–65.

Kaplan, B. (1958). Further remarks on compensation for ideas in California. *California Law Review, 46*(5), 699–714.

Machlup, F., & Penrose, E. (1950). The patent controversy in the nineteenth century. *The Journal of Economic History, 10*(1), 1–29.

Merges, R. P. (1999). As many as six impossible patent before breakfast: Property rights for business concepts and patent system reform. Berkeley Technology Law Journal, *14*, 577–615.

Merges, R. P., & Reynolds, G. H. (2000). Proper scope of the copyright and patent power. *Harvard Journal on Legislation, 37*, 45–68.

Mezrich, B. (2010). *The accidental billionaires: The founding of facebook: A tale of sex, money, genius and betrayal*. New York: Doubleday.

New York Times (2009). *ConnectU's 'Secret' $65 million settlement with facebook*. 10 Feb, by Stone, B. http://bits.blogs.nytimes.com/2009/02/10/connectus-secret-65-million-settlement-with-facebook/. Accessed 02 March 2013.

Rajan, R., & Zingales, L. (2001). The firm as a dedicated hierarchy. *Quarterly Journal of Economics, 116*(3), 805–851.

Roberts, E. B. (1988). Managing invention and innovation. *Research Technology Management, 31*(1), 11–29. reprint 2007 50(1):35–54.

Scotchmer, S. (2004). *Innovation and incentives*. Cambridge: MIT Press.

Shulman, S. (1999). *Owning the future: Staking claims on the knowledge frontier*. Boston: Houghton Mifflin.

The New Yorker (1993). *The flash of genius: Bob Kearns and his patented windshield wiper have been winning millions of dollars in settlements from the auto industry, and forcing the issue of who owns an idea*. 11 Jan, by Seabrook, J. www.newyorker.com/archive/1993/01/11/1993_01_11_038_TNY_CARDS_000363341. Accessed 02 March 2013.

Udell, G. G., Bottin, R., & Glass, D. D. (1993). The wal-mart innovation network: An experiment in stimulating American innovation. *Journal of Product Innovation Management, 10*(1), 23–34.

Ueda, M. (2004). Banks versus venture capital: Project evaluation, screening, and expropriation. *Journal of Finance, 59*(2), 601–621.

Yosha, O. (1995). Information disclosure costs and the choice of financing source. *Journal of Financial Intermediation, 4*(1), 3–20.

Narrowing the Gap 9

What has been done so far to make it possible for inventors and businessmen to work together to realize ideas? Many interesting approaches exist that work around Arrow's paradox. The most prominent and promising so far is to offer prizes. A prize is promised for the best idea to solve a given problem. This has proven a successful method. Many platforms have sprung up offering innovation prizes, creating a multi-billion dollar industry. But innovation prizes have some severe limitations. They only apply to ideas that address given problems. Many ideas are novel in nature, or improve on something already quite good. No prize exists for them. Prizes help, but do not solve the paradox.

Travel, trade, and transport in the eighteens century were risky and dangerous. Forces of nature, but also the inability of men to determine their position at sea posed a serious threat. While the stars easily revealed the position of ships in terms of latitude – how far north or south of the equator – determining longitude – how far east and west you are, for example, where on the Atlantic you are between London and New York – was a magnificent challenge. For the better part of human history, this problem eluded a sensible solution. As exploration, global empires, and international trade became increasingly important, the need for a solution became more and more pressing. "Lacking the ability to measure their longitude, sailors throughout the great ages of exploration had been literally lost at sea as soon as they lost sight of land. Thousands of lives, and the increasing fortunes of nations, hung on a resolution."[1] After explorers, entire armies, and vast riches of trade were lost because of it, England, as well as France, Spain, and the Netherlands offered large prizes for solving this problem and demonstrating a practical method to determine longitude at sea. The English Longitude Act of 1714 alone promised a fortune of £20,000 (equivalent to maybe $12 million today). The prizes spurred the greatest minds of their time, and many quacks of the age to pursue and propose a range of solutions. Luminaries such as Galileo Galilei and Isaac Newton had failed to solve

---

[1] Sobel (1995), cover

the problem. Everyone was vying for a celestial solution. Just as the stars were able to tell latitude through measuring the elevation angle of the Pole star, such a method was sought to also reveal longitude at sea. John Harrison built a clock instead. This unlikely candidate, an unknown carpenter and clockmaker, managed to find a workable solution to the problem of longitude by building a highly resilient clock able to keep time at sea, which allowed determining longitude by calculating the difference in local time to the time in London.[2] Similarly the problem of food preservation that plagued the supply of the Napoleonic armies scattered across Europe. Extensive wars and colonial expansion required long and far transportation of food to supply troops. Food spoilage and impasses in resupplying troops was a major problem, leaving armies 'travelling on their stomachs'. A vast cash prize was offered for a ready solution to preserve food. In 1810, the fairly unknown chef and confectioner Nicolas Appert claimed the 12,000-franc prize, proposing, without understanding the science of sterilization, to heat food and seal it in glass jars.[3] Such prizes seem to have a particular tradition in aviation. The 1919 Orteig Prize for the first solo-crossing of the Atlantic (non-stop from New York city to Paris) spurred Charles Lindberg to be the first to achieve this in 1927; similarly the Kremer as well as the Sikorsky Prizes established in 1959 and 1980 offering large cash prizes for human powered flight, or the Ansari X Prize offering $10 million for the first privately funded reusable manned spacecraft to fly into space. Small but highly popular prizes include the Windows-on-a-Mac Prize to boot Windows on a Mac. Many more such examples exist. Today Prizes seem to be mushrooming, offering rewards from building a computer program that predicts the location of potholes, to designing efficient toilets, to designing driverless vehicles, to creating better, cheaper, and faster ways to sequence genomes.[4]

These quaint anecdotes have a point. Prizes demonstrate a powerful mechanism to incentivize invention. As the examples show, they can very successfully elicit innovation. What is more and more important here, prizes suggest an elegant way to work around Arrow's Fundamental Paradox of Information. Instead of solving the

---

[2] For an intriguing and entertaining description of the 'Problem of longitude' see Sobel (1995)

[3] Incidentally, the same issue seems to be acute again today – though in a different context: space exploration (see example of NASA Prize of space storage below)

[4] For details on Nocolas Appert, see appert-aina.com (02.03.2013). On the Orteig Prize, see, for example, charleslindbergh.com/plane/orteig.asp (02.03.2013) or innovationinthecrowd.com/2011/03/01/charles-lindbergh-and-the-orteig-prize/ (02.03.2013); for the Kremer Competitions see haerosociety.com/About-Us/specgroups/Human-Powered/Kremer (02.03.2013), for the Igor I. Sikorsky Human Powered Helicopter Competition see vtol.org/hph (02.03.2013), and for the Ansari X Prize see space.xprize.org/ansari-x-prize (02.03.2013). For driverless vehicles, see Defense Advanced Research Projects Agency (DARPA) Grand Challenge (darpa.mil/About/History/Archives.aspx 02.03.2013). On the prediction of the location of potholes, see InnoCentive (innocentive.com/ar/challenge/9932752 02.03.2013), on designing toilets see Gates Foundation (gatesfoundation.org/media-center/press-releases/2012/08/bill-gates-names-winners-of-the-reinvent-the-toilet-challenge 02.03.2013), and on sequencing genomes see the Archon Genomics X PRIZE (http://genomics.xprize.org/ 02.03.2013). For more examples, see Krohmal (2007) and Love (2008), also Davis and Davis (2004) and Maurer and Scotchmer (2004).

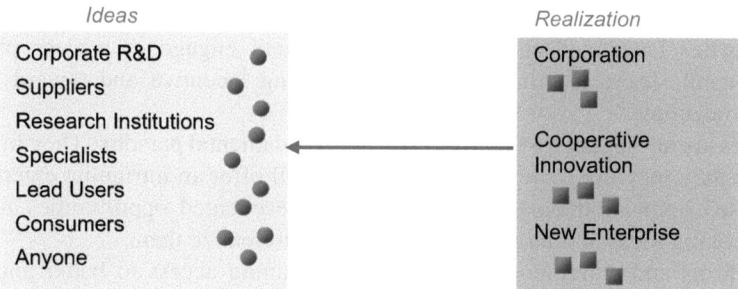

Fig. 9.1 Prizes: looking for ideas to solve a problem

paradox, prizes bypass it. Instead of trying to devise solutions for those with ideas to approach the ones able to realise them, prizes work the other way around: How can the ones seeking ideas best approach those with ideas (see Fig. 9.1).

This is not the same thing. It is not an idea seeking ways to realize it, it is a realization path seeking specific ideas. Someone with a bright idea can still not approach an entrepreneur or firm, this, as the fundamental paradox of information has shown is not that easily possible. It would require negotiation. The one bidding for the idea can only determine a reward knowing the idea. Once it is shared, there is no need to reward it any longer. Without knowing the idea, no value can be assessed. Using prizes, this is different. Prizes look for ideas. It is not ideas looking for realization. The ones announcing the prize already know what they would want the idea to do. They already have a plan, a vision, a strategy. They know very well what the economic return for the idea would be; know what the solution is worth to them. They are the ones intending to realize this idea. This can make a huge difference. A prize is promised. Detailed conditions of success in solving the challenge are spelled out up front. Any workable idea able to solve the challenge can be proposed. The value of the idea is already known. What is missing is the matching idea. It cannot be anything. The cause has to fit its effect. The impact is known. How to get there is not. This offers a clear assessment of the value, though without knowing the actual idea. No negotiation is required. No haggling over the ideas is needed. Clearly the one awarding the prize is interested in the solution. Ideas have been explicitly invited that are obviously sought. From the side of the inventors it is a self-selection process. If someone has a specific solution they offer it if the reward is appealing enough. As long as the idea is able to fulfil the predefined criteria, the prize is theirs. There is no real negotiation. The prize has been determined ex ante in light of any fitting idea. With additional safeguards such as employing independent intermediaries to guarantee or handle the prize, or mediate a clearly predefined reward and the associated evaluation of the solutions, little conflict potential arises. The mechanism can be made reasonably robust. The one looking for a solution knows what it is worth to them and what it would cost them. The ones offering up ideas know the potential reward for their idea up front. Both parties have a clear reward structure. Both engage in a clear planning environment. There is little uncertainty regarding incentive and reward and no need to negotiate.[5]

Prizes are an elegant way to avoid Arrow's fundamental paradox. They bypass it. Though they may not solve the paradox, they still offer an intriguing extension to the closed innovation model. Prizes offer unprecedented opportunities for both those seeking ideas and those with ideas unable to realize them.

For firms and entrepreneurs it may mean gaining access to better and faster solutions. Someone faces a specific challenge. They could task their researchers to solve it, could ask some expert to solve it. This would be a closed form on innovation. Possibly the available in-house resources have no ready answer, have no good idea. Many solutions are possible. It may not be clear which one is the most promising to pursue.[6] Some else, somewhere out there may already have a solution, have a solid hunch. Someone could have a new perspective to solve it faster and cheaper. Someone could simply be smarter. When opening up the invention process to anyone able and willing, anyone with a good idea to come forward and overcome the challenge. Typically, the firm or entrepreneur will specify a problem to be solved offering a prize for the winning contribution.[7] Those able to solve it put forward their ideas. The ideas are vetted and a solution is selected and rewarded. Instead of narrowing down the solution range by approaching a specific expert, or relying on internal R&D, a broader scope of potential ideas is tapped into. Challenges can, and often are solved by people unknown in the respective field, people that would never have been targeted directly, never been commissioned to solve the challenge. Certainly the unknown carpenter and clockmaker John Harrison would not have been commissioned to solve the problem of longitude, but rather, as indeed they were, leading astronomers and other famous academics. Often problems are solved specifically by those less involved, those approaching the problem from expressly different angles.[8] This can lead to better and faster solutions – and solutions to problems that would otherwise not be solved. Prizes offer an easy way to tap into the vast pool of possible solutions in a decentralised

---

[5] Clearly the value of the prize often extends far beyond the pecuniary. It may include the prestige, etc. (see especially also Brunt, Lerner, and Nicholas 2008. For an intriguing example also see Nelson 1993). Should this however become less the exception, but rather the norm, the non-pecuniary aspects (standing out, acquiring prestige, etc.), now of considerable value, will diminish fast.

[6] See Abernathy and Rosenbloom (1969). Also Sommer and Loch (2004), Terwiesch and Ulrich (2009) and Girotra et al. (2010), Kornish and Ulrich (2011).

[7] Blue-sky prizes, as sometimes proposed, fall victim to Arrow's Fundamental Paradox in the sense of adverse behavior and conflict potential. Where such prizes are indeed effective typically more subtle mechanisms are at work (see discussion in Chap. 10).

[8] See Page (2007), also Jeppesen and Lakhani (2010). Also see Granovetter (1973). Also Abernathy and Rosenbloom (1969). Related also Bonawitz et al. (2011).

way. It could be characterized as 'decentralised commissioned research'. More commonly, and certainly more eloquently and pithy, it is labelled 'crowdsourcing'.[9] The assumption is that plenty of experts are located all over the world. Few work for one specific firm. They are widely dispersed. They work for other companies, often in unrelated industries; many are retired; many independent; plenty are bright, eager, and creative students; some are simply bored looking for a challenge, looking to profit from their creativity and brilliance.[10] NASA tried out such crowdsourcing processes. It may very nicely illustrate this approach. NASA surely employs plenty of experts, plenty of the best and the brightest minds of our time. But it also faces an incredibly wide range of challenges. Why reinvent the wheel if solutions to some of the challenges exist? Why devote scarce resources to problems that others have solved or could solve much quicker and easier, possibly also better, cheaper, and more elegantly? Why bare the risk of failing if others are often willing to dare? With this in mind and to try out how open innovation could work for NASA several challenges were posted: among them a $30,000 challenge on forecasting solar activity, one for $15,000 on finding a material for a space food packaging, and a third for $20,000 for an exercise device to reduce the bone and muscle loss astronauts suffer in weightlessness.[11] Proposals were received from all over the world, many from people outside of the narrow realm of space technology researchers, many that have nothing to do with NASA or for that matter even the particular research area in question. More than 1,300 eager inventors participated from over 60 countries.[12] The first, the solar activity forecasting challenge was solved by Bruce Cragin a retired radio frequency engineer from New Hampshire (though with an extensive space research history). As he reports: "Though I hadn't worked in the area of solar physics as such, I had thought a lot about the theory of magnetic reconnection. Also, the image analysis skills I acquired in the 80s, while looking into something called the 'small comet hypothesis,' turned out to be very useful."[13] Yuri Bodrov a scientist from St. Petersburg, Russia, won a partial award for his proposal of a new material for food packaging. While reading the challenge description he immediately had an idea for a solution. As he claims, "his wide and varied work experience has provided him with many different perspectives, which

---

[9] See Wired Magazine (2006), also Howe (2009). Sometimes also other expressions, implying very similar concepts but from a slightly different perspective are used: e.g. 'broadcast search' (Lakhani 2006 and Jeppesen and Lakhani 2010), or 'distributed innovation' (Lakhani and Panetta 2007), or 'Tournaments for Ideas' (Morgan and Wang 2010)

[10] See for example the analysis of participants in the NASA Challenges (InnoCentive 2010, p. 23).

[11] (i) Data-Driven Forecasting of Solar Events (gw.innocentive.com/ar/challenge/9059496 02.03.2013); (ii) Improved Barrier Layers ... Keeping Food Fresh in Space (gw.innocentive.com/ar/challenge/9050426 02.03.2013); (iii) Mechanism for a Compact Aerobic and Resistive Exercise Device (gw.innocentive.com/ar/challenge/9051616 02.03.2013).

[12] See InnoCentive 2010. NASA has published some excellent maps locating the various project teams and submitters.

[13] InnoCentive 2010, p. 30–31

he believes helped him see the solution immediately." "He had heard of using graphite-based materials for holding volumes but never for this particular application"[14] The third, the astronauts exercise device was won by Alex Altshuler a mechanical engineer from Massachusetts, usually working on laser beam scanning technology. "I've never worked on exercise devices! I knew nothing about them – I had to look on the web to learn the basics. I didn't want to re-create the wheel!"[15] Yet the result as NASA attests was 'outstanding'! As the White House professed, Altshuler "had never before responded to a government Request for Proposal (RFP), let alone worked with NASA. And NASA may never have found him or benefitted from his winning insight were it not for the open innovation approach"[16] Unlike internal innovation or 'directly commissioned research', problems are opened up for anyone to solve. This increases not only the likelihood of it being solved, but also to find among the many possible solutions the best one (in terms of costs-benefit, in terms of future potential, etc.). The risk of invention, the chance that valuable time and money leads to no, inferior, or belated solution, is 'outsourced' to all willing to engage in this invention process. The risk of betting on the wrong horse, the risk of failing to come up with a solution despite investing huge sums is reduced. Only if successful is the prize awarded. Possible incentive issues of unmonitorable inputs, etc. are avoided. Search costs are reduced. Firms only have to administer the process and adequately frame the challenge. They have to pay a prize. Yet they only pay for a result. They only pay for a ready solution, not for trying, not for failing, only for the best and most effective and efficient solution. To companies such an approach promises to be more successful as well as possibly considerably more cost effective than mere internal innovation – getting the biggest bang for the buck.[17]

It offers equally appealing opportunities for inventors, for anyone bright and creative. Prizes pose an alluring challenge just as a promising outlet for ideas. Clearly, the prize is a motivation. Anyone able and willing could participate. It may spur people into participating in research, being faced with challenges they may never have considered otherwise. It may spur research by focussing highly creative and bright experts to solving a particular issue, engaging a variety of experts and thus tapping into a variety of fields. Often those who otherwise would not have considered applying their expertise in a particular area are inspired to contribute. Now that they see a possible application, they might be inspired to try. In 2007 the Oil Spill Recover Institute (OSRI), set up in response to the Exxon Valdez oil spill in Alaska, offered a prize of $20,000 to break the viscous shear of crude oil under cold weather conditions – simply put, to reduce the 'thickness' of oil to allow it to

---

[14] InnoCentive 2010, p. 27–28

[15] InnoCentive 2010, p. 29

[16] whitehouse.gov/blog/2010/07/13/nasa-open-innovation-competition-delivers-three-winning-solutions (02.03.2013). Also nasa.gov/centers/johnson/news/releases/2010/J10-017.html (02.03.2013).

[17] Also see Morgan and Wang's decision tree when and when prizes are appropriate (Morgan and Wang 2010, p. 81. Also see Terwiesch and Xu 2008 and Boudreau et al. 2011).

be pumped off more easily. John Davis, a chemist from Illinois, with no connection to the oil industry, realized that the same problem applies to pouring concrete. As he had observed when pouring cement one summer, "concrete vibrators are used to allow the concrete to easily flow into fine cracks and crevices, and also are used to restore liquid flow to concrete that has begun to set-up prematurely"[18] With minor modification, pneumatic concrete vibrators would do the same to oil, winning him the prize.[19]

Apart from inspiration, prizes offer access. Prizes often offer an outlet for existing ideas that would otherwise not be realized. In the early 1990s Dr. Hoffman, a Professor of Environmental Chemistry at the California Institute of Technology, developed a solar-powered portable toilet. As the obvious partner, he offered it to NASA for use on the space station. His idea was met with little interest. It remained shelved. Until now. In 2011, the Bill and Melinda Gates Foundation launched a $100,000 prize, Reinventing The Toilet Challenge. Dr. Hoffman won. His once lost invention now is hailed as a major contribution to "bring safe, affordable and "sustainable" loos to the 40 % of the world's population who lack access to basic sanitation. This could help prevent many of the 1.5 m childhood deaths from diarrhoea that now occur each year."[20] A prize made it possible. It offered an outlet for the idea that would otherwise have likely remained shelved. How to find out who may be in need for such invention? Prizes signal interest. They signal demand for a solution – often from areas the inventor may not have considered.

More importantly still, prizes offer access. They enable access where there was none. Many people may have brilliant ideas, but little or no access to those able to realize them. Incumbents often have insurmountable advantages precluding competitive entry. It would not pay to set up a car manufacturer just because you have a brilliant idea to improve fuel efficiency. The idea might be highly valuable to the car company, but you may have no way of collaborating with them on this. As long as there is no solution to Arrow's paradox, you cannot safely offer your idea to them. Were they to offer a prize for improving fuel efficiency this would offer you access and a reward for your idea. An extreme case would again be the example of NASA. Solar-flare prediction would probably find no alternative use. Who else would employ it? Even if you had an idea, why would you give it to NASA? Whom to contact? How to be rewarded? It would fall victim to Arrow's paradox. Now that NASA offers a hefty reward for solving such challenge, access is suddenly provided for. This is the beauty of prizes. Since it is not possible for the ones with ideas to approach the ones that may or may not be in need of such idea, or else would fall victim to Arrow's fundamental paradox of information, firms may get access to ideas by offering rewards for specific solutions to challenges they face – and inventors get access to firms they would otherwise not have. This offers an outlet for ideas and/or inspires them to look for new solutions. Anyone, from whichever

---

[18] See John Davis blog on InnoCentive (blog.innocentive.com/2008/07/15/john-davis/ 02.03.2013).
[19] See New York Times (2008). Also see Wired Magazine (2006) for the example of Ed Melcarek.
[20] The Economist 2012c. Also see caltech.edu/article/13432 (02.03.2013).

perspective, and whatever background is called on to contribute to highly innovative and unexpected solutions. Whoever has a suitable idea to solve it has an outlet to be rewarded for their ideas, offering direct access to those often most able to realise the idea. For existing solutions, this increases the appropriation of the returns to innovation as it is applied more broadly (also in areas where they may not have thought of applying it originally). Prizes may inspire new solutions – and offer a ready reward for them.

Prizes seem a win-win.[21] To firms they offer faster, cheaper and better innovation. To inventors it offers a reward opportunity, offers inspiration and access to realization paths. No wonder the potential of such prizes has long been widely recognized.[22] Many such prizes already exist. The 'innovation prize' sector is estimated to be in the billions of dollars.[23] Naturally a variety of services have sprung up, connecting those looking for and those offering solutions (see box 9.1). These mostly offer structured platforms to process innovation challenges and match those seeking solutions with those able and willing to solve the 'problem'. A 'coming boom' of such facilitators is prophesised. They are enjoying a remarkable growth, and, according to the providers, have a considerable success rate in finding solutions (or better, those able to propose solution) to the challenges faced by their clients. More and more challenges are posted and participation rates appear to increase, indicating a general satisfaction with the approach. They offer an increased idea pool to be tapped into by those seeking solutions to challenges they are facing, and through it a better utilisation of ideas, more broad based and open innovation.

---

[21] Bar one aspect: Prizes transfer the risk of failing to come up with the winning idea to the inventors. They may invest time and effort, often resources into the invention process – and not win. They typically receive nothing for their efforts. Openness may stifle participation and reduce effort if too broad or too appealing even. The chance of solving it is reduced; the risk increases and many will not partake. This may be dismissed with reference to informed choices and calculated risk for the participants, but may raise macro-inefficiency concerns (see Scotchmer 2004; Taylor 1995; Che and Gale 2003; Moldovanu and Sela 2001; Garcia and Tor 2009; Konrad and Kovenock 2010. Also Boudreau et al. 2011).

[22] The literature on the use of innovation prizes can be traced to a lengthy tradition (e.g. Polanyi 1943; Wright 1983; de Laat 1996; Kremer 1998, 2000; Shavell and van Ypersele 2001; Gallini and Scotchmer 2002; Scotchmer 2004; Maurer and Scotchmer 2004; Davis and Davis 2004, etc.). Discussions typically focused on comparing prizes to patents systems, discussion efficiencies and the use of knowledge in the public domain. Analysis often was based on fairly restrictive assumptions (see for example discussion in Shavell and van Ypersele 2001). Still, many insights can be gained from this literature. The question however differs to the one here as none of the literature considers cooperation and efficiency of implementation in a constraint realization environment, but rather equates invention to innovation. From a more applied business perspective, also see McKinsey (2009), Lakhani and Jeppesen (2007), and Kalil (2006). A more journalistic overview: The Economist 2007, 2010a, 2010b. Also see the related literature on research contests (Terwiesch and Xu 2008; Taylor 1995; Che and Gale 2003), though, again, mainly focused on the optimal design.

[23] Rigby and Zook (2002, p. 83). McKinsey (2009, p. 16) shows a fast growth in 'innovation prizes' estimating the total prize sector to be up to 2 billion dollar in 2007.

Box 9.1 Examples of Innovation Intermediaries
Several Approaches exist, making use of prizes, or passive approaches in general. They are becoming increasingly sophisticated and refined. Some of the more prominent examples of such 'Innovation Intermediaries' include:
i. *InnoCentive (innocentive.com)*
   InnoCentive, spun off 2001 as a venture from pharmaceutical giant Eli Lilly and Company, could be considered a cliché of passive intermediary approaches – in a positive sense. Through its 'global network of millions of problem solvers' it connects those looking for a solution to those with ideas to solve important business, social, policy, scientific, and technical challenges.[24] Companies, or for that matter others, such as non-profit organizations, aptly called '*seekers*', submit a challenge. This challenge is then posted on the InnoCentive website, with a prize, a 'significant financial award' promised for the winning solutions. At the other end there are a self-reported more than 270,000 registered solvers, typically engineers, scientists, inventors, business people, and research organizations in more than 200 countries, and through its strategic partners a reported reach of over 12 million solvers (see Fig. 9.2).[25] These so called '*solvers*' need to be registered, though anyone can easily register, with no verification needed. Simply set up an account and off you go. It is fully open. Anyone can participate. '*Solvers*' can access the different challenges and submit solutions. InnoCentive helps the '*Seekers*' evaluate the submitted solutions. The '*seeker*' selects the winning solution along some criteria stated in the posted Challenge. There's a guarantee that at least one '*Solver*' will win an award. As a rudimentary form of a code of conduct ad hoc 'Lessons Learned' are provided to refine the posting, submission and collaboration mechanism and address several of the issues encountered on both sides, *seekers* and *solvers*.[26] Several services are provided to support this, and assist either side.
   InnoCentive is highly successful. Since 2001 more than a 1,500 challenges have been posted with more than 37 million in prizes. More than 34,000 submissions were made; some 1,300 prizes were awarded with a combined value of over 10 million dollars.[27]
   This is not a selfless effort. It has high ideals – and a business model. InnoCentive Inc. takes a cut of the prize and it charges for posting ideas (and of late offers consultative services to increase returns). Participation of

(continued)

---

[24] See innocentive.com/about-innocentive (02.03.2013)
[25] See innocentive.com/about-innocentive/facts-stats (02.03.2013)
[26] See innocentive.com/faq (02.03.2013)
[27] See innocentive.com/about-innocentive/facts-stats (02.03.2013) and innocentive.com/for-solvers/why-solve (02.03.2013).

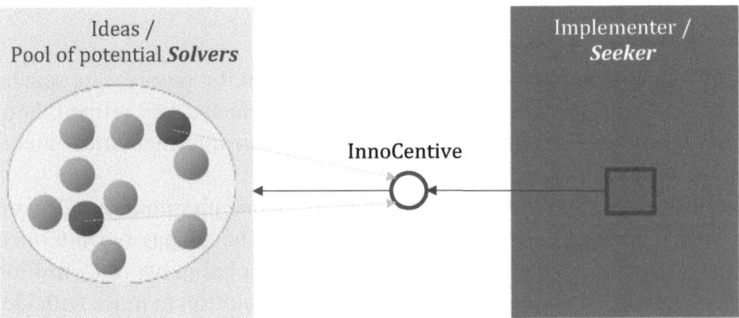

Fig. 9.2 Intermediaries: InnoCentive

> **Box 9.1** (continued)
> solvers is free. The ones looking for ideas, seekers, pay the intermediary. This allows for broad and open access, no barriers, no deterrence. For seekers this is a small price to pay, and makes it easy to assess the costs up front, to weigh the benefits and costs. It is a fixed price model. Everyone knows what to expect, knows what rewards are pending, or costs to be incurred.
>
> The appeal of such approach is surely the breadth of possible solutions, the access to a wide range of possible ideas. Ideas can come from any corner, from unexpected sources. No filter is in place to target the most likely, to approach the path dependent, the solvers to-be. Thus the appeal of such an 'online middleman'. It helps seek out solutions that are either not that apparent in the initial field of inquiry, or to enhance the chances of finding a better solution than the most apparent.[28]
>
> ii. **NineSigma** *(NineSigma.com)*
>
> Preceding the canonization of the Open Innovation movement in 2000, MIT trained Professor of Engineering Dr. Mehran Mehregany founded NineSigma. Its claim of being the largest, leading, the most experienced, the pioneering, and the most mature approach to open innovation is presumptive, but not necessarily unjust.
>
> Most interesting in the current context are its intermediary services such as technology search. Firms looking for a technical solution to a problem can approach NineSigma to find an adequate solution for them. They offer to "Reach, Find, Filter and Focus – expertly articulating the need to expand the pool of potential solutions, searching broadly across geographies, industries

(continued)

---

[28] See especially the finding in Jeppesen and Lakhani 2010, drawing on a large dataset of InnoCentive Challenges

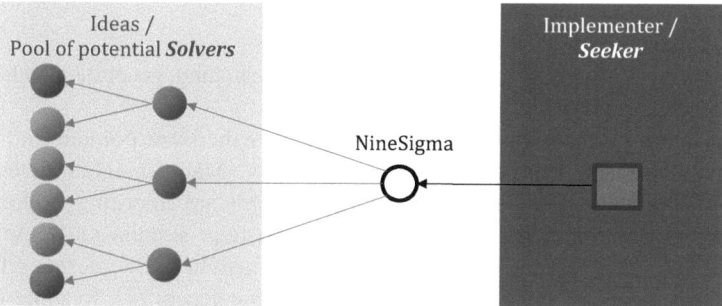

Fig. 9.3 Intermediaries: NineSigma

> **Box 9.1** (continued)
> and technical disciplines – and filtering responses for quality and fit which allows you to focus on the solution(s) that best match your need."[29]
>
> In principle, it is similar to InnoCentive, with a slightly different approach. It takes a more active role in matching solutions to searches. It is not an open self-selection.[30] NineSigma actively targets only (which is still fairly broad) those it deems best suited to solve the challenge. However, these may forward it to others they consider able to solve it. The advantage is a more focused search, targeting a sub-network of the most promising solvers, avoiding being cluttered with proposals (see Fig. 9.3). It lends itself to more complex and broader challenges often across disciplines. An interesting extension of this approach is its system of offering rewards those identifying possible solvers.[31]
>
> The most striking difference is the remuneration approach. Prizes are not set in advance, but negotiated specific to the solution provided. This may consist of 'supply agreements, joint development agreements, co-development or licensing arrangements, research contracts, and consulting agreements'. Cost, for searching, and larger fees if a proposal is accepted, are born by the searching party. NineSigma offers an array of supportive services such as consulting, technology

(continued)

---

[29] ninesigma.com/File%20Library/Product%20Slicks/Datasheet-Tech-Search-final.pdf (02.03.2013)

[30] Though NineSigma Grand Challenges come close, though with a more PR component to it (see ninesigma.com/open-innovation-services/grand-challenge 02.03.2013)

[31] Not an idea, but connectivity and networking is rewarded. Spotting ideas, not just solving! This incentivizes more active matching, delivering the idea to the solver. It rewards knowledge intermediaries able to identify potential solutions even in the most unexpected areas. Though serious concerns of strategic behavior, pricing and potential abuse may arise. See especially also the findings from DARPAs Red Balloon and Tag Challenge (archive.darpa.mil/networkchallenge/ 02.09.2012, or web.mit.edu/press/2009/darpa-challenge-1210.html 02.03.2013). Also The Economist 2012d.

Box 9.1 (continued)

mapping, etc. to augment the mere passiveness of matching, generating additional income streams.

These are just two examples. They are surely the most popular and most widely acclaimed. Both have come a long way. Many insights have been gained, lessons learned, and rudimentary standards are emerging, addressing the intricacies, the challenges of collaboration (e.g. submission templates, codes of conducts of handling ideas, etc.). More insights will emerge, refining these approaches and making them more and more accessible. In some ways, these two are indeed the pioneers, archetypes, and poster children of late of passive approaches. They have enabled and inspired many emerging providers (e.g. InnovationExchange (innovationexchange.com), Inpama (inpama.com/), skipso (skipso.com), Crowdspring (crowdspring.com), ChallengePost (challengepost.com), Zooppa (zooppa.com), IdeaConnection (ideaconnection.com), Big Idea Group (bigideagroup.net), and many, many more).[32] The concept has also been used on an individual basis. Companies have discovered it partly as marketing, partly as a true innovation tool, bypassing the intermediaries. For example, in 2000, GoldCorp, a Canadian gold mining company launched the widely noticed Goldcorp Challenge, offering prizes totalling more than half a million dollars for suggestions of locating gold on their 55,000-acre Red Lake stake.[33] In 2006, Netflix launched the highly popularized Netflix 1 Mio Dollar Prize (netflixprize.com) for finding a better suggestion algorithm for its movies. Political interest in the use of such prizes for social innovation is growing too. Such approaches are increasingly explored and taken up by non-profit organizations, philanthropic initiatives, even national governments and international institutions (e.g. challenge.gov, onebillionminds.com, etc.).[34] Many variations in approaches can now be found. Some incentivize participation though prizes, some though reputation, fame and good will. The principle, of seeking solutions to specific defined challenges remains the common thread.[35]

---

[32] For a broader mapping, see Diener and Piller (2009). They also provide a brief overview and introduce more rigorous classification methodology and structure. For further discussion, also see Chesbrough (2006a). For a structured overview of history, concepts and the relevant literature on the diversity of intermediary functions also other than the passive approaches addressed here, see Howells (2006).

[33] Also see Tapscott and Williams (2006). Also Articles in FastCompany (2002), and in Forbes (2000).

[34] The Economist 2010b. Also Kay (2011), or Goldsmith et al. (2010). Also see National Research Council (2007), Stine (2009), as well as Ziens (2010).

[35] The approaches still vary, though everyone still claims to be the incarnation of open innovation. No clearly dominant approach has yet emerged. Many adjustments are made along the way, signals of a fast evolving, but also not yet mature concept.

This has potentially far-reaching implications for the innovation process. Prizes are an appealing way to work around Arrow's information paradox. In achieving this, prizes contribute to more and better innovation. They improve the use of existing knowledge. Where solutions exist, outreach and identification can help find them. Prizes may inspire participation in the innovation process, and draw in different fields of inquiry to best use their abilities to tackle a specific challenge, thereby overcoming considerable path dependence by broadening the potential sources of ideas to any other field of inquiry not just the most relevant one. Often better and cheaper solutions can be found as the possible source of ideas is broadened. It may also elicit solutions to challenges not otherwise solved. More innovation! It encourages a division of labour, it brings the firms closer to ideas, and their role as implementers. If efficient and practical, prizes can contribute to a better and broader utilization of ideas, thus contribute to the innovation process, increasing innovation and making it more efficient.[36]

Unfortunately, prizes do not solve the information paradox. They avoid it. To some extent they neglect it. They merely confine the problem, which in turn creates issues of its own. The most severe limitation of such an approach is the scope of ideas it encompasses. It only targets a very narrow portion of possible ideas. It does not span the divide for all ideas. It applies only for a small sub-segment of ideas. It only captures those 'ideas' already thought of by the implementers. Only such challenges will be posted to problems the ones posting them have already thought of; products and services they have already envisioned and have identified the bottlenecks. It becomes a question of technical expertise rather than broad ideas, new market opportunities, and clever ideas for new products and services. Prizes do not address the vast range of practical ideas that no one is asking for, precisely because they may not yet have thought of it. Ideas that are not asked for remain without a realistic realisation path. Your idea for a completely novel design of vacuum cleaners may not be thought of by others. They do not see the need, nor may they have your vision to see the potential. Yet, your idea would highly improve the product, would offer a completely new product. As it is not asked for, as no prize is set for it (as it has not been thought of by the incumbents yet), you cannot trade your idea to them. Even more problematic for new business ideas, spotting market opportunities, or improving existing processes. If not asked for, no outlet exists. Asking simply for any new product to be proposed and offering a prize for it would subvert the prize mechanism of avoiding negotiation. It may not be clear how much value such improvement would bring. Negotiation would become necessary – and Arrow's paradox would apply.

---

[36] There may be some concern of macro-inefficiency. The more people participate in the discovery process, the more redundancy of activity is likely to occur. Duplications of efforts and costs occur. If this is greater than the benefits of additional discoveries resulting from this, and the efficiency gain in having access to the invention earlier, this could be deemed inefficient. (See Tandon 1983; Gallini and Kotowitz 1985). Also Maurer and Scotchmer (2002), and Scotchmer (2004). Similarly Che and Gale (2003), Fullerton and McAfee (1999), or Taylor (1995) who argue for a restriction in participants.

Moreover, prizes typically target specialised experts not anyone with an idea. Many ideas are intuitive and very practical; once you have them, there is little of a deep rooted challenge. Those that do are typically highly technical in nature. In addition, very practical issues of handling ideas are often not addressed.[37] How are submissions handled? Who decides what to do with the ones that have not won? The ideas are revealed, to at least those judging them – and if this is the recipient corporation, who is to say they may not still use it, combine it with the winning one, use it in other contexts, etc. without fair compensation. Either the challenges are defined in such a narrow manner that this becomes unlikely because only one, the most efficient solution is needed, or other means of handling ideas need to be sought. Legal mechanisms are sometimes employed, but they are hardly reasonably enforceable in such a context. If they were, this would be just another IP process in disguise. It would be an IP market or product matching exercise. Not a market for ideas. For more open challenges that involve more original ideas such as requests for product ideas, legal provisions would not apply in any case. But with it also many more questions of handling ideas arise, prizes are ill equipped to deal with (how to determine a prize up front? How to handle complementary or competing ideas?). Arrow's Fundamental Paradox of Information would apply more and more the more 'open' such challenges become. It would not get very far. This leaves another severe shortcoming of prizes. The offered remuneration is typically predefined and absolute. The use of blue-sky prizes where the reward is determined only after the idea is revealed are unrealistic and would fall victim to Arrow's fundamental paradox.[38] Thus, prizes are typically defined ex ante. This, by design, encourages those to submit ideas that see this prize worth their effort, but also only those to submit it that do not consider alternative uses of the idea (e.g. setting up a company with a rival product) more lucrative. Take the example of John Wesley Hyatt, the inventor of the first plastic, celluloid. "Hyatt had originally invented celluloid in order to claim a prize posted by a manufacturer of billiard balls who wanted to replace ivory. However, when Hyatt realized that his invention had wider applicability, he chose patent protection instead, apparently judging the value of the patent to be greater than the prize."[39] Also, other strategic behaviour may render the use of prizes inefficient. Sometimes the potential promise of a future prize may deter revealing it, biding time when such prizes or more lucrative prizes are offered, postponing rather than spurring innovation. All of this implies the submission of ideas is confided to a specific spectrum of ideas – a narrow spectrum. In addition posting a specified challenge to be solved will be a revelation mechanism for competitors, offering insights into the specific problems being faced, revealing the state of research, its directions, and may be a source of good ideas for competitors. Such broadcasting of challenges may highly mitigate first mover advantages and other forms of economic benefits to those

---

[37] Also see McKinsey (2009, p. 46)

[38] See for example Scotchmer (2004, p. 40), who acknowledges: "the innovator must trust that an appropriate prize will be given." Also Maurer and Scotchmer (2004).

[39] For an example see Maurer and Scotchmer (2004, p. 9). To be correct, Hyatt really simplified the production of celluloid, Alexander Parkes invented it.

posting the challenge – which is often the most valuable part of the innovation. In a larger sense this also may give rise to issues of research alignment between competitors, may create undue market noise or incentives to abuse such mechanisms for strategic misdirection.[40]

In light of these issues, it would not be unfair to say, such reversed innovation approaches, still confine the 'open', broad based innovation process to a somewhat modified form of closed innovation. In practice, they focus almost exclusively on intellectual property considerations, they typically target highly specialised experts, and are by design confined to preconceived internal strategies and ideas. More visionary ideas, original ideas from outside, disruptive ideas, etc. all those not directly sought by the implementers are neglected. It remains confined to predetermined mostly path dependent innovation processes, in terms of thought-out requests. Bigger, more out-of-the-box ideas, still remain limited to a more disruptive, entrepreneurial approach – which is not addressed through such mechanisms. Challenges will remain narrowly defined, and thus only slightly increasing the involvement of the idea pool. The paradigm of prizes remains: ideas are commissioned, not invited.[41] It may narrow the gap somewhat by enabling a linkage between certain kind of ideas, and for a certain range of innovations, but it is unable to address the underlying challenges or come close to bridging the gap. It is a complementary approach, not a solution.

References

Abernathy, W. J., & Rosenbloom, R. S. (1969). Parallel strategies in development projects. *Management Science, 15*(10), 486–505.

Bonawitz, E., Shafto, P., Gweon, H., Goodman, N. D., Spelke, E., & Schulz, L. (2011). The double-edged sword of pedagogy: Instruction limits spontaneous exploration and discovery. *Cognition, 120*(3), 322–330.

Boudreau, K. J., Lacetera, N., & Lakhani, K. (2011). Incentives and problem uncertainty in innovation contests: An empirical analysis. *Management Science, 57*(5), 843–863.

Che, Y. K., & Gale, I. (2003). Optimal design of research contests. *The American Economic Review, 93*(3), 646–671.

Chesbrough, H. W. (2006a). *Open business models: How to thrive in the new innovation landscape*. Cambridge: Harvard Business School Press.

---

[40] This has been realized by many passive providers: e.g. InnoCentive. Additional mechanisms have been put in place (e.g. anonymity of seekers). "Managed intermediary process" are becoming the norm. These still do not address the issue fully. They still reveal plenty of information and can be an important source of information to competitors, and even without the details can inspire competitors to pursue similar research. See for example Lakhani and Panetta (2007).

[41] Without getting into semantics, the difference would be that in these cases a specific task to be solved is posted. Ideas may be submitted that address this challenge. This is not an open invitation for any kind of idea. In the extreme – if the request for solutions is phrased so broadly as to include any kind of ideas – of course the two may coincide. In this case however the same issue of the information paradox would apply, and the notion of a passive approach, identified through its restrictive and targeted use to confine or avoid these issues, is lost.

Davis, L., & Davis, J. (2004). *How effective are prizes as incentives to innovation? Evidence from three 20th century contests*. Paper to be presented at the DRUID summer conference 2004 on industrial dynamics. Innovation and Development Elsinore, Denmark. commercialspace. pbworks.com/f/Davis+%26+Davis+2004.05.07.pdf. Accessed 02 March 2013.

de Laat, E.A.A. (1996). Patents or prizes: Monopolistic R&D and asymmetric information. *International Journal of Industrial Organization, 15*(3), 369–390.

Diener, K., & Piller, F. (2009). *The market for open innovation: Increasing the efficiency and effectiveness of the innovation process* (Open innovation accelerator survey 2009). Aachen: RWTH-TIM Group.

FastCompany (2002). *He struck gold on the net (Really)*. By Tischler, L., 31 May. www.fastcompany.com/44917/he-struck-gold-net-really. Accessed 02 March 2013.

Forbes (2000). *Gold rush*. By B Coffey, 3 July. www.forbes.com/forbes/2000/0703/6601126a.html. Accessed 02 March 2013.

Fullerton, R. L., & McAfee, R. P. (1999). Auctioning entry into tournaments. *The American Economic Review, 107*(3), 573–605.

Gallini, N. T., & Kotowitz, Y. (1985). Optimal R and D processes and competition. *Economica, 52*(207), 321–334.

Gallini, N.T., & Scotchmer, S. (2002). Intellectual property: When is it the best incentive system? *Innovation Policy and the Economy, 2*, 51–77.

Garcia, S. M., & Tor, A. (2009). The N-effect: More competitors, less competition. *Psychological Science, 20*(7), 871–877.

Girotra, K., Terwiesch, C., & Ulrich, K. T. (2010). Idea generation and the quality of the best idea. *Management Science, 56*(4), 591–605.

Goldsmith, S., Georges, G., & Burke, T. G. (2010). *The power of social innovation: How civic entrepreneurs ignite community networks for good*. New York: Wiley.

Granovetter, M. S. (1973). The strength of weak ties. *The American Journal of Sociology, 78*(6), 1360–1380.

Howe, J. (2009). *Crowdsourcing: Why the power of the crowd is driving the future of business*. New York: Random House.

Howells, J. (2006). Intermediation and the role of intermediaries in innovation. *Research Policy, 35*(5), 715–728.

Jeppesen, L. B., & Lakhani, K. R. (2010). Marginality and problem solving effectiveness in broadcast search. *Organization Science, 21*(5), 1016–1033.

Kalil, T. (2006). Prizes for technological innovation. Brookings Institution Discussion Paper 2006–08, http://www.brookings.edu/research/papers/2006/12/healthcare-kalil, www.brookings.edu/ research/papers/2006/12/healthcare-kalil. Accessed 2 Mar 2013.

Kay, L. (2011). *Managing innovation prizes in government*. IBM Center for The Business of Government, Washington, DC. www.businessofgovernment.org/report/managing-innovation-prizes-government. Accessed 02 March 2013.

Konrad, K. A., & Kovenock, D. (2010). Contests with stochastic abilities. *Economic Inquiry, 48*(1), 89–103.

Kornish, L. J., & Ulrich, K. T. (2011). Opportunity spaces in innovation – empirical analysis of large samples of ideas. *Management Science, 57*(1), 107–128.

Kremer, M. (1998). Patent buyouts: A mechanism for encouraging innovation. *Quarterly Journal of Economics, 113*(4), 1137–1167.

Kremer, M. (2000). Creating markets for new vaccines. Part II: Design issues. *Innovation Policy and the Economy, 1*, 73–118.

Krohmal, B. (2007). *Prominent innovation prizes and reward programs*. KEI research note 2007:1, http://www.keionline.org/misc-docs/research_notes/kei_rn_2008_1.pdf. Accessed 02 March 2013.

Lakhani, K.R. (2006). Broadcast search in problem solving: attracting solutions from the periphery. *Technology Management for the Global Future, 4*, 2450–2468. doi:10.1109/PICMET.2006.296842.

References

Lakhani, K. R., & Panetta, J. A. (2007). The principles of distributed innovation. *Innovations Technology Governance Globalization, 2*(3), 97–112.

Lakhani, K. R., & Jeppesen, L. B. (2007). Getting unusual suspects to solve R&D puzzles. *Harvard Business Review, 85*(5), 30–32.

Love, J. (2008). *Selected innovation prizes and reward programs.* KEI Research Note 2008:1, http://www.keionline.org/misc-docs/research_notes/kei_rn_2008_1.pdf. Accessed 02 March 2013.

Maurer, S.M., & Scotchmer, S. (2002). The independent invention defence in intellectual property. *Economica, 69*(276), 535–547.

Maurer, S. M., & Scotchmer, S. (2004). Procuring knowledge. *Advances in the Study of Entrepreneurship Innovation and Economic Growth, 15*, 1–31.

McKinsey. (2009). And the winner is…Capturing the promise of philanthropic prizes. http://www.mckinsey.com/App_Media/Reports/SSO/And_the_winner_is.pdf, www.mckinsey.com/App_Media/Reports/SSO/And_the_winner_is.pdf. Accessed 2 Mar 2013.

Moldovanu, B., & Sela, A. (2001). The optimal allocation of prizes in contests. *The American Economic Review, 91*(3), 542–558.

Morgan, J., & Wang, R. (2010). Tournaments for ideas. *California Management Review, 52*(2), 77–97.

National Research Council. (2007). *Innovation inducement prizes at the national science foundation.* Washington, DC: National Academies Press.

Nelson, K. E. (1993). Dow's energy/WRAP contest- A 12–Yr energy and waste reduction success story. *Proceedings from the fifteenth national industrial energy technology conference*, Houston, 24–25, Mar 1993. http://repository.tamu.edu/bitstream/handle/1969.1/92057/ESL-IE-93-03-03.pdf?sequence=1. Accessed 02 March 2013.

New York Times (2008). *If you have a problem, ask everyone.* 22 July, by Dean, C. www.nytimes.com/2008/07/22/science/22inno.html. Accessed 02 March 2013.

Page, S. E. (2007). *The difference: How the power of diversity creates better groups, firms, schools, and societies.* Princeton: Princeton University Press.

Polanyi, M. (1943). Patent reform. *The Review of Economic Studies, 11*(2), 61–76.

Rigby, D., & Zook, C. (2002). Open market innovation. *Harvard Business Review, 80*(10), 80–89.

Scotchmer, S. (2004). *Innovation and incentives.* Cambridge: MIT Press.

Shavell, S., & van Ypersele, T. (2001). Rewards versus intellectual property rights. *Journal of Law and Economics, 44*(2), 525–547.

Sobel, D. (1995). *Longitude: The true story of a lone genius who solved the greatest scientific problem of His time.* New York: Walker and Company.

Sommer, S. C., & Loch, C. H. (2004). Selectionism and learning in projects with complexity and unforeseeable uncertainty. *Management Science, 50*(10), 1334–1347.

Stine, D. D. (2009). *Federally funded innovation inducement prizes.* CRS report for congress, Congressional Research Service, Washington, DC. www.fas.org/sgp/crs/misc/R40677.pdf. Accessed 02 March 2013.

Tandon, P. (1983). Rivalry and the excessive allocation of resources to research. *Bell Journal of Economics, 14*(1), 152–1650.

Tapscott, D., & Williams, A. D. (2006). *Wikinomics: How mass collaboration changes everything.* New York: Penguin.

Taylor, C. R. (1995). Digging for golden carrots: An analysis of research tournaments. *The American Economic Review, 85*(4), 872–890.

Terwiesch, C., & Xu, Y. (2008). Innovation contests, open innovation, and multiagent problem solving. *Management Science, 54*(9), 1529–1543.

Terwiesch, C., & Ulrich, K. T. (2009). *Innovation tournaments: Creating and selecting exceptional opportunities.* Cambridge: Harvard Business Press.

Wired Magazine (2006). *The rise of crowdsourcing.* 14 June 14, by Howe, J. www.wired.com/wired/archive/14.06/crowds.html. Accessed 02 March 2013.

Wright, B.D. (1983). The economics of invention incentives: Patents, prizes, and research contracts. *The American Economic Review, 73*(4), 691–707.

Ziens, J. D. (2010). *Guidance on the use of challenges and prizes to promote open government.* Executive office of the president: Office of management and budget, Washington, DC. www.whitehouse.gov/sites/default/files/omb/assets/memoranda_2010/m10-11.pdf. Accessed 02 March 2013.

Bridging the Gap

10

If innovation prizes are not the solution, how can Arrow's paradox be solved? In some niches such as the movie industry or venture capital, trading ideas is somewhat possible. Identifying the underlying principles can help draw up a solution to make trading possible in general. Trust and fairness between the partners are essential. Research has made significant progress in analysing and finding mechanisms to support both. Establishing a reputation mechanism and norms of fairness are the most promising elements to make cooperative innovation a reality.

Prizes are good. They narrow the gap. They allow some more to partake in the innovation process. Solutions are better. To enable true trading of ideas passive approaches such as prizes do not suffice. They cannot bridge the missing link. They do not overcome or compensate for the information paradox. Though they may indeed expand the spectrum of ideas to be traded, this spectrum remains mostly confined to a still narrow segment of the idea pool. It is no broad based solution. To offer a solution for a broader spectrum of ideas, and for those generating these ideas to be included and ultimately to enable the realization of their ideas, different approaches are required. More active solutions to trading ideas are needed. Solutions are required where those with ideas can freely approach the ones able to realize them – without being confined to a specific predefined challenge. This is, if at all, currently mostly only feasible where clearly defined intellectual property rights apply. But the scope of IP is limited to but a fraction of ideas; its applicability is exclusive and often prohibitively expensive. Most ideas fall outside. Your idea may not be patentable or not patentable yet while you are in the initial phase. No challenge fitting your idea may exist. You may have no access to appropriate means of realising the idea, or the entrepreneurial skills required. Without family, friends, and fools (the innovator's three F) to help out and possibly a befriended and trusted entrepreneur, what are you to do? Similarly on the other side: There may be a strong reluctance of those able to realize ideas to fully open up to idea proposals. How are they to be handled? How evaluated? How would cooperation work? Much conflict potential arises here and possibly high costs are associated with such opening up to any ideas.[1] In addition, the existing corporate innovation management tools are still unrefined to handle this so that many ideas fall victim to still outdated and too narrow knowledge filters. Mechanisms and tools are needed to address these issues on both sides in order to bridge this gap. Many, potentially vastly valuable ideas are lost to such lack of a realistic trading mechanism – and without practical solutions the *Idea Economy* cannot advance much further.

Solutions exist. Active approaches are not that uncommon. They are often just not that obvious. If you look around carefully you will find that such active approaches are already commonplace in several niches and on a smaller scale. Here the ones with ideas do indeed approach those able to realize them. Here often some seemingly refined markets for ideas exist. To some extent it is often secured through latent intellectual property law – as it could apply to several other sectors – but more importantly, a variety of subtle mechanisms are at play that may hint at underlying general concepts. The most common examples of such markets for ideas are the venture capital industry, the film industry, and several others: pitching ideas, filtering ideas, realising ideas. Some invent, others realize, or make realization possible. A division of labour exists. Ideas are practically traded. Some come up with them and offer them to others, others better suited to realize them. Such markets for ideas build on often subtle mechanism, that, when identified properly, can be scaled up and opened up for a wider audience, across different industries, and beyond some confined valleys, villages or clusters.[2]

Some firms have tried to tap into such market for ideas. They have tried to open up their innovation process in a similar manner, fully opening up to outside inventions on a broad basis. The most prominent and possibly the first and most successful is Procter and Gamble's 'Connect and Develop' (pgconnectdevelop.com) (see box 10.1). Several other firms have established similar access points (e.g. General Mills (openinnovation.generalmills.com), Dell (ideastorm.com). These invite inventors to approach the firms and offer their ideas. Anyone with a clever idea can approach them and offer it to them. The hope is to cooperate in realizing these inventions – and to both profit from them. In some cases, such cooperation is the only option for the inventor to see their idea realized. Often it is the simpler and more lucrative option to pursue than to attempt an entrepreneurial innovation path.

---

[1] See the example of Disney (in Rigby and Zook 2002, p. 86–87: "when Disney decided to seek ideas from the outside, it got hit with some eye-popping lawsuits. In August 2000, a jury awarded $240 million to All Pro Sports Camps, which had accused Disney of entertaining then stealing its ideas for a sports-themed entertainment complex. That kind of legal threat, which resulted from Disney's ad hoc approach to the situation, can lead to protocols that discourage employees from even listening to proposals from outsiders.")

[2] The most striking analysis that hints at a more general solution as suggested in the following is surely Shane and Cable (2002). Here the emphasis in on social ties, both direct and indirect and the role of reputation. This study however does not go further to identify a more general principle. It rather builds on the proximity and use of close social ties for information.

> **Box 10.1 Connect and Develop**
> Procter and Gamble's active approach to accessing Ideas.[3]
>
> > Do you have a game-changing product, technology, business model, method, trademark, package or design that can help deliver new products and/or services that improve the lives of the world's consumers? Do you have commercial opportunities for existing P&G products/brands? If so, we'd like to consider a partnership.[4]
>
> This is how you are welcomed when visiting Procter and Gamble's (P&G) Connect and Develop Website (pgconnectdevelop.com). Captivating, promising, succinct. It sounds like an opportunity. It is. For you and P&G. This is what true openness sounds like, offering a partnership to the benefit of both.
>
> P&G set itself an ambitious goal: "We want to partner with the best innovators everywhere, which is why Connect + Develop is at the heart of how P&G innovates", in order "to acquire 50 % of our innovations outside the company."[5] This is a truly active approach to cooperative innovation. They have not only pioneered its own interactive outreach for active submissions, but are at the same time offering passive approaches such as prize challenges, to guide innovation and are participating in existing passive initiatives while also employing a network of their own technology scouts. Together this extends their outreach to a potential target audience (see Fig. 10.1).
>
> The process seems straightforward. You have an idea. You go to the Connect and Develop website and submit your idea. "Each submission is reviewed by a Connect + Develop team member who determines how to route it to ensure it reaches the relevant business and/or technical personnel within our Company. In many cases, this means review by multiple P&G organizations to evaluate technical merit or strategic business fit for P&G. We are committed to reaching a decision as quickly as possible. However, a thorough initial review, especially of a technology innovation, may take as long as 8 weeks to complete."[6] Simple; reasonably transparent; intuitive;

(continued)

---

[3] For more information see especially Huston and Sakkab (2006) and Sakkab (2007) and pgconnectdevelop.com (02.03.2013).

[4] This was in 2009. It has since changed. It still reads "Could your innovation be the next game-changer?", but the enthusiasm has faded and it offers: "Is there a new product you'd like to see, or a change to an existing one? Our Consumer Relations Team welcomes your input and is happy to answer product questions." "If you are the owner or representative of an innovation that might help improve the lives of consumers, this site can help you submit your innovation for review by our Connect + Develop Team." (ipgconnectdevelop.com/home/home.html 02.03.2013)

[5] Huston and Sakkab (2006, p. 61) – as opposed to 10 % in 2000.

[6] pgconnectdevelop.com/home/frequently_asked_questions/about-submitting-to-pg.html (02.03.2013)

Box 10.1 (continued)

successful![7] So far, C + D has led to more than 2,000 successful agreements with innovation partners around the world.

This approach is born out of visionary pragmatism: "The strategy wasn't to replace the capabilities of our 7,500 researchers and support staff, but to better leverage them. Half of our new products, [...] would come from our own labs, and half would come through them."It was, and still is, a radical idea. As we studied outside sources of innovation, we estimated that for every P&G researcher there were 200 scientists or engineers elsewhere in the world who were just as good–a total of perhaps 1.5 million people whose talents we could potentially use. But tapping into the creative thinking of inventors and others on the outside would require massive operational changes. We needed to move the company's attitude from resistance to innovations "not invented here" to enthusiasm for those "proudly found elsewhere." And we needed to change how we defined, and perceived, our R&D organization–from 7,500 people inside to 7,500 plus1.5million outside, with a permeable boundary between them.[8] Brilliant! So simple – seemingly. Others have made the same calculation but failed to draw the same pragmatic conclusions.[9] P&G's approach remains at the forefront of openness. And for a good reason. P&G believes "that connect and develop will become the dominant innovation model in the twenty-first century [...] Companies that fail to adapt to this model won't survive the competition."[10]

Though promising, it is not perfect. Many practical issues remain to be addressed still.[11] Most limiting: it remains highly IP-centric, confining it to a very particular set of ideas. While originally also non-IP ideas were invited this is no longer the case. It seems, too many challenges arose that did not yet

(continued)

---

[7] See pgconnectdevelop.com (02.03.2013). Other sources complement this number:

"•P&G works proactively with more than 85 network and 120+ universities. Seventy five percent of their searches within these networks result in viable leads.

•P&G's "open door" for unsolicited innovation submissions – Connect + Develop – generated nearly 4,000 leads last year. The site is now available in 5 languages (English, Chinese, Japanese, Portuguese, and Spanish).

•P&G invests in relationships over time to become the preferred partner for open innovation. Forty percent of their relationships results in repeat deals.

•More than 50 % innovation is sourced externally (up from < 10 % in 2001)."

15inno.com/2010/06/02/criticalfactslessons/ (02.03.2013, originally published June 2, 2010)

[8] Huston and Sakkab (2006, p. 61)

[9] For example Merck (see anrpt2000.com/innovation2.htm 02.03.2013, also see Chesbrough 2006a, p. 53).

[10] Huston and Sakkab (2006, p. 66)

[11] Also see Sakkab (2007)

> **Box 10.1** (continued)
> have a clean solution.[12] While once a pioneer of more collaborative innovation, others now seem to be carrying the torch.[13]

There is a problem with such examples. What works for Procter and Gamble, what works for Hollywood and Silicon Valley may very well be confined to specific circumstances. They cannot be generalized that simply. What may work in a special case, and has evolved over a long time, cannot simply be generalized. The basic principle of how it works first needs to be identified. What may work for the film industry may not work for ideas in general. The industry has matured and the mechanisms have been refined to a considerable degree. They are not without problems. How often have you seen strikingly similar movie ideas come out of different studios, that make you wonder who copied from whom. A number of high profile legal quarrels exist, authors suing studios over stealing their ideas.[14] Its principles however may still hold the key to establishing market for ideas. The same holds for other examples such as the venture capital industry, or attempts by consumer firms to open up their innovation process. Procter and Gamble's attempt is on-going. Identifying its basic principles of how it works can still help inform more general solutions. More importantly, as it is still a young concept, raw and somewhat unrefined, the difficulties it faces may help point out challenges important to be taken into account when considering more general solutions. Indeed, when analysing the underlying mechanisms of such examples and by drawing on the current state of research, solutions to more broadly applicable active approaches to cooperative innovation seem within reach. Much progress has been made in the understanding of many of the underlying principles to address the information paradox and overcome, or at least work around it. Drawing on a wide range of research and prevailing practices some very pragmatic solutions can be found. From such analysis, two broad underlying principles can be identified. Cooperative innovation requires both: trust and fairness.

---

[12] "For your protection and ours, we may decline to review your submission if it appears to lack intellectual property protection." pgconnectdevelop.com/home/frequently_asked_questions/about-submitting-to-pg.html (02.03.2013) and further: "Idea submission guidelines now read: "Innovations should also include protectable intellectual property, typically in the form of a granted patent or published patent application ... Please do not submit the following: Innovations which you do not own or have the legal right to represent, or about which you are unable to provide information on a nonconfidential basis. ... Ideas, suggestions or thoughts that do not include protectable intellectual property." pgconnectdevelop.com/home/submission_criteria.html (02.03.2013)

[13] For example, Vileda (vileda.co.uk/uk/innovations 02.03.2013) who appear suspiciously similar to the early stages of Connect and Develop, using very similar language and tone.

[14] See Desny v. Wilder, 299 P. 2d 257 – Cal: Supreme Court 1956: Gunther-Wahl Productions, Inc. v. Mattel, Inc., 128 Cal. Rptr. 2d 50 – Cal: Court of Appeals, 2nd Appellate Dist., 8th Div. 2002; Grosso v. Miramax Film Corp., 383 F. 3d 965 – Court of Appeals, 9th Circuit 2004; Montz v. Pilgrim Films & Television, Inc., 606 F.3d 1153, 1158 – Court of Appeals, 9th Circuit 2011.

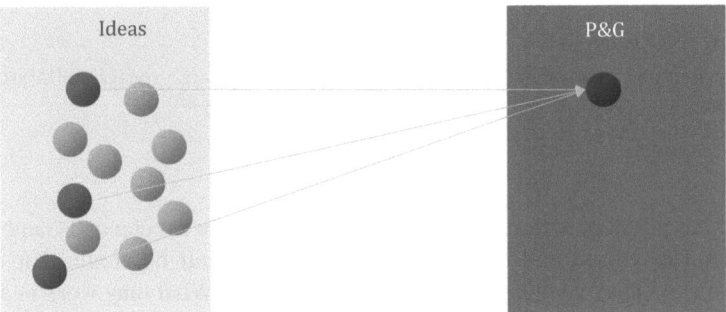

Fig. 10.1 Connect and develop

10.1 Trust

When it comes to cooperative innovation, trust is essential. Cooperation is based on the notion of each contributing to a common objective. It requires that people trust one another to have a common interest in making the cooperation succeed and make the most of the idea. You need to be able to trust your partners not to cheat you, steal your idea or sabotage your success. But how can you trust your partner. He/she may be a decent person, honest and cooperative, eager to realize the idea jointly, benefitting from the idea together with you. Or not. He/she may be scrupulous. You sharing the idea shifts the power over to your partner. He/she is typically better positioned to realize the idea – that is the reason why you chose him/her. Now knowing, now in the possession of the idea, what would keep him/her from running off with it, realize it, and exclude you from any proceeds? Nothing. Indeed, the incentive structure is rather in favour of him/her to cheat. He/she could make much more profit from not sharing. Your contribution other than the idea itself is probably replaceable. And what is there to fear? No real legal sanctions apply here. Only a sense of moral may ensure a fruitful cooperation. This is a problem, a serious one. To ensure that such uncooperative incentive is overruled a robust mechanism is sought, able to incentivise cooperation. Then, even strangers you can trust in their intention, you can trust in their cooperation.[15]

Trust is a very basic notion underlying many social processes just as it does existing approaches to trading ideas. Certain mechanisms are at play that, when identified, are very similar to those that enable trust in many of the existing approaches to cooperation. Exposing them can help draw up a more general mechanism that can incentivize cooperation also when it comes to ideas. When looking carefully at social interaction and at functioning cooperation in the innovation process, they become more obvious: Assume you tell a secret to a

---

[15] It is important to note that this need neither imply that all would indeed cheat (see Bolton and Ockenfels 2009), nor that the least common denominator should be the basis for the following discussion (see Bowles and Hwang 2008)

friend. This friend tells it to others you did not want it to know. You feel betrayed. You will stop sharing your innermost thoughts with this person. Knowing someone gossips will presumably keep you from divulging too much information in the future. The friend will be kept out of the loop in the future. You may shun him/her on occasion. You may want to distance yourself a little. You may tell your friends that he/she gossips and that they should be careful. Very similar applies to ideas. Imagine you are a screenwriter with a brilliant idea for a movie. You pitch the idea to a producer. He recognizes the value of your idea, takes it up, realises it, and makes a handsome profit. You get nothing. He steals it from you without compensating you for it. Would you ever pitch him another idea? Presumably not. More importantly yet: had you known he steals, you probably would not even have offered your idea to him in the first place. Now you are upset. You may confront him, write angry letters complaining. Certainly, you are sure to tell anyone who will listen not to trust this producer. He steals. The same could be said about venture capital, even, in a very similar sense, about Procter and Gamble. If they were to steal, you would probably not trust them again.

It may not be that obvious here, how this may overcome the information paradox. The examples merely seem to illustrate the problem rather than offer solutions. Knowing what you are looking for these examples make a lot of sense. The basic insights: people you do not trust you will not engage, you will not cooperate with. If someone steals ideas, you will no longer cooperate with him/her. Their reputation is ruined. No one who knows they steal will cooperate with them. This can severely hurt them. Your former friend will likely slowly be excluded from the group if he/she does not change; the film producer will not receive any more proposals from you or anyone you know, or who has heard of your story. It becomes even more obvious in fact when you think of ideas directly. Assume a large corporation invites you to submit ideas for new products. You have a great idea that fits very well with the company's profile, their competence and ambition. You offer it to them in the hope of being rewarded for it. You are not. Your idea is taken up, realised, and the corporation generates significant profits with it. None is for you. Clearly, this would be expected from such corporation if it had the possibility (not saying all would). If it can access the invention for free, why not? It probably has no legal or legally enforceable obligation to reward you. So why should it not take advantage of the situation and profit from it. You surely would be upset. But so what? One disgruntled customer more or less. This is where the logic falls short. Stealing ideas would be a myopic approach. You are unlikely ever again to submit any ideas to them, and certainly none that they would profit from. This, again, may not matter to them this much, unless of course you are such a brilliant mind that their fortunes may depend on it. Others have good ideas too. But they may take your experience to heart. They will not want to cooperate and share ideas with them either, knowing this company is not trustworthy: 'they steal ideas'. Their reputation among the idea community is ruined. No one is likely to share any more ideas with them. In a market where innovation is the core competitive advantage, this would have significant fallout. Drawing on the ideas of outsiders would no longer be an option. Merely the old model of trying to generate ideas within the company could still be an option. Even more unfortunate for them: Inventors will offer the ideas to others, to other competitors. Competitors, those more cooperative and more trustworthy would still be offered ideas and increasingly compete away more and more of the market. While they might prosper, the untrustworthy firm suffers. This is no tale of moral and good corporate behaviour. It is one of survival and incentive. The fallout of failing to access the idea pool outside the firm in the future is most likely to be much greater than the gain of freeriding on a very limited number of ideas before people find out and stop offering up ideas – and, to even greater dismay, rather submit their ideas to a more trustworthy competitor. Given the vast scope of the idea spectrum outside a corporation today, and the phenomenal economic potential of tapping into these ideas, a continued access is more likely to succeed and more lucrative than to ever freeride and default on the offered cooperation of those with ideas.

Generalizing this concept leads to what is known as repeated games and reputation (see box 10.2). Building on the common notion of trust and interaction can help address the information paradox.[16]

> **Box 10.2 Theory of Cooperation**
> Repeated games and reputation.
>
> The theoretical intuition of cooperation and incentive structures is fairly complex, but its basic intuitions can be shown in a simple example.[17] Assume Merrill, an attentive consumer and skilled tinkerer, has a brilliant idea. He contacts Albert, a gifted serial entrepreneur, to help him realise it. It is an idea, amorphous, nothing legally binding. Albert could just take up the idea, realize it and not share any profits with Merrill. Why should he? There is no incentive for him to reward Merrill for his idea. On the contrary! Albert would be better off not having to share any spoils with Merrill. Albert has,
>
> (continued)

---

[16] Interestingly enough Kenneth Arrow drew a very similar conclusion, but failed to connect it in a specific setting of information, of ideas (see Arrow 1972, p. 367).

[17] As small tribute to Merrill Flood (and Melvin Drescher) as well as Albert W. Tucker. In a more formal setting, the example could be interpreted as a prisoner's dilemma: not only Albert could cheat Merrill, but also the other way around. Merrill could use Albert's help to realise the idea and then exclude him from the profit (a less likely scenario, and more easy to remedy by legal provisions). Risk of deviation seems to be greater with the one realising than having the idea. Merrill needs someone to realise the idea, or he would not have cooperated with Albert in the first place. Once he reveals the idea to Albert, Albert could still realise the Idea, while Merrill would probably have a more difficult time (lacking the required skills, Albert's first mover advantage, etc.). Hence, the risk of defection is stronger on the side of the realisation partner than the one having the idea. But by no means exclusive. Merrill may offer the idea to several partners at once, and see who realises it quickest and to greatest advantage – thus Albert would invest in realising it and plan with it, realising only later a competitor has beaten him to it.

Box 10.2 (continued)

what is called, 'an incentive to defect'. So why would Merrill take the risk? Knowing he might, or likely will be cheated, he would not risk sharing his idea with Albert. Indeed, Albert might be a virtuous and fair cooperation partner. But how can Merrill know, or be sure. It would still be tempting for Albert to cheat. The incentives would work against cooperation. Cooperation is unlikely. It would be a huge risk for Merrill to cooperate, and a good opportunity for Albert to cheat. Without adequate mechanism to overcome these adverse incentives, cooperation is unlikely to be a sustainable solution. Binging a dog may not suffice (although it has been suggested the presence of a dog will significantly increase cooperative behaviour).[18] Many approaches exist. For the current context, one in particular is relevant: reputation.

This is a very static perspective. It holds if this only happens once and has no other implications. If this game is repeated, more dynamic incentives arise, often overturning the initial non-cooperative incentives. Assume Merrill is a brilliant person who comes up with highly promising ideas frequently. Albert knows this. Merrill and Albert cooperate frequently in realising new ideas. Both could make a handsome profit from it (see Fig. 10.2, scenario 1). But this assumes Albert would be cooperative. Why would he? Would he not be better off cheating? Assume at some point Albert would defect (see Fig. 10.2, scenario 2, at t_5). He takes Merrill's idea, realises it and keeps the profit – substantially more than if he had to split it with Merrill. Merrill would be upset. He would lose out on the profit. Why would he keep on cooperating with Albert? Albert cheats. Merrill would simply walk away and never again cooperate with Albert.

This is more than a stubborn reaction. This clearly would hurt Albert.[19] All possible future profits from cooperating with Merrill would be lost. Taken together, all the profits from future cooperation would most likely be much larger than profiting once from a defection. The aggregate payoff would most likely be considerably larger in a lasting cooperation. Merrill on the other hand may find someone else to cooperate with, or try to realize his ideas on

(continued)

---

[18] See The Economist 2010c.

[19] In fact the mechanism could indeed be strengthened, and the punishment of defection increased by destroying the actual value of the idea ex post. If someone steals the idea, instead of letting him/her walk away and actually profiting from it, the other party could reveal the idea publically or at least to competitors, often severely mitigating the potential payoff of such an idea. The one stealing would thus be left with less, possibly even less than in the case of continued cooperation. This rests on well-established analysis especially by Anton and Yao (1994, 2004, 2005). Indeed if it were possible (though it may strongly vary with the nature of the idea, the relative position of the partners, etc. not least also possibly the fact that the one with ideas would have likely chosen the one best to realize the idea, and revealing it, only ex post would not mitigate its value too much) this would alter the analysis somewhat. It certainly would strengthen any such reputation mechanism. It may even sometimes be a mechanism in its own right. But it is not general, nor is it without severe drawbacks. It raises a number of additional questions and concerns.

Box 10.2 (continued)

his own. Even if this is not as successful, it is better than being frequently cheated by Albert. And yet, still, better for both it seems, would be to cooperate on a continuous basis. The lesson drawn from such simple example is: cooperation in repeated games may be superior for both than to cheat. In the longer term, both have a strong incentive to cooperate.[20]

This of course is a two people example. You could easily argue: so what? Albert could simply look for someone else with ideas to cooperate with – and do it all over again: Take the profit from defecting on Merrill; do it again to the third person and then move on; and so on (see Fig. 10.3). This is surely better than to always having to split the proceeds. Why should he care if Merrill never cooperates with him again? Others may have good ideas too. They don't know him. They could be cheated just the same. Thus, the system would easily break down. Anyone able to realise the idea could think like Albert. They could cheat all those with ideas. They would learn the hard way, but eventually will learn and adapt. They too will stop sharing. Everyone will stop sharing. In effect, no cooperation, no trading of ideas would take place.

True. But the principle can easily be generalised. It holds just the same. Admittedly, additional complications with transparency arise, but they can be overcome.[21] In the two people scenario, as in small social systems the track record/reputation is generally known. In large and anonymous systems, it is absent and adverse selection poses a serious threat to cooperation. Not everyone is known, not everyone's reputation can be determined. However, small systems can be simulated by making available the relevant information on reputation accessible to all (i.e. publishing track records and feedback evaluation).[22] This way there is little difference between the two people and

(continued)

---

[20] The intuition is much more complex and the analysis highly technical, especially when considering uncertain payoffs, variable length and intensities of cooperation, etc. The basic insight still holds though. A variety of strategies could be thought out of how to maximise the disincentive to defect and/or increase the return for either side also in a continuous cooperation between just the two. The effectiveness will depend on the commitment and credibility of threat of any such strategy and the dynamics it will entail. For an introduction and overview, see Mailath and Samuelson (2006). More accessible see Axelrod (1984), and (1997). Also see Schelling (1960).

[21] See especially Bolton and Ockenfels (2009) on the discussion of perfect reputation systems and the various issues.

[22] "Economists have long recognized reputation as an effective means of enforcing cooperation when an institution exists to track and disseminate such information (e.g., credit agencies; Milgrom et al. 1990), or within a small group where people are intimately familiar with of one another's history (Fudenberg and Maskin 1986). In contrast, the effectiveness of reputation in circumstances where players are essentially strangers, knowing about one another only through word-of-mouth, is far less certain." Bolton et al. (2005, p.1458) (for the references see: Milgrom et al. 1990; Fudenberg, and Maskin 1986).

Box 10.2 (continued)
the many people example. Even if no one would ever cooperate with the same person more than once, it would still hold. Indirect reputation still may be available. If everyone knows Albert cheats, no one would cooperate with him. And maybe more importantly, he knows that if he cheats all other will know and he would forgo future profit. The threat of negative reputation disciplines him. He will have an incentive to cooperate even in a multi-people setting. Hence, if the third person knew that Albert cheats, he would likely also ignore him just the same as Merrill would. Of course, other factors play a role, such as personalities, adjustment, special aspects of cooperation, etc. And the mechanism could be argued to death with the intricacies of game-theoretical analysis and the inclusion of endless alterations and additional assumptions.[23] The concept however still holds.

Two principles are at work here: reputation and deterrence. Reputation is a basic principle of social interaction – your past behaviour will reflect on you. A track record is often considered a good indicator of trustworthiness. The idea is that by knowing the previous behaviour it is possible to infer on the personal characteristics and project or expect similar future behaviour. Clearly, this may not always hold. This clearly does not mean we can simply categorize people this easily. They differ. They may change. Even if they have a good track record, if opportunity comes along, they may still cheat. Maybe even strategically generate a good track record for a better opportunity. Also in reverse: there might have been good cause for defecting, it might have been a desperate exception, etc. while typically they are reliable. Still, despite this, it is how we typically would behave. And it mostly remains a good indication. It certainly is a typical factor in deciding between options. If you had to choose between otherwise identical candidates, would you not choose the one with the better reputation? If it becomes the most important factor it easily also outdoes many other good attributes a candidate may have. A track record, even if sometimes flawed, is an often-used variable. The better the track record, at least in a relative sense, the better chances

(continued)

---

[23] Many variations in the game exist (rationality constraints, forgiveness, payoff size, communication, etc.) and special aspects apply in the current case (variable payoff, payoff size, uncertainly of payoff, etc.). (See Mailath and Samuelson 2006. As a small sample also see Klein and Leffler 1981; Shapiro 1983; Kreps et al. 1982; Kreps and Wilson 1982; Benoit and Krishna 1985; Habourian 1990; Bendor et al. 1991; Milgrom et al. 1990; Smale 1980; Fudenberg and Maskin 1986; Fudenberg et al. 1998; Compte 1998; Kandori and Matsushima 1998; Cripps et al. 2004; Bolton et al. 2005; Bolton and Ockenfels 2009). Though backward induction seems a logical problem in finite games, due to the nature of the game (infinite participants, variable ends, and behavioural components) this seems to apply only with severe caution (see, for example, Selten and Stocker 1986; Pettit and Sugden 1989; Aumann 1995; Dow and da Costa Werlang 2008). Many complementary experiments amend these findings (see especially trust games: Berg et al. 1995; Ostrom and Walker 2003. Also Wilson and Eckel 2006).

> **Box 10.2** (continued)
> are to be cooperated with.[24] This leads directly to the second principle: deterrence. Knowing they will tarnish their record and forfeit future gains from cooperation, participants will refrain from defecting. The partners know, that exploiting cooperation will mean that it gets more and more unlikely to be selected to cooperate with again, or at a much less promising level. This will act as an adverse incentive to defect in the first place. Even if it is worth it in the individual case, in the long run it will not be worth foregoing follow-up benefits of repeated or increased and/or increasingly valuable cooperation in the future.
>
> Therefore, by enabling a reputation mechanism it becomes possible to incentives cooperation. Partners no longer have to trust in the goodwill of the other not to defect, but it becomes an incentive for both to conduct themselves properly, to be cooperative.

The economic theory of cooperation sounds like a nice story, concocted by some defunct economist in an ivory tower somewhere. Indeed, it was. The principles matter though. If you look around, you will find many examples where this applies, with your friends, at work, in daily life. Reputation mechanisms are almost ubiquitous – they are just not explicit. We all place trust in people and companies on a daily basis – more often than not based on their reputation. When we go to the local grocery market, we expect the goods to be of a certain quality. They cannot afford to cheat us. It would quickly spread and ruin their reputation. They would lose customers in droves. That is why we trust them also not to dupe us in the future. They could not afford it. We expect our physicians to do a diligent job. We cannot check their work. We hardly understand what they do. We trust in their knowledge and experience. They live of such reputation. We expect them to do a good job, if only for the reason they know they would have no future customers if they were sloppy. [25] Reputation may be the most significant reason why many large firms are more trusted than unknown ones, might be the reason why their valuation commands a significant mark-up. It also may be a reason why so many behave exemplary. Some effects are of course due to legal deterrents. Many though are based on the risk of reputation. With public relations nightmares such as the Brent Spar debacle for Shell (when they got caught and punished by consumers for trying to dispose an old oil platform at sea), Firestone tires (when several series of tires exhibited high failure rates, risking serious accidents) or Toyota's sticking accelerator pedal fiasco (when sticking gas pedals feared to cause unintended-acceleration crashes) and the recent Apple's iPhone 'Antennagate', or Sony's exploding

---

[24] At worst, one may choose to act on the reputation as punishment and/or to simply uphold the effectiveness of the deterrence mechanism.

[25] Also see discussion on credence goods. For an overview, see Dulleck and Kerschbamer (2006).

10.1 Trust

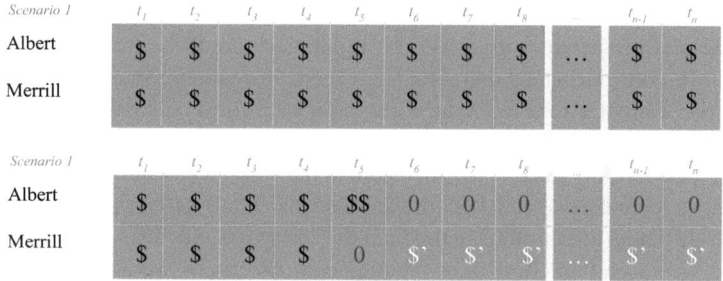

Fig. 10.2 Working together versus cheating one another

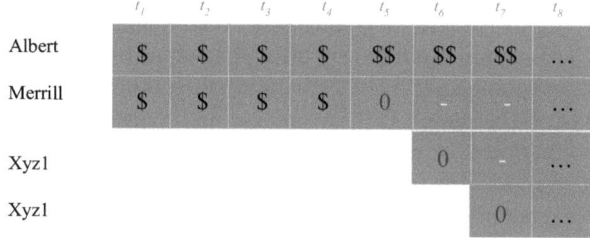

Fig. 10.3 Cheating one, then the others

batteries, trust as a key component in daily commerce become painfully obvious.[26] Even where there are no legal remedies, and even if there are some, loss of consumer trust can have immense economic repercussions.[27]

But it is not that simple. Typically, a reputation mechanism applies directly in fairly small social systems where such information on the reputation of others is freely available. In groups or organizations where people exchange information, gossip if you will, such information on the reputation of group members is readily available. This may equally apply to specialised networks or medium sized communities where word-of-mouth networking works effectively. For example if a venture capital firm in Silicon Valley would start stealing ideas, news would very quickly spread and no one in Silicon Valley would any longer pitch ides to them.[28] It sometimes also works on a larger scale where such information spreads widely, through the media, grass root initiatives, etc. If Procter and Gamble as the most prominent proponent of a more open innovation were to suddenly 'steal' ideas, it is very likely the media would quickly pick up on this and report it widely, even, as they

---

[26] As Arrow put it: "Virtually every commercial transaction has within itself an element of trust, certainly any transaction conducted over a period of time. It can be plausibly argued that much of the economic backwardness in the world can be explained by the lack of mutual confidence;" Arrow (1972, p. 367)

[27] See for example Shapiro (1983), Jarrell and Peltzman (1985), Mitchell and Mahoney (1989), or Karpoff and Lott (1993), Banerjee and Duflo (2000)

[28] For a general example on this see Atanasov et al. (2008).

often do, blow it way out of proportion. It is a fair bet that many would no longer participate and cooperate with P&G. Its reputation would be tarnished and ruined among those considering cooperating with them in realising ideas. If only for that reason, P&G cannot afford to cheat. If all other firms however would offer a similar open approach, and cooperative innovation becomes ubiquitous, attention to the particular case would fade, and diffusion of misconduct would be minimal. The reputation mechanism would become ineffective. If all offer open access no one would get sufficient attention from media if they were to steal ideas, with the exception perhaps of the largest and best-known companies. Overview would be lost. Transparency would be lost. Successive freeriding would become possible. Thus, if all do it, it would be hard to keep track of culprits and know whom to trust and whom not. Anonymity enables fraud. Thus, on a larger scale and especially where many such actors are involved this is far more challenging. The reputation mechanism typically breaks down as information is either very difficult to come by, or more importantly, if quality, validity, and comparability of such information is hard to assess. Cooperative innovation builds on the notion that anyone can participate. In fact, the larger the potential range of partners the more likely the most suitable partner will be selected, making innovation more effective and efficient. This however would require a large scale and reliable reputation mechanism to work in practice.

The phenomenal spread of information and communication technologies (ICT), especially the internet, has given such information exchange an entirely new dimension and opened up fascinating new possibilities.[29] Cooperation is no longer location bound but much more readily available to anyone, anywhere. Many more possibilities to cooperate open up and come within reach. This also means more chances to cheat. While this implies on the one hand more anonymity for those wanting to cheat as the network of potential victims gets larger, these technologies also enable more transparency and comparability. Information on potential partners is more mobile and ubiquitously available, to anyone, anywhere, anytime. Those stealing can be identified much more easily. Since such large networks also offer those able to realize ideas access to a much larger pool of ideas to potentially cooperate with, this means the costs of cheating, and forgoing any access to all these ideas are increasing, the cost of defection become enormous. They can no longer pack up and relocate and start anew, can no longer exploit one network after another. But it is not that simple. The internet is no ready solution. The biggest problem is the quality of such reputation information. It is overwhelming and unreliable. Think of yelp, Google places, eBay, Amazon, Angies List, etc. On the internet, you can 'feedback' almost anything these days: medical doctors, drugs, professors, childcare, hair salons, mechanics, nightclubs, teachers, colleagues and people.[30] No wonder it is sometimes jokingly referred to as 'institutionalized

---

[29] See, for example, Bakos and Dellarocas (2005), Dellarocas (2003, 2005, 2006, 2007), Ba and Pavlou (2002), Resnick and Zeckhauser (2002), Melnik and Alm (2002). Also Güth and Kliemt (2004), Bolton et al. (2004), Brennan and Petti (2004), Hardin (2004), Pettit (2004), McGeer (2004), Bierhoff and Vornefeld (2004).

[30] See, for example, the Rating and Review Professional Association (rarpa.org).

10.1 Trust

gossip'.[31] If you look up a firm's reputation on the internet, the information may have been posted by a disgruntled customer, a laid off employee, a vindictive competitor, etc. seeking to defame and deliberately damage the company – as Facebook's attempted smear campaign against Google memorably underscores.[32] Identities can seldom be verified, and much abuse exists of such systems. Network hopping, aggregation problems, self-selection bias, etc. complicate the matter further. You cannot know whom to trust and verify the validity of their claim.[33]

For many such reasons much work has been done on better understanding and refining reputation mechanisms. Often even to avoid many such issues, reputation mechanism have often been institutionalised and/or pooled. Various examples readily exist where such well-functioning mechanisms are in place. Most prominently: credit histories, rating agencies, elaborate platform-rating systems. Credit histories are the closest example where such reputation mechanism readily applies. Most countries have it in one form or the other. If you default on your loan, pay late, or generally have a very bad track record of financial conduct, your credit rating will deteriorate, and the costs or even the possibility of renting an apartment, getting access to credit, and much more may be impaired. You may want to have a good credit history to maintain access to credit, or at low costs, to be able to get an apartment, a mobile phone contract, etc. You wanting to have a good credit rating may deter you from doing something now that would affect it negatively, knowing the fallout in the future will be likely higher and more costly than your gain at this moment. Mostly your misconduct now will hurt you later, and will hurt you much more than the gain right now.[34] It acts as a deterrence effect. The striking feature of the credit history reputation mechanism is that is works in a decentralised setting. It is not bound to one partner. You may default on your car loan, but this will not only hamper your chances of buying a car from that particular vendor or brand, but will affect your general credit history, and thus also your credit card payments, your mortgage, student loans, etc. This is serious. And this deterrence is relatively mild, as it often only determines a risk mark-up, and mostly not a binary system of full access to credit or none at all. Imagine if you were to default anywhere once, you would never get any credit ever again, no loan, no credit card, no mortgage, no credit of whatever kind. You would seriously reconsider.[35] This may be a crass illustration, but it merely

---

[31] Bolton and Ockenfels (2009, p. 16)

[32] The Daily Beast (2011)

[33] See Dellarocas (2003), Resnick and Zeckhauser (2002), Friedman and Resnick (2001), Cook et al. (2009), Nissenbaum (1999), Tadelis (1999), Yamagishi (2003), Yu et al. (2008). Also Bolton and Ockenfels (2009).

[34] A good example, especially in the current setting is defaulting on government bonds. A country failing to repay their bonds now faces severe penalty in the future through higher interest rates for often a considerable time and scope, putting often a hefty fine on default, by far outweighing the initial gain. See for example Zivney and Marcus (1989).

[35] For various reasons this is not easily possible. Punishment has to fit the crime. Too severe a punishment may not be optimal either (see Gary Becker 1968, Hamilton and Rytina 1980, and Andreoni 1991).

exemplifies the reputation mechanism by degree of sanction. The principle is the same. This is a negative sanction. An alternative way of looking at this mechanism, and to spin it in a positive manner: Good behaviour is rewarded through more and better chances in the future. It is a virtuous loop. If you have a good credit history, you will have easier access to credit, get better conditions, and draw on the full array of financial services. Having a good credit history thus is worth a lot.

This applies directly to trading ideas. Here the feedback is even more, much more important. Trust is the most important aspect of cooperative innovation. Reputation may be a convenient additional consideration when it comes to buying goods and services. It helps to better inform you of potential issues, and may help raise the quality, timeliness, etc., but the most important aspect still remains the price. When you go to Amazon.com to buy a book, most likely you will go by price rather than paying a hefty premium for a highly regarded trader. Many legal remedies and often many integrated safeguards and mechanisms of resolving issues with your purchase place little premium on the feedback mechanism. The reputation score remains secondary at best.[36] When it comes to ideas, trust, and with it reputation is your only safeguard. There is no price up front. You choose to engage in cooperation with someone because you think you can trust them, and you hope that the reputation mechanisms will deter them from stealing your idea. Surely, also here a multitude of attributes play a role apart from trust (competence, experience, success rate, etc.), but the larger the pool of possible sources, and the more similar the competitors, trust will likely remain by far the most important criteria of choice. Precisely because trust is so essential when it comes to ideas, the deterrence effect is likely to apply fully here. If someone is seen as trust unworthy because he/she has a bad track record, and possibly stole ideas even only once, they are unlikely to be considered ever again to be cooperated with. This also means no future possibilities of participating in the profit of realizing the ideas of others. The future loss will likely by far outstrip the current gain.[37] For example a large corporation: were it to steal an idea and ruin its reputation in the market for ideas, it would likely not get access to any future ideas again. The profits from realizing the one idea on its own is sure to be much less that the potential profit it could make by cooperating to realize many, many more ideas. It will likely fail in the market where innovation is the key driver of growth.[38]

---

[36] This may still mean higher prices, especially for risky transactions or high prices. See Dellarocas (2003), Resnick et al. (2006), as well as Güth et al. (2007).

[37] This is not that simplistic, as many other factors need to be taken into account, discount rates, risk of failure, opportunity costs, etc.

[38] Again, it is not that simplistic. Consider the argument made in part one where companies become implementers and the innovation markup will be competed away more and more and remain with the idea rather than the implementer. Then also the incentive to default becomes larger. This may complicate the argument and require some form of modification, if at all, but the principle still holds.

Additional pressure comes from public perception. With a rising acknowledgement of the value of ideas, social and socially induced economic pressure can be very powerful supplementary deterrents to stealing ideas. If you steal an idea, and it is publically known, the more successful you are, the more you may be socially penalized. People may shun you – knowing all too well how awful stealing of ideas is, even though you are not punished for it directly. They may boycott your products to some degree and jeopardize your success. This will at least reduce the payoff. The more acknowledged and treasured ideas become, the more likely and the more forceful such supplementary social effects become. This strengthens the deterrence effect further as it increases the fallout and reduces the payoff from stealing ideas.[39]

Establishing a practical reputation mechanism seems the single most promising approach to bridging the gap, to trading ideas, and to enable an open innovation process. It may help to better find and choose partners, and especially promote cooperative behaviour in realizing ideas. Operationalizing it is far from easy. A crucial aspect still needs to be addressed for any reputation mechanism to work: fairness. Reputation relies on the notion that norms exist that provides a clear criterion of what makes up a reputation. Think again of Amazon: If someone does not deliver the good, you provide negative feedback; if someone stalls and delivers very late, you provide negative feedback; if someone provides low quality products despite claiming certain standards, you provide negative feedback. These are already a range of different aspects of feedback that are not easily compared. Does being late really warrant someone ruining your online reputation? Perhaps. These and many more questions arise when it comes to Amazon. When it comes to ideas, it becomes more complicated still. In some instances, it may be obvious. Stealing an idea would be a breach of trust. A clear and unambiguous negative feedback seems appropriate. Most aspects of cooperation are far less black and white. Even the notion of 'stealing' an idea may have a range of interpretations. Some initial insight that leads to a better idea – is that stealing? Applying an idea to a completely different sector, a completely different aspect, never thought of by the initial inventor – is that stealing? Some will say yes, some will say no. This is where norms come in. It becomes much more complicated when it comes to the intricacies of cooperation. What if someone delays implementation because he/she is too busy causing you to lose out on your profits? Is that 'failing' cooperation? You very much dislike the style of your partner. Does that mean you can provide bad feedback? Many, many more such questions arise.

---

[39] Another argument that has been put forward in this context is the notion of sunk costs to build reputation (see Hellmann and Perotti 2011, also Kreps, 1990). First you have to build up a reputation to be considered a trustworthy partner. You have to do some simple things first. Maybe advertise, demonstrate you are investing money in your reputation. These sunk costs may deter you from defaulting to some extent. If the projects you are working on offer a smaller additional return on stealing the ideas than your reputation is worth there would be no incentive to steal. In any case, even where this is not the case, such sunk costs reduce the payoff of stealing as they would then be lost.

The reputation mechanism builds on such a notion of norms. Who is to judge their behaviour as being cooperative or not? More importantly, what constitutes such behaviour? This is where the notion of fairness plays a fundamental role. Any such reputation mechanism is inseparable from the notion of fairness.

10.2 Fairness

Arrows paradox, curiously, in some ways may work in favour of companies. Indeed, some are exploiting this lack of solutions. In the absence of realistic avenues for many inventors to realize their ideas, some clever companies lure them into revealing their ideas to them for the promise of 'fun', 'reputation', a 'sense of belonging' and the benefits of delivering on long longed for products to them as a consumer – only to make away with a healthy profit themselves.[40] Threadless.com, an online t-shirt vendor, offers a platform for virtual t-shirt design competitions where anyone can submit a design and compete. About 2,000–3,000 designs are uploaded each week. More than two million users can vote on the designs. After several rounds of user voting, a handful of winners emerge each week. Winners get a prize of $2,000. The company sells more than 120,000 T-shirts a month and revenues are estimated at $30 million last year, pegged to continue to grow rapidly still.[41] "Simple. Brilliant. Most importantly: Ridiculously cost-effective."[42] With profit margins of 30 % and higher the company is essentially freeriding on peoples willingness to give away their ideas – in the face of little to no alternatives. "The expected hourly wage for an individual submitting a design to Threadless can be calculated as being a small fraction of the federal minimum wage in the US"[43] Unsurprisingly among businessmen it has become a celebrate business approach to bait unsuspecting inventors to share their ideas with companies – for free, or almost free.[44] Given the lack of alternatives, many indeed have to make due with a warm handshake and a virtual star next to their screen name. Deprived of any reasonable rout to profiting from their idea, they have to settle for peanuts or opt out. Firms like Threadless are grateful. Yet, while such business models are still thriving, discontent is mounting. Not everyone is willing and happy about giving away their ideas for free and let others profit from them. Users are

---

[40] See for example Wu and Sukoco (2010) or Füller (2006). See also Franke et al. (2010).

[41] See Chicago Magazine (2012). Also see threadless.com (02.03.2013). For more details, also see Forbes (2010); also Howe (2006); also see Brabham (2010).

[42] Howe (2006), also see Howe in Wired Magazine (2006).

[43] Franke et al. (2012, p. 5)

[44] On a much smaller scale – modification rather than idea sharing – this has also been identified as a business opportunity to allow customers to participate in the process, typically by customizing the product to some extent. This would raise their utility and customers would be more willing to pay higher prices. In areas with high heterogeneity of need this promises to hold significant value added (e.g. von Hippel and Katz 2002, Franke and von Hippel 2003).

increasingly demanding more profit participation. Especially high performers, those with many ideas are increasingly disgruntled. They are less and less willing to continuously be exploited. Many do opt out. They rather not see their idea realized than be exploited for it.[45]

This pinpoints a critical aspect of cooperation: fairness. In principle, fairness should not matter. Even if cooperation would only yield little return, or even if it failed, as long as the gain, be it money or 'fun', exceeds 'costs', however small, anyone should be happy. But this is not the case. People typically harbour an inherent notion of fairness. This often leads to odd behaviour. People may forego a profit rather than be treated unfairly. If someone were to offer you an electric toaster for your brilliant idea that you deem highly valuable, would you take it? Presumably not – despite the simple fact: a toaster is better than no toaster. This may be due to the fact that you may think you may have alternative routes of realising your idea, that will yield more than just a toaster, or, and this is often the case, even where no such alternative path exists you would deem the offer unfair, and would rather forego the toaster than to be exploited. You would rather dwell on pride and principle than to see someone else, a corporation no less, make away with a hefty profit from your idea. You would rather forgo any return than to accept a modest, nay, minute reward compared to the 'real' value of your idea. Imagine a very simple example: you identified a market niche in your area and have a plan to profit from it – a quality coffee shop. It offers a great opportunity. But, sadly, you do not have the means or skills to realise it on your own. You share this idea with a local businessman. He implements it, sets up a coffeehouse, and, indeed, makes a killing. Occasionally he may thank you by giving you a free coffee. Are you happy? Maybe. You too like coffee, and are grateful you no longer have to walk so far, and really appreciate the free coffee once in a while. More likely than not, you think it unfair. Someone else profits highly from your idea.[46] The benefit is in no comparison to the reward the other person receives. In retrospect, and likely in the future you will rather not share it with others and see them benefit from it than to enjoy such marginal benefits. You expect more. You would rather wait, in hope for a better opportunity (though in the meantime it is likely a Starbucks will have popped up), or simply forgo the benefit than to see you being screwed over. It would probably be even worse if it was not a local businessman, but some huge, to you anonymous, multinational corporation that you entrusted your idea.

This is far from trivial. It overrules basic notions of incentive. And in this case, it raises critical questions about what then cooperation means (see box 10.3). Surely, if someone steals your idea, you would not make a profit, likely a loss, and forgo considerable opportunities. But even if you were to be rewarded, if any kind of reward does not suffice, how much is sufficient, what is fair, who deems it fair, etc.[47]

---

[45] See Franke et al. (2012)

[46] You may debate whether this is truly an 'idea' or not. Generally it is. Innovative – maybe. Think of any other high-tech example if this is more comfortable. The point here is rather: You spotted the profit opportunity. Someone else reaped the benefit.

[47] It would be a mistake to think of a reward only in terms of monetary profit. It may have aspects of recognition, and many other kinds of perceived rewards other than money. See for example also Nelson (1993).

Box 10.3 Fairness
On how fairness matters.

Experimental economics has shown repeatedly that people show a remarkable inherent notion of fairness. This is best exemplified in what has come to be known in experimental economics as *'ultimatum games'*.[48] To demonstrate this, consider two players. Again, think of Albert and Merrill. Albert is given an amount of money – say 100 Euro. It is up to him to split it up between himself and Merrill. Merrill can only either accept the proposed splitting of the money or reject it. Should Merrill reject, no one will receive any money. If Merrill accepts the offer, each one can keep their share of that was offered (see Fig. 10.4).

In principle, any amount should be preferable to none at all. Thus whatever Albert offers him, whether 99/1 or 50/50, it should be acceptable to Merrill. 1 Euro is better than no Euro. Albert could thus offer any amount. Merrill would be assumed to accept anything that is more than 0 (and even at 0 he should be indifferent). Anything is better than nothing, be it 1 cent, 1 euro, 10 euro, 50 or a 100. Experiments regularly show that this is not the case. A typical result rather shows an acceptance at around 70:30. Below this, the second person would rather end up without any money at all than be treated unfairly. This is a profound insight. People would often rather forgo a certain amount of money than to be treated unfairly. The typical conclusion drawn is thus: people have inherent notions of fairness. This needs to be taken into account. The classical notion of rationality fails here. Anything is not better than nothing. Only a certain anything is better than nothing. How much this 'certain' amount is, is a big question. Yet it is immediately clear, such behaviour can highly distort economic outcomes.

This of course is a very simple experiment, and has received wide criticism and many additional factors influencing the outcome have been discussed. E.g. if larger sums are taken the outcome will differ; if it is based on a perception of relative status of the two people this will alter the result; repeated or altering cooperation will also change the outcome; even things as the attractiveness of the bargaining partner or the testosterone level of the players may have an influence; etc. But none the less, the principle still holds. Fairness matters. The result may only be indicative. The exact split is

(continued)

---

[48] Güth et al. (1982). Also Thaler (1988), Schotter et al. (1996).

Much research has been done on this. For an overview, see, for example, Camerer and Thaler (1995), Güth (1995), Camerer (2003), Murnighan (2008), and Hoffman (2008).

Research has been conducted in many adjacent fields further discussion the ultimatum game, some with intriguing conclusions (e.g. Sanfey et al. 2003; Wallance et al. 2007; Burnham 2007).

See especially also publications by Ernst Fehr and Armin Falk for details and a more elaborate analysis of fairness and reciprocity (e.g. Fehr and Schmidt 1999; Falk and Fischbacher 2006).

Box 10.3 (continued)

circumstance dependent and can certainly not be universalised. The insight that fairness matters, however, can.

This notion applies directly and particularly well at that, to the innovation process: the reluctance to realise an idea, and the conditions for successful cooperation. Such inherent notion of fairness may pose a strong disincentive to participate in the realisation process. People will rather not attempt to realise their ideas, fearing that if they reveal the idea, it will be taken up by someone else without a fair remuneration. This goes beyond the typically rational incentive problem, but rather builds on the notion of relative income. The typical question: 'Why should I make someone else rich with my idea? – Even though I gain from it he/she gains disproportionably more from it. I would rather not see it realised than be exploited.'

From an analytical perspective, this seems remote, unemotionally abstract. What does this mean in practice? Consider a case of cooperation. You have an idea. But you are no expert or entrepreneur. You need a partner to cooperate with for your idea to materialize and thrive. Your idea is brilliant, and it will work simply because it is visionary. Your idea creates the extra benefit. The entrepreneur is more of a means to realize it. Success, you are convinced, is your credit. You deserve a large chunk of the profit. The entrepreneur may see it differently. Operationalizing the idea, making it work in practice, sweating the details is what counts. He is investing time and effort, lending expertise, making something out of the idea. Success depends on him more than just on the idea. He too believes a sizable part of the profit should be his. More likely than not, both claims overlap (see Fig. 10.5).

A conflict arises. The outcome depends on the bargaining position and skill. In any case, it seems questionable if the outcome will be considered 'fair' by both. You should be happy to get anything for your idea at all. There are often hardly any costs associated with 'production'. As long as any such costs are covered you should be fine. So should your partner. He/she should be happy to be selected, should acknowledge the lack of own ideas. As long as it covers the opportunity costs, anything should be fine. Even add a sizable risk premium for both. As long as the reward is bigger, both should be happy. Yet they are typically not. Distribution matters. Fairness matters.

This claim goes beyond mere distribution of profit. This concerns relative input intensity, feedback levels, etc. – most importantly also the way of handling ideas. Fairness is fundamental to cooperation. Cooperative innovation cannot work without finding some way of accounting for such behaviour. Fairness in cooperation, also beyond the distribution of profits, can be a significant source of conflict. It may be the cause of failure, at least of dispute and discontent – undermining the principles and workings of any

(continued)

> **Box 10.3** (continued)
> reputation mechanism. For cooperation to succeed a clear and mutual understanding of what fairness means, in terms of the distribution of profits as well as the handling of ideas, is essential. The sense of fairness may highly distort this. Rather successful cooperation may fail, and be considered a failure because of fairness disputes.

Fairness is not just a question of stealing ideas. That is unfair. No question about it. The notion of fairness is much broader. It entails any aspect of cooperation. Not to steal the idea is but one of many such normative expectations. What about the distribution of profit? What about credit for the success? What about risk taking, management decisions, teamwork, etc.? Any aspect of cooperation builds on some notion of inherent norms of fairness. This is a vague notion, and nowhere more so than when it comes to something as blurry and intangible as ideas. Yet, still, for a reputation mechanism to work, a clear notion of fairness is required. It needs to be clear from the start what is expected and what everyone will be judged against. Even the best of intended cooperation may fail due to great differences in mutual expectations and claim, harbouring much conflict potential. General norms are needed to establish an acceptable common ground for all parties engaged in cooperation – how to share, how to handle ideas, how to treat cooperation. But especially when it comes to ideas, what exactly does fair mean? What are the different expectation and claims? Can an acceptable agreement be found? Without such a notion of what can be expected from each other, the often highly diverging claims are likely to lead to huge conflicts, threatening not only the success of a reputation mechanism, but the entire concept of cooperative innovation. Imagine you were to cooperate with someone. In good faith, you have entrusted them with your idea. What is it that you expect of your partner to accomplish? Where to you draw the line of what constitutes failed cooperation? The extreme case of course is clear: someone runs off with your ideas and you don't see any of the proceeds. But what would you consider a fair remuneration? What would you consider a fair behaviour? If someone makes an honest mistake, jeopardising your return on your idea? What if someone simply is arrogant and rude to you? The question is: What is reasonably acceptable; what is cooperative, and what is not; what is fair and what is not? The same can be asked by the partner. What is fair behaviour on your part? If your partner is too busy realizing the idea in a timely manner, can you default on the cooperation and change partner? But even though you default, maybe rightfully so, would you consider it fair if your partner would still pursue the idea? He/she knows your idea and has invested in it; can he/she still pursue it without being seen as stealing ideas? These and many more issues need to be addressed. It needs to be clear what everyone can expect from their partners and what is expected of them; what is the criteria cooperation is judged against; what makes up the reputation mechanisms, and what does it tell you about your partners.

10.2 Fairness

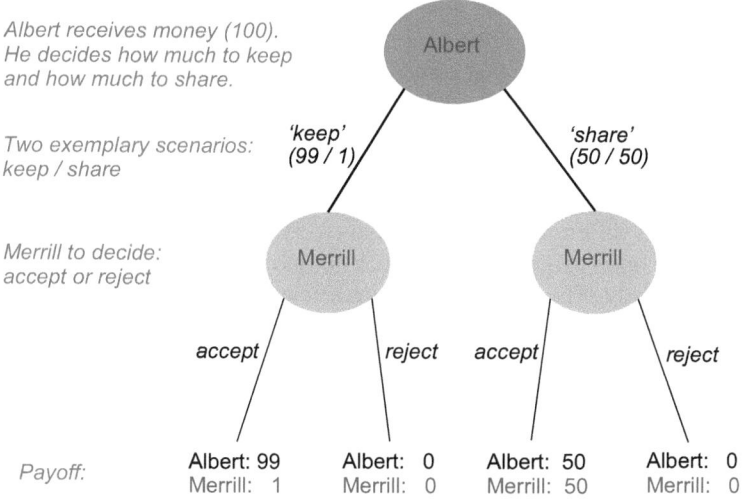

Fig. 10.4 Calculated greed: fairness versus rationality

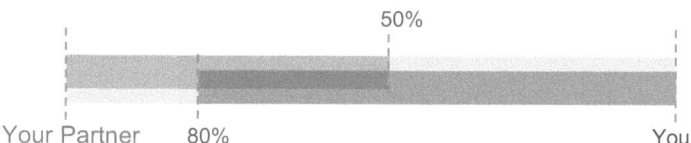

Fig. 10.5 Expectations and conflict

When you default on a loan, are too steeped in debt, repay late, or you try to get credit too often, your creditworthiness suffers, i.e. your 'credit reputation' suffers. Such aspects weigh in on your credit score. The criteria are well defined and available to you and anyone else. The norms that determine your credit reputation are clearly spelled out and are clear-cut. You know exactly what is expected of you. In other cases, norms can be much softer. They are not spelled out clearly nor are they binary, but rather vague and implicit – and yet often they are quite generally accepted. Think of the rather curious case of couchsurfing. This is a phenomenon with millions of so called 'couchsurfers' that travel the world and crash on strangers guest beds and literally 'couches' – for free. Online platforms have sprung up (e.g. CouchSurfing International at couchsurfing.org) that connect travellers to locals, surfers to couches. When a host chooses whether to let someone into their house, they look at the applicant surfer's profile. This typically contains references from previous experiences that help guide their decision. Very bad profiles typically would dissuade hosts from welcoming the traveller. 'Surfers' therefore desire a favourable profile to be able to rely on such free accommodation opportunities also on future travels. The criteria used to provide a reference are fully undefined. The most important is: is the other party trustworthy? He/she stole you flat screen – obviously, he/she is not. But many other criteria come into play as well. You may think it unacceptably rude not to have received a host gift, or that someone arrived too late at night, much later than what was agreed on. Or someone did not clean up after him/her, or did not thank you when parting. Even though norms are not clearly defined, a typical core exists of what is acceptable and what not, what is fair and what is not. With ideas, neither are the norms clearly spelled out, nor are they implicitly accepted. No such norms yet exist. People have very different expectations, have very different notions of what fairness means when it comes to ideas. Without such norms cooperative innovation is doomed to fail spectacularly. The entire array of intricacies of cooperation, as banal as they often may seem, are of utmost importance here.

Only if addressed jointly can trust and fairness enable cooperative innovation, can enable the successful trading of ideas and can enable the evolution towards the *Idea Economy*. Cooperative innovation builds on both these notions. Only if addressed satisfactorily will active approaches to trading ideas become practically feasible. No doubt they will be overcome eventually. But for the sake of bringing about such solutions, much needs to be done to tackle these issues of trust and fairness. Once they are, active approaches where those with ideas can approach those able to help realise ideas will become the new mantra of innovation, the new modus operandi in the emerging *Idea Economy*.

References

Andreoni, J. (1991). Reasonable doubt and the optimal magnitude of fines: Should the penalty fit the crime? *The RAND Journal of Economics, 22*(3), 385–395.
Anton, J. J., & Yao, D. A. (1994). Expropriation and inventions: Appropriable rents in the absence of property rights. *The American Economic Review, 84*(1), 190–209.
Anton, J. J., & Yao, D. A. (2004). Little patents and big secrets: Managing intellectual property. *The RAND Journal of Economics, 35*(1), 1–22.
Anton, J. J., & Yao, D. A. (2005). Markets for partially contractible knowledge: Bootstrapping versus bundling. *Journal of the European Economic Association, 3*(2–3), 745–754.
Arrow, K. J. (1972). Gifts and exchanges. *Philosophy and Public Affairs, 1*(4), 343–362.
Atanasov, V. A., Ivanov, V. I., & Litvak, K. (2008). The effect of litigation on venture capitalist reputation. *EFA 2009 Bergen meetings paper*. www.efa2009.org/papers/SSRN-id1343981.pdf. Accessed 02 Mar 2013.
Aumann, R. J. (1995). Backward induction and common knowledge of rationality. *Games and Economic Behavior, 8*(1), 6–19.
Axelrod, R. (1984). *The evolution of cooperation*. New York: Basic Books.
Axelrod, R. (1997). *The complexity of cooperation*. Princeton: Princeton University Press.
Ba, S., & Pavlou, P. A. (2002). Evidence of the effect of trust building technology in electronic markets: Price premiums and buyer behavior. *Management Information Systems Quarterly, 26*(3), 243–268.
Bakos, Y., & Dellarocas, C. (2005). *Online reputation and litigation as mechanisms for quality assurance: Strategic implications for profits and efficiency*. www.rhsmith.umd.edu/faculty/cdell/papers/replit.pdf. Accessed 02 Mar 2013.
Banerjee, A., & Duflo, E. (2000). Reputation effects and the limits of contracting: A study of the Indian software industry. *Quarterly Journal of Economics, 115*(3), 989–1017.

References

Becker, G. S. (1968). Crime and punishment: An economic approach. *Journal of Political Economy, 76*(2), 169–217.
Bendor, J., Kramer, R. M., & Stout, S. (1991). When in doubt... cooperation in a noisy prisoner's dilemma. *Journal of Conflict Resolution, 35*(4), 691–719.
Benoit, J. P., & Krishna, V. (1985). Finitely repeated games. *Econometrica, 53*(4), 905–922.
Berg, J., Dickhaut, J., & McCabe, K. (1995). Trust, reciprocity, and social history. *Games and Economic Behavior, 10*(1), 122–142.
Bierhoff, H. W., & Vornefeld, B. (2004). The social psychology of trust with applications in the internet. *Analyse und Kritik Zeitschrift fuer Sozialtheorie, 26*(1), 48–62.
Bolton, G. E., Katok, E., & Ockenfels, A. (2004). Trust among internet traders: A behavioral economics approach. *Analyse und Kritik Zeitschrift fuer Sozialtheorie, 26*(1), 185–202.
Bolton, G. E., Katok, E., & Ockenfels, A. (2005). Cooperation among strangers with limited information about reputation. *Journal of Public Economics, 89*(8), 1457–1468.
Bolton, G. E., & Ockenfels, A. (2009). The limits of trust in economic transactions: Investigations of perfect reputation systems. In K. S. Cook, C. Snijders, V. Buskens, & C. Cheshire (Eds.), *eTrust: Forming relationships in the online* (pp. 15–36). New York: Russell Sage Foundation.
Bowles, S., & Hwang, S. H. (2008). Social preferences and public economics: Mechanism design when social preferences depend on incentives. *Journal of Public Economics, 92*(8–9), 1811–1820.
Brabham, D. C. (2010). Moving the crowd at threadless. *Information, Communication & Society, 13*(1), 1122–1145.
Brennan, G., & Petti, P. (2004). Esteem, identifiability and the internet. *Analyse und Kritik Zeitschrift für Sozialtheorie, 26*(1), 139–157.
Burnham, T. C. (2007). High-testosterone men reject low ultimatum game offers. *Proceedings of the Royal Society B, 274*(1623), 2327–2330.
Camerer, C. F. (2003). *Behavioral game theory: Experiments on strategic interaction*. Princeton: Princeton University Press.
Camerer, C. F., & Thaler, R. H. (1995). Anomalies: Ultimatums, dictators and manners. *The Journal of Economic Perspectives, 9*(2), 209–219.
Chesbrough, H. W. (2006a). *Open business models: How to thrive in the new innovation landscape*. Cambridge: Harvard Business School Press.
Chicago Magazine (2012). *How Jake Nickell built his threadless empire*. July 2012, by Froelke Coburn, M. www.chicagomag.com/Chicago-Magazine/July-2012/How-Jake-Nickell-Built-His-Threadless-Empire/. Accessed 02 March 2013.
Compte, O. (1998). Communication in repeated games with imperfect private monitoring. *Econometrica, 66*(3), 597–626.
Cook, K. S., Snijders, C., Biskens, V., & Cheshire, C. (2009). *eTrust: Forming relationships in the online world*. New York: Russell Sage Foundation.
Cripps, M. W., Mailath, G. J., & Samuelson, L. (2004). Imperfect monitoring and impermanent reputations. *Econometrica, 72*(2), 407–432.
Dellarocas, C. (2003). The digitization of word of mouth: Promise and challenges of online feedback mechanisms. *Management Science, 49*(10), 1407–1424.
Dellarocas, C. (2005). Reputation mechanism design in online trading environments with pure moral hazard. *Information Systems Research, 16*(2), 209–230.
Dellarocas, C. (2006). Strategic manipulation of internet opinion forums: Implications for consumers and firms. *Management Science, 52*(10), 1577–1593.
Dellarocas, C. (2007). Reputation mechanisms. In T. Hendershott (Ed.), *Handbooks in information systems, volume 1: Economics and information systems* (pp. 629–660). Amsterdam: Elsevier.
Dow, J., & da Costa Werlang, S. R. (2008). Nash equilibrium under Knightian Uncertainty: Breaking down backward induction. *Journal of Economic Theory, 64*(2), 305–324.
Dulleck, U., & Kerschbamer, R. (2006). On doctors, mechanics, and computer specialists: The economics of credence goods. *Journal of Economic Literature, 44*(1), 5–42.
Falk, A., & Fischbacher, U. (2006). A theory of reciprocity. *Games and Economic Behavior, 54*(2), 293–315.

Fehr, E., & Schmidt, K. M. (1999). A theory of fairness, competition, and cooperation. *Quarterly Journal of Economics, 114*(3), 817–868.
Forbes (2010). *Need to build a community? Learn from threadless*. By Burkitt, L., 7 January. http://www.forbes.com/2010/01/06/threadless-t-shirt-community-crowdsourcing-cmo-network-threadless.html. Accessed 02 March 2013.
Franke, N., & von Hippel, E. (2003). Satisfying heterogenous user needs via innovation toolkits: The case of apache security software. *Research Policy, 32*(7), 1199–1215.
Franke, N., Schreier, M., & Kaiser, U. (2010). The "I Designed It Myself" effect in mass customization. *Management Science, 56*(1), 125–140.
Franke, N., Keinz, P., & Klausberge, K. (2012). *"Does this sound like a fair deal?" Antecedents and consequences of fairness expectations in the individual's decision to participate in firm innovation*. Forthcoming in Organization Science, doi:10.1287/orsc.1120.0794.
Friedman, E., & Resnick, P. (2001). The social cost of cheap pseudonyms. *Journal of Economics and Management Strategy, 10*(2), 173–199.
Fudenberg, D., Levine, D., & Pesendorfer, W. (1998). When are nonanonymous players negligible. *Journal of Economic Theory, 79*(1), 46–71.
Fudenberg, D., & Maskin, E. (1986). The folk theorem in repeated games with discounting or with incomplete information. *Econometrica, 54*(3), 533–554.
Füller, J. (2006). Why consumers engage in virtual new product developments initiated by producers. *Advances in Consumer Research, 33*(1), 639–646.
Güth, W. (1995). On ultimatum bargaining experiments – a personal review. *Journal of Economic Behavior and Orpranization, 27*(3), 329–344.
Güth, W., Schmittberger, R., & Schwarze, B. (1982). An experimental analysis of ultimatum bargaining. *Journal of Economic Behavior and Organization, 3*(4), 367–388.
Güth, W., & Kliemt, H. (2004). The evolution of trust (worthiness) in the net. *Analyse und Kritik – Zeitschrift fuer Sozialtheorie, 26*(1), 203–219.
Güth, W., Mengel, F., & Ockenfels, A. (2007). An evolutionary analysis of buyer insurance and seller reputation in online markets. *Theory and Decision, 63*(3), 265–282.
Habourian, H. (1990). Anonymous repeated games with a large number of players and random outcomes. *Journal of Economic Theory, 51*(1), 92–110.
Hamilton, V. L., & Rytina, S. (1980). Social consensus on norms of justice: Should the punishment fit the crime? *The American Journal of Sociology, 85*(5), 1117–1144.
Hardin, R. (2004). Internet capital. *Analyse und Kritik Zeitschrift für Sozialtheorie, 26*(1), 122–138.
Hellmann, T. H., & Perotti, E. P. (2011). The circulation of ideas in firms and markets. *Management Science, 57*(10), 1813–1826.
Hoffman, E. (2008). Reciprocity in ultimatum and dictator games: An introduction. In C. R. Plott & V. L. Smith (Eds.), *Handbook of experimental economics results* (Vol. 1, pp. 432–453). Amsterdam: North-Holland.
Howe, J. (2006). *Crowdsourcing. Pure, Unadulterated (and Scalable) crowdsourcing*. [Blog post] 15 June. http://crowdsourcing.typepad.com/cs/2006/06/pure_unadultera.html. Accessed 02 March 2013.
Huston, L., & Sakkab, N. (2006). Connect and develop: Inside procter & gamble's new model for innovation. *Harvard Business Review, 84*(3), 58–66.
Jarrell, G., & Peltzman, S. (1985). The impact of product recalls on the wealth of sellers. *Journal of Political Economy, 93*(3), 512–536.
Kandori, M., & Matsushima, H. (1998). Private observation, communication and collusion. *Econometrica, 66*(3), 627–652.
Karpoff, J. M., & Lott, J. R. (1993). Reputational penalty firms bear from committing criminal fraud. *Journal of Law and Economics, 36*(2), 757–802.
Klein, B., & Leffler, K. B. (1981). The role of market forces in assuring contractual performance. *Journal of Political Economy, 89*(4), 615–664.

Kreps, D. (1990). Corporate culture and economic theory. In J. E. Alt & K. A. Shepsle (Eds.), *Perspectives on positive political economy* (pp. 90–143). Cambridge: Cambridge University Press.
Kreps, D., & Wilson, R. (1982). Reputation and imperfect information. *Journal of Economic Theory, 27*(2), 253–279.
Kreps, D., Milgrom, P., Roberts, J., & Wilson, R. (1982). Rational cooperation in the finitely repeated prisoner's dilemma. *Econometrica, 27*(2), 245–252.
Mailath, G. J., & Samuelson, L. (2006). *Repeated games and reputations: Long-run relationships*. New York: Oxford University Press.
McGeer, V. (2004). Developing trust on the internet. *Analyse und Kritik Zeitschrift fuer Sozialtheorie, 26*(1), 91–107.
Melnik, M. I., & Alm, J. (2002). Does a seller's ecommerce reputation matter? Evidence from eBay auctions. *The Journal of Industrial Economics, 50*(3), 337–349.
Milgrom, P., North, D., & Weingast, B. (1990). The role of institutions in the revival of trade: The law merchant, private judges, and the champagne fairs. *Economics and Politics, 2*(1), 1–23.
Mitchell, M. L., & Mahoney, M. T. (1989). Crisis in the cockpit? The role of market forces in promoting air travel safety. *Journal of Law and Economics, 32*(2), 329–355.
Murnighan, J. K. (2008). Fairness in ultimatum bargaining. In C. R. Plott & V. L. Smith (Eds.), *Handbook of experimental economics results* (Vol. 1, pp. 411–416). Amsterdam: North-Holland.
Nelson, K. E. (1993). Dow's energy/WRAP contest- A 12-Yr energy and waste reduction success story. *Proceedings from the fifteenth national industrial energy technology conference*, Houston, 24–25, Mar 1993. http://repository.tamu.edu/bitstream/handle/1969.1/92057/ESL-IE-93-03-03.pdf?sequence=1. Accessed 02 March 2013.
Nissenbaum, H. (1999). The meaning of anonymity in an information age. *The Information Society, 15*(2), 141–144.
Ostrom, E., & Walker, J. (2003). *Trust and reciprocity: Interdisciplinary lessons for experimental research*. New York: Russell Sage Foundation.
Pettit, P. (2004). Trust, reliance and the internet. *Analyse und Kritik Zeitschrift für Sozialtheorie, 26*(1), 108–121.
Pettit, P., & Sugden, R. (1989). The backward induction paradox. *The Journal of Philosophy, 86*(4), 169–182.
Resnick, P., & Zeckhauser, R. (2002). Trust among strangers in internet transactions: Empirical analysis of eBay's reputation system. *Advances in Applied Microeconomics, 11*, 127–157.
Resnick, P., Zeckhauser, R., Swanson, J., & Lockwood, K. (2006). The value of reputation on eBay: A controlled experiment. *Experimental Economics, 9*(2), 79–101.
Rigby, D., & Zook, C. (2002). Open market innovation. *Harvard Business Review, 80*(10), 80–89.
Sanfey, A. G., Rilling, J. K., Aronson, J. A., Nystrom, L. E., & Cohen, J. D. (2003). The neural basis of economic decision-making in the ultimatum game. *Science, 300*(5626), 1755–1758.
Sakkab, N. Y. (2007). Growing through innovation. *Research Technology Management, 50*(6), 59–64.
Schelling, T. C. (1960). *The strategy of conflict*. Cambridge: Harvard University Press. Reprint 2006.
Schotter, A., Weiss, A., & Zapater, I. (1996). Fairness and survival in ultimatum and dictatorship games. *Journal of Economic Behavior & Organization, 31*(1), 37–56.
Selten, R., & Stocker, R. (1986). End behavior in sequences of finite prisoner's dilemma supergames a learning theory approach. *Journal of Economic Behavior and Organization, 7*(1), 47–70.
Shane, S., & Cable, D. (2002). Network ties, reputation, and the financing of new ventures. *Management Science, 48*(3), 364–381.
Shapiro, C. (1983). Premiums for high-quality products as returns to reputations. *Quarterly Journal of Economics, 98*(4), 659–680.
Smale, S. (1980). The prisoner's dilemma and dynamical systems associated to non-cooperative games. *Econometrica, 48*(7), 1617–1634.

Tadelis, S. (1999). What's in a name? Reputation as a tradeable asset. *The American Economic Review, 89*(3), 548–563.

Thaler, R. H. (1988). Anomalies: The ultimatum game. *The Journal of Economic Perspectives, 2*(4), 195–206.

The Daily Beast (2011). *Facebook busted in clumsy smear on google.* 11 May, by Lyons, D. www.thedailybeast.com/articles/2011/05/12/facebook-busted-in-clumsy-smear-attempt-on-google.html. Accessed 02 March 2013.

von Hippel, E., & Katz, R. (2002). Shifting innovation to users via toolkits. *Management Science, 48*(7), 821–833.

Wallance, B., Cesarini, D., Lichtenstein, P., & Johannesson, M. (2007). Heritability of ultimatum game responder behavior. *Proceedings of the National Academy of Sciences of the United States of America, 104*(40), 15631–1563.

Wilson, R. K., & Eckel, C. C. (2006). Judging a book by its cover: Beauty and expectations in the trust game. *Political Research Quarterly, 59*(2), 189–202.

Wired Magazine (2006). *The rise of crowdsourcing.* 14 June 14, by Howe, J. www.wired.com/wired/archive/14.06/crowds.html. Accessed 02 March 2013.

Wu, W. Y., & Sukoco, B. M. (2010). Why should I share? Examining consumer motives and trust on knowledge sharing. *The Journal of Computer Information Systems, 50*(4), 11–19.

Yamagishi, T. (2003). *The role of reputation in open and closed societies: An experimental study of internet auctioning.* Working paper, Hokkaido University. http://citeseerx.ist.psu.edu/viewdoc/download?doi=10.1.1.7.204&rep=rep1&type=pdf. Accessed 02 March 2013.

Yu, H., Kaminsky, M., Gibbons, P. B., & Flaxman, A. D. (2008). SybilGuard: Defending against Sybil attacks via social networks. IEEE/ACM Transactions on Networking, *16*(3), 576–589.

Zivney, T. L., & Marcus, R. D. (1989). The day the united states defaulted on treasury bills. *The Financial Review, 24*(3), 475–489.

Cooperative Innovation   11

How can trusted and fair cooperation work in practice? Making cooperation work in practice is much more difficult than the theory suggests. Many details need to be addressed to make it happen. This includes (i) a good matching between those with ideas and those able to realise them; (ii) a commonly acceptable code of conduct to trading ideas to establish a benchmark for fairness and thus minimise friction and foster better cooperation; (iii) a transparent feedback mechanism to support building up a reputation and thus incentivize cooperation; and (iv) clear support and guidance throughout the innovation process. It is a new approach. To jumpstart it must be simple yet refined enough to make it work.

Cooperative innovation is the most promising approach to offer anyone a fair chance to pursue innovation and profit from his/her ideas. Imagine once more, this time in more detail: you have an idea. What next. You may be able to identify someone or some firm that could realise your idea. They are open to ideas from outside; they invite the submission of ideas; they pursue an active approach. Knowing a reputation mechanism is in place, you need not fear being exploited. Clear codes of conduct are in place offering guidance and means for trading ideas. Now both, you and your cooperation partner have a strong incentive to cooperate and make the most of the idea. Both profit from its success, no one from trying to freeride or default. If they would default, they would lose out in the long run as no one else will cooperate with them thereafter. In short: now trading ideas is becoming feasible, and realistic. Now there is a clear incentive to cooperate. For both fair and cooperative innovation is the default option. You need no longer be an entrepreneur, need no longer bear all the risk, need no longer rely on the benevolence and often unfounded trustworthiness of others. Cooperative innovation now is becoming commonplace. Many possibilities exist beside the classical internal or entrepreneurial path. Anyone now has easy access to realising and profiting from his or her idea. Your ingenuity is rewarded fairly. This is what the *Idea Economy* is all about.

This is still seemingly a far cry from what it is today. But it will come. It is underway. To accelerate its arrival, and to promote the effectiveness of such approach, more specific considerations to employ a reputation mechanism in practice are needed. It is easy to make grand claims. Operationalizing them is often the much harder part, far more messy, encountering a range of unexpected challenges that may make or break the entire concept. Having identified the basic mechanism that can enable, and the bottlenecks that may impede the development towards cooperative innovation, more applied considerations need to be addressed. This is challenging. The devil is in the detail. Carful and diligent weighing of different options is required. Often this means very pragmatic trade-offs between what is optimal and what is realistic, between what is feasible and what is sound. Such mechanisms are never perfect. The aim is to make them as good and as efficient as possible. The following tries to flesh out such practical approach in more detail. For convenience different modules can be identified that, if designed properly can make a significant difference in operationalizing and truly enabling cooperative innovation.

11.1 Matching

Cooperative innovation builds on a division of labour. Different partners cooperate that best complement each other to make the most of an idea, to best realise an idea. Realisation is more efficient and effective the better the partners are matched. The most obvious approach is thus to offer better information and access to a large pool of potential partners. Both are often a problem. Who would be best suited to realise a specific idea? Where to find them, how to find them, how to connect to them, how to contact, in what manner, etc.? Having a large pool of partners vying for your idea gives you the choice of identifying the most promising partner to make the most of your idea. Knowing who they are and what they do is key. Offering transparency enables such matching. Comparability may also foster competition among the different partners, further improving the services and range of partners, thus making the innovation process more efficient. The more transparent the competition for ideas, the better the innovation process.

Such matching approaches exist, and are becoming more and more prevalent and refined. Universities are matching their technical students with business students; researchers are matched with firms; ideas are matched to venture capitalists; challenges are matched to ideas. Many claim this domain. Clusters, allegedly facilitating such matching, are mushrooming everywhere. Especially online systems such as Xing (xing.com), LinkedIn (linkedin.com), or Viadeo (viadeo.com), also job-search-engines such as Monster (monster.com) or Experteer (experteer.de), even add-on modules in social networks such as Facebook (facebook.com), etc. seem, and often claim to be, natural candidates. Existing passive approaches could easily adapt (e.g. InnoCentive (innocentive.com), NineSigma (ninesigma.com)). Currently this does not capture the complexity of such matching needs nor does it account for the particularities that come with focussing on ideas. When it comes to ideas, more and different information is needed than some generic CV publishing, business card exchange, or photo sharing. Information needs to be more specific and far more idea-focussed. This holds both for the information required and for the design of the matching methodology.

11.1 Matching

Relevant variables include: skills, experience, capacity, etc. – the typical business attributes. Additional skills are required, and/or the relevant qualities better explicated, e.g. project management, entrepreneurial skills, proven access to resources, etc. But also less obvious variables may be highly important if attainable: for example location often matters in longer lasting cooperation. This mostly also implies culture, language, etc. aspects that can highly influence the success of cooperation. It also matters in terms of legal implications, access to markets, tax rules, etc. Many others matter. Such variables are sure to evolve further and become much more specific, adapting to the emerging needs and practices.[1]

Most importantly of course is the inclusion of a feedback mechanism and cooperation variables. This makes or breaks any such approach. Without the proper design and meaningful integration of trust and fairness into the cooperative innovation mechanism, matching will not prosper beyond a very narrow scope where limited legal provision apply or existing social mechanism are still sufficiently active to mimic the broader reputation mechanism. Trust is the decisive variable that needs to be included. True, it is a trade-off between competence and trust. But while in many approaches trust is just another, mostly secondary variable (for example with eBay or Amazon, where price is likely to be the more important variable, as also many alternative means of enforcing your rights and penalising misconduct exist), when it comes to cooperative innovation, trust is key. Competencies matter, no doubt. But with increasing participation, and increasing competition between highly able and competent partners, the distinguishing variable is reputation (read: trust). Still, in many cases a trade-off remains. It often seems a calculated risk that needs to be considered in any matching approach.

Practical questions are equally pressing. A careful trade-off is required between what is needed and what can be a burden. Too much data is overwhelming to those using it; too much data is considered unreasonable by those providing it. But the more data, the better and more refined the matching can be. But especially in times of data protection concerns, in time of risks of too much data pooling, reluctance can be high. If the convenience and potential benefit of participation is high enough, much of this reluctance can be overcome, but a strong participation bias, by age group, culture, etc. may occur that could severely limit broad based cooperation.

Matching also requires engagement. Many networks quickly clutter. They carry much dead weight. This can severely impede its effectiveness. Some mechanisms need to be included to deter inactiveness, to encourage and incentivise active participation. This can range from entry costs deterrence, to activity rules, to simple participation reminders.

This leads to another serious question: scope. Innovation is a process that can be segmented into numerous stages, and many different paths, sectors, themes, etc.

---

[1] Research has often identified a range of such variables that might be highly relevant: e.g. *'personal similarity'* Bengtsson and Hsu (2010); *'Social ties'*; Shane and Cable (2002); *'geography'* Sorenson and Stuart (2001); etc. Also see de Jong and Freel (2010); Boschma (2005); Mohr and Spekman (1994).

It is a broad spectrum. To make it manageable some categorization needs to be devised. This may include simple data classifications. Discussions may range from fully decentralized tagging and self-categorization (you may give yourself a classification, create a new one, anyone as they see fit, use keywords, etc.) to stringent classification standards (e.g. UN International Standard Industrial Classification (ISIC)).

More interesting though are questions of competing attributes and abuse potential due to different roles and tasks. Matching does, and should, go far beyond a simplistic notion of idea-to-realisation. Many different combinations and sequences are possible. An idea may be matched to an expert to develop the idea, and then together matched to an entrepreneur, then a competent partner firm. Someone may want to go at it all alone, merely draw on some supportive services along the way. Some may want to directly trade it to a firm or broker, be involved little to none. The pressing question then: if there are different classes of participants, do their tasks compete, interfere, pose a threat and abuse potential? If so indeed, segmentation may be prudent. Initially this may apply crudely thus: services, partners, traders (see box 11.1). These categories have different levels of engagement, different commitments and challenges to face in cooperating, in handling ideas. This is not per se a logical split, and the demarcation line may often be blurry. It may seem unfair to some. It is pragmatic. It helps to minimize conflict of interest. This makes it necessary for practical reasons.

> **Box 11.1 Different Kind of Partners: Services/Partners/Traders**
> Partners could be segmented any which way. The trick is to find the most meaningful, while still accessible segmentation. 'Meaningful' means finding categories that do indeed differ in terms of engagement, differ in the need to share ideas, differ in their incentive structures to handle ideas, categories where, if not segmented, conflicts of interest would arise. 'Accessible' mostly implies a limited number of clear-cut categories, where demarcation follows along some intuitive lines.
>
> An initial segmentation could look something like the following: service providers, partners, traders. They differ mainly in the degree of participation in the realisation process. *Service providers* only offer a specific service but do not really get involved in the realisation process of any particular idea. Instead they will be rewarded specifically for their services. A *partner* does indeed get involved, shares the risk and the spoils of realising an idea with the one that has the idea. A *trader* basically purchases the idea and realises it on his/her own terms. This sounds somewhat intuitive. But it is also meaningful, almost necessary. Clearly any of these categories allows for a wide range of participants. The demarcation line may not always be that clear. There are certainly many forms of partnership, many degrees of involvement, and many different kinds of services with different levels of cooperation intensity. But,
>
> (continued)

Box 11.1 (continued)

all in all, it allows for a simple differentiation by somewhat intuitive characteristics. This does not yet explain its need. The following might:

Service providers: A large variety of services may be provided. It could be at any stage along the realisation path. Services could address and support any which aspect. Services could be free, could be for a fee. Such segmentations may be useful to classify them. But they are not necessary. One aspect however might be: whether services are active or passive. The criteria of demarcation would be how much an idea would need to be revealed to the service providers. Asking for a market analysis may not require sharing an idea. Contracting someone to produce a critical element or even assemble the components mostly would. Assessing the viability of an idea, or providing assistance with the business plan or implementation strategy, surely requires sharing of the idea. From the very nature of such a service provision this might be problematic. Typically you would use an expert, the best you could find. So would your competition. Hence, this person will likely receive many very similar ideas. Provisions of what this means could be made, and some form of idea-neutrality could be established. But, were this person also allowed to participate in the realisation process, thus have a vested interest in the success of a particular idea, a strong conflict of interest would arise. To avoid this, as a prudent measure to minimize the likelihood of such instances it would be reasonable to limit participation to service provision alone. This risk applies to a lesser extent to passive services, where the idea needs not be revealed, but may still apply to a certain degree. Risks could for example include provision of outdated information, or deliberate misinformation or concealment of opportunities, or the inference of ideas through requests. As a preventive measure a clear cut seems adequate. Service providers can only be just that, service providers, not partners, not traders. This may seem unfair to many. You might be an expert, biding your time until the perfect idea comes around. In the meantime you are happy to help others. Often their ideas are not in competition even to yours. Again, this may be the case, but in some, if not likely many, a conflict could arise. To avoid this, and to avoid the uncertainty this creates also in your partners, it is best to keep these functions separate.

Partners: Similar argument holds for a partner. As a partner you should not at the same time be involved in competing activities. As a service provider you may be conflicted to provide sound services, or if you provide sound advice, conflict with your obligations to your partners in realising an idea. You might participate strategically. You might want to get access to other's ideas are out there to best use them for your own endeavours, or even worse, misinform and stall their development. Creating a separate

(continued)

Box 11.1 (continued)

group where partnership is exclusive reduces this conflict of interest somewhat.

Traders: The distinction between services and partners seems straightforward. The same logic holds for traders and services. Further differentiating between partners and traders is subtler. Traders could simply be seen as ambitious partners. Not quite. A trader differs from a partner insofar as the needed skill set, the handling of ideas, and the incentive structure is concerned. A partner is likely to be highly involved in the implementation, sharing in decisions and personal stakes with the inventor. It is a continuous engagement where the outcome only defines the prospect for the idea. With trading ideas they are likely to be positioned differently (as intermediaries of corporate access points), likely to be handling many more ideas. A different kind of decision-making takes place. Ideas are handled differently (pooling, combining, reusing, trading, segmenting, etc.). Valuation becomes a crucial question. Trading offers a different form of implementation. To avoid conflict between trading and partnering, a differentiation seems sensible.

Such segmentation also implies a serious restriction: everyone may only participate in one category. To some this may be unfair. Some, while waiting to be selected as a partner, may want to benefit from providing services to others; some would like to become a partner as well as being able to trade ideas. The restriction is indeed problematic, but necessary. Different categories require different commitments and most important of all, they require different handling of ideas. Such segmentation has to be placed within the larger picture. It needs to be addressed in the mechanism design, remunerations strategies, the code of conduct, etc. In principle it should not matter. The reputation mechanism still holds in either case. Any kind of abuse and misconduct would ruin a reputation. But to minimize abuse incentive and conflict potential, for pragmatic purposes such segmentation seems sensible. But such a strict division of services may also have additional benefits: it increases the specialisation and division of labour, it shows an additional commitment to position oneself in one or the other category, and may lead to a more complete market for services, partners and traders.

Another major challenge for any such matching approach is the question of inclusion of legal entities: firms. How can they be integrated, made comparable? With individuals different characteristics apply. Different consideration (not least age and standing in life) play a role. Companies have different attributes. They can be short lived or be disposable, or show a much longer time horizon than individuals. How can institutional competencies be included? Companies can offer a much wider range of competencies, offer capacities individuals cannot. Still, some form of comparison is needed. The inventor would want to explore different options and paths. Categories need to adapt or clear and transparent algorithms devised to make legal entities comparable to natural persons. Additional questions of reliability, of capture, of sustainability, and of multiple participation arise (see also discussion below on feedback mechanism).

Yet another issue is overuse. This sounds odd. Participation and usage is good – to a certain degree. Think of email or mail. The more you are spammed the less valuable it is to you. It is tedious, it takes time and you cannot properly absorb or filter the information. Typically you discard it altogether. It is annoying. It diminishes its usability and may let it fail. The question is where spamming starts. Some very eager, call them hyperactive participants, may use it in good conscience. Others see it as spamming. This must be addressed. It gets more complicated still when it comes to highjacking such matching systems for alternative use. Matching is intended for innovation. You could seek partners, could seek assistance, etc. But what if you are looking for a firm to produce certain parts? Does this still fall within the innovation path? Perhaps. What about head hunting? Like many other networks, it can be used for multiple purposes. Headhunting would be an appealing use. The potential to innovate is known. Your profile reveals it. It is innovation specific. This could be highly appealing to headhunters. Does this pose a threat, a deterrence to participate? What about corporate stalking – if you are checked up on every time you apply somewhere or engage with others? These are not just privacy concerns. These are also questions of appeal of participation. Overuse can be a threat.

Many more such issues exist when matching partners. These are but some of them. To cooperate, first adequate partners need to be found. Enabling this, and making any such approach as robust, easy to use and innovation specific as possible is the initial key to enable cooperative innovation.

11.2 Feedback Mechanism

The reputation mechanism is at the heart of cooperative innovation. It is an abstract concept. Operationalizing it is challenging. Little things matter. It needs to be designed carefully. It often requires a delicate trade-off. Some aspects are absolutely vital, some are highly desirable – and some are practically unfeasible. Often very pragmatic solutions are required for it to work, in order to strike a balance of what is needed and what is possible.

Identity is essential. If identity cannot be established, any such mechanism is destined to fail. The feedback mechanism builds on a clearly attributable history. Any deterrence effect to incentivize cooperation depends on this. Without it cooperation seizes to be a dominant strategy. Freeriding likely prevails. If someone cheats, the mechanism assumes a clear assignment of a negative feedback to that person, and adverse consequences for future cooperation opportunities. Freeriding thus is costly. If it is not possible to clearly attribute such feedback, it cannot act as a deterrent. Freeriders could simply change their identity; start off anew with a clean slate. And repeat this process over and over again. Start off, engage, cheat; start off, engage, cheat, and so on. In effect, cooperation would seize, as the mechanism fails. Ensuring identity is key. This is particularly tricky in larger, and especially in online networks. In small social settings ensuring identity is fairly easy. Strong social ties, clear identification characteristics, etc. enable it. In large networks with mostly unfamiliar or anonymous counterparts this is unlikely. Other means of identification are needed. Many possibilities exist, and various options are offered. Some are better than others, but also vastly more expensive. Whichever is chosen, at least a reasonable identification needs to be ensured to avoid manipulation and systematic and strategic abuse of such networks that undermine the feedback mechanism (in academic parlance this is often referred to as 'Sybil attacks' and 'cheap pseudonyms'[2]).

The inclusion of non-natural persons, i.e. legal entities: firms, funds, and associations, adds considerable challenges. In many countries setting up such entities is trivial compared to the potential gains of abuse. Spending a 100 lb Sterling to register a Limited company (Ltd) is 'peanuts' compared to the abuse potential of setting up a front for an identity to steal ideas. Though identity of the entity could be ensured, it can easily be used as a front for others.[3] Some very direct and personal association to key representatives would be needed (to avoid a 'hit and run', that is, to set up a company, freeride, and then move on, setting up a new, clean slate firm). But then the questions of frontmen, changing leadership, etc. arise (and with it a problem of 'hit and fire', that is, hiring someone to take the fall, then remove this person so as not to be associated any longer, which may be especially problematic in a very hierarchical organisation with strong economic dependence). Thus a combination of both, legal entity and personal responsibility of a member/members seems necessary. Additional issues arise when it comes to large firms. This means additional complexity. It poses serious questions about the accuracy of cumulative feedback and the multiplicity of engagement. In case of high volume engagement, cumulative data may conceal strategic abuse of high value cooperation as opposed to negligible or alibi cooperation. In large organizations personal responsibility may differ from the actual cooperation partners on which feedback is provided on. Indeed a large number of such partners in an organization are likely to exist, making clear attribution an issues. Only if strong accountability and unambiguous enforcement to ensure long-term returns can be ensured, can this work.

When it comes to the details of the *feedback mechanism design* several very practical challenges need to be considered (see also box 11.2). By themselves they might not appear primary, but in their consequence, especially when taken together, can seriously unbalance the approach and undermine its credibility on a larger scale. Who can evaluate the quality of cooperation – an impartial judge, the parties

---

[2] See Cheng and Friedman (2005), Friedman and Resnick (2001), Kollock (1999), Levine et al. (2006), Nissenbaum (1999), Tadelis (1999), Yamagishi (2003), Yu et al. (2008).

[3] See especially the discussion in van der Does de Willebois et al. (2011).

themselves or a mixture thereof? What design can inhibit abuse and subversion of such reputation mechanism? Given the range and volume of possible cooperation, any centralized approach seems destined to fail. Even with a detailed code of conduct evaluating each cooperation would be immensely costly and unlikely to be feasible. On the other hand, designing a fully decentralised reciprocal feedback mechanism is complicated and still harbours inherent conflict and abuse potential. If each partner evaluates the quality of cooperation either periodically throughout cooperation or at least in retrospect, the challenges of cooperation often appear too demanding to build on such direct feedback alone. The mechanisms designed to support any such direct feedback system need to be highly refined to incentivise fair conduct and accord, also in the mutual evaluation process, while at the same time be pragmatic, intuitive and realistic.

Box 11.2 Feedback Reputation: Designing Decentralised Feedback Systems: Some Aspects

Designing a proper feedback system in practice if challenging to say the least. The intricacies of human behaviour, of mechanisms, law, technology, etc. all need to be taken into account and play together to form a coherent whole. Many very practical constraints exist, some conceptual ones. Many different approaches have been tried – many have failed. Some aspects to consider:

Quantification: Pressing feedback on cooperation into quantifiable data is not easy. It is always a story, with many facets, situations, details, particularities, special circumstances. Rating it is difficult. It can never capture the heterogeneity of cooperation, but squeezing it into a rather narrow format. In order to compare, sort, rank feedback, specific predefined categories are needed. A considerable spectrum exists in-between fully open feedback essays and a single binary feedback questions. One is too open, the other too narrow. A reasonable balance is needed that allows to capture different aspects and provide a richer picture, to provide quantifiable data while capturing some of the underlying variation and nuances.

Aggregation: Detailed data is good. Summaries are convenient. Given the overwhelming breadth of potential partners, choosing between them is important. Comparing different aspects, or even the same one but with different data points can be taxing, can be overwhelming. At least for an initial screening some form of aggregation is needed. One could assign a summary variable to fairness per se. It could be disaggregated into many different variables. Each variable could consist of many data points. E.g. someone could have engaged in multiple cooperation. Some were good, some bad. How to build an average, over what time period, how are the different outcomes weighted, captured, etc. This is not trivial. The manner of representation, the way they are aggregated distorts the data to some

(continued)

Box 11.2 (continued)

extent. It is biased. Hopefully this bias is designed in such a way that it is acceptable, and helpful to the user. In detail such considerations face plenty of challenges.

History: History maters. This is an essential premise of the feedback mechanism. It helps to decide whom to cooperate with. Its availability generates the deterrence effect. If everyone would get a clean slate, forgive and forget as soon as it occurred, the feedback mechanism would be empty. Nothing to interpolate, nothing to deter. But how long, and to what extent does history matter? A radical position: Fail once and you are out. A more conciliatory: forgive; give second chances; allow for redemption. This poses some interesting questions in themselves. The true difficulty lies with the not so clear, the not so binary. The quality of cooperation can have a large scale: very good, good, quite good; 1–100, etc. but in this case averages, aggregates, etc. matter. Time is a factor. Does the most recent one count more than older ones? The other way around even? This is not so trivial. For example, in a system where the entire history matters, with more and more cooperation the positive effects to an average decreases. Any additional cooperation adds less benefit to the reputation, or, more interestingly, defection does no longer distort it very much. The way in which history matters is thus important. Any rule will motivate a certain strategic behaviour. The design of such rules needs to be chosen carefully.

Segmentation: How to cut and slice, how to aggregate and weigh cooperation, its different aspects, different stages, different success rates, different combinations, different sequences, etc.? Is it less of a problem failing in the early stages of cooperation than in the later stages when more time, money, etc. has been invested? Does it make a difference if a large group of partners fail, or whether it is just two? Is there a difference in cooperating with a service provider helping to write a business plan, or a big firm trading the idea to? Many options exist of how to segment the data, and weigh it in different manners. The choices made will codetermine the information the data is able to convey.

Sequence: Sequence can be very important. Imagine you cooperated with someone. You thought it went ok. Not well, but ok. Now you see the feedback your former partner provided. It is bad. She thought you performed poorly. Many would no longer be objective, but equally downgrade their feedback – be it for reasons of retaliation, perceived blame assignment (if it is also the others fault it may be easier to explain). If on the other hand feedback would be provided at the same time without the other seeing the feedback this problem could be avoided. Surely this can always be a problem if you suspect you will get a negative feedback. It may even cause a generally lower feedback level (which should not matter as it is essentially a relative selection process more than an absolute one).

(continued)

Box 11.2 (continued)

But it can also lead to abuse using feedback as a strategic tool, postponing feedback, etc. Fairly sophisticated feedback mechanisms designs are required to make this work in a least distortive manner.

Selection/Sampling bias: Giving feedback requires some, though small, effort. You cannot force someone to provide it. How? Even if this was feasible, those feeling forced are likely to do it sloppily. In a voluntary feedback system, a self-selection bias prevails so that often only those participate that have sufficient incentives to promote or denounce. In other words, outliers dominate in voluntary systems. This can highly distort feedback. Look at online rating platforms. If it does get sufficient traction, people will try to encourage only those with exceptional feedback, hide from those disgruntled. The voluntary effort is typically only undertaken by the most upset. Very few make the effort to praise. Most are indifferent enough to avoid even a fraction of effort. As long as potential feedbacks can be identified 'neutral'/reasonably positive feedback defaults could be used. For example, now, instead of just one negative ten feedbacks exist, one bad and nine fairly positive default ratings. This helps the aggregate, but may also be the least interesting for the ones assessing potential cooperation. The manner of accounting for the participation incentive in the feedback evaluation matters to cater to the notion of fairness and to minimise abuse. Both are needed to make such approach work.

Mediation: Things can go wrong. Not everyone is fair, is honest or plays by the rules. What to do in such cases? Mechanisms need to be in place to address such cases. Though likely exceptions, they matter as they pose a risk anyone could face, deterring participation, increasing the risk and costs of cooperation. Also, they attract attention and form opinions, based on often crude and unwarranted extrapolations about the entire approach. Some form of mediation is required. To what extent, and in what format this can be integrated is a challenge. Can there be 'objective standards'? If so, what mechanisms best trade off overuse and costs with adequate inclusion of worthy cases? Some form of mediation can help. Typically this is costly. It needs to be individual and flexible to accommodate the particularities of the different cases. The mechanism design in when and how mediation processes apply should include deterrents and frame the incentives to call on such mechanisms in a way that attract only warranted cases. This may for example include cost sharing as a function according to deviation of new to previous feedback value. Other designs are possible, and likely more refined designs will emerge. Other questions include issues of strategic abuse of such mechanisms. What happens in the meantime, while a feedback is being contested? Is it excluded – thus will be strategically contested, or does it count, and may thus be used for manipulation. Many more such questions apply and will need to be taken into account when designing a mediation process.

(continued)

Box 11.2 (continued)

Abuse and Manipulation: No feedback system is perfect. If attractive enough, people will find ways to work around any constraints, and discover often highly creative ways to manipulate or abuse it. Simple situations may include: cooperation fails. Either one would give a bad feedback. They agree to avoid this, and rather give mutually positive feedback. Alternatively, blame is clear; one will abstain from negative feedback in return for adequate compensation. Or more calculating: to avoid getting a deservedly bad feedback you threaten the other person to do the same and rather settle on a mutually beneficial feedback. More complex and presumable more serious abuse may for example include the emergence of a market for reputation – offering good reputation feedbacks for sale. Someone will fake a positive cooperation for a fee; or someone offering to be the fall-guy for failed cooperation; someone offering to pose as a front for cooperation abuse. A range of possibilities can be devised to address these issues. Some preventive, some punitive. Some are outside of control of any such approach. It needs to be addressed with vigour, as this can seriously undermine cooperation, the trust in the system, the functioning of cooperative innovation per se.

A broader question is that of scope of the feedback mechanism, whether an *integrated or a specific* approach is more appropriate. A highly targeted feedback system can best identify the crucial and often highly content specific characteristics of cooperation. There could be one for some specific technologies, different stages in the innovation process, different sectors, etc. But such fragmentation can have serious drawbacks, for innovation in general (given the assumption ideas come from anywhere and often unrelated sectors) and the effectiveness of the reputation mechanism in particular. The most serious threat is 'network hopping'. Particularly if the networks are too similar, or offer parallel systems, this can hamper the effectiveness, and undermine the feedback mechanism. Someone could participate, even with proper identification, could free-ride, leave the network, do the same in other networks, and so on. The mechanisms would fail. Hybrid systems of mutual sharing or platform solution with a common underlying core and specific add-ons could be an option to address this (but then questions of interchangability and consolidation of data, categories, etc., arise). Even if not by initial design, some intermediate form of broad based coordination between more dedicated and targeted networks will surely evolve.

Again, these are but some, by far not all challenges facing the design of a practical feedback mechanism. Many of these details can undermine the entire approach. Many more issues exist. Pragmatic solutions to address them are needed to make the reputation mechanism work. It may not need to be perfect – it never is. Sometimes it may even be better not to over-engineer, to keep it simple and accessible, to allow for flexibility, and to allow for intrinsic notions of trust.

Focussing too much on the least common denominator, pursuing a view that stubbornly focuses on design optimality in an otherwise imperfect setting with an unimaginable complexity of intertwining factors, can often be suboptimal itself.[4] A feedback mechanism needs to be effective. It also must be pragmatic. It may never be perfect, but sufficiently precise.

11.3 Code of Conduct

Fairness is crucial. But what does that mean in practice? Clearly it is a broad and often very heterogeneous concept. It is constantly evolving. Thus any practical approach can only be indicative and transitory. Worse, when it comes to ideas, cooperation falls on utterly unchartered territory. Only some initial codes of conduct can be proposed. Practice will show their validity and feasibility, likely to be altered and amended as the concept evolves. Surely, with practice, research, and increasing competition, eventually good practices will emerge. Initially, to initiate and nurture its development the task is to develop a commonly acceptable skeleton of a code of conduct. It may help minimise initial friction and initial backlash against cooperative innovation.

Such considerations of fairness are very real, very practical. By far this is no vanity exercise of idealism, of what could, should, might. Its impact is real. The feedback mechanism is built on such very notion of fairness. You are judged against what the other considers fair, and you judge others according to what you deem fair. Without specific outlines of what constitutes fair behaviour people would be unsure of what such fair behaviour includes when searching and selecting partners and how to conduct themselves properly so as to avoid a negative feedback. You need to know what the other considers to be fair, and vice versa. Clearly, standards may differ. You may have a different view on fairness than your partner. All the more important to establish a common basis, so as for all to know, for all to judge, and for all to be judged against. Such Code of Conduct somewhat defines fairness when it comes to cooperating in realising ideas. It is a playbook for the different players involved that lays the foundation of cooperation and feedback, the criteria for cooperation and for providing feedback against. Any deviation of it may be included in the feedback evaluation accordingly, just as it will be considered the standard for any mediation process. Therefore a code of conduct has to go beyond process questions, and address the very essence of cooperative innovation – how to treat ideas; what to do with complementary ideas; what with substitutive; etc.? (see box 11.3 for more details) A code of conduct of course goes both ways. It should not be seen as a norm for the realisation partner. It applies to all partners. All need to know what to expect, and what is expected of them. This reduces uncertainty, reduces discord, and provides some yardstick to identify deviation and noncompliance.

---

[4] For an intriguing discussion, see Bowles and Hwang (2008). Also see Ostrom (2000). In the current context specifically also see Bolton and Ockenfels (2009) and Bolton et al. (2004).

Box 11.3 Code of Conduct: Outlining Key Aspects
Both inventors and implementers need to behave and be treated fairly for any cooperation to succeed. The one submitting an idea needs to follow a certain code of conduct, as does the one receiving ideas. The challenges they face differ somewhat. For the latter it is somewhat more complicated. Each will be considered in turn.

Ideas: How to Engage in Cooperation
The one with the idea yields considerable power. He/she determines the initial realisation path, approaches potential partners, etc. Once the idea is shared however, this situation typically shifts. This raises many questions. It brings with it certain rights as well as certain obligations that are needed for a feedback mechanism to work in practice.

How refined does the idea need to be? A hunch is good. A well thought out idea is better. A hunch can lead to many things. Some may be thought of, others neglected. If an idea is shared, it is very important to frame its scope. It can spark many new ideas. Often ideas apply to a range of industries, and, modified in various forms, can be reused in many instances. These can go much beyond the original. Then the question becomes: who's idea is it? What is included, and what is not? Where can the boundaries be drawn? Defining the idea in some detail, conducting some basic investigation into the details of the idea, its application, the potential market, etc. not only helps to convey the idea, but also helps frame its scope. This can help place a certain claim on the idea by providing a framework of applicability, of what is included, what not (the sector, the product, the process, the customers, etc.). Only what is defined, and defined explicitly, can be included. This incentivises clarity and initial work on part of the inventor. It also minimises ambiguity and transaction costs of conveying the idea, as well as reducing possible clutter and overwhelming of recipients with half-baked ideas.[5] What this may look like in detail likely varies from a general to a very specific set of aspects and level of detail.

How will offering an idea work in practice? Blurting out an idea would be a mistake. The question is, what would such a process of contacting a potential partner look like, and what rights and obligations does it bring with it? Are certain processes important? Does the sequence matter? Can an offer be taken back? Certain procedures are needed of how establishing contact and offering an idea can work in practice. (i) As a basic aspect, ideas will have to be invited. The potential for strategic abuse, for

(continued)

---

[5] This is a very practical concern often raised by those trying out open innovation systems (e.g. Huston 2007).

11.3 Code of Conduct

Box 11.3 (continued)

safeguarding ideas, and for establishing certain obligations, some format will be needed to establish contact and agree to take a look at the idea and possible cooperation. Uninvited submissions are to be considered void. No claim on the idea can be made or else strategic abuse would be rampant. This also forces some discipline and compliance with format and processes. Once such contact is established and a general interest is signalled, cooperation can start official cooperation. It will have to start prior to submission or else it would be problematic to have a basis for what feedback can be provided on. Otherwise anyone could provide feedback on anything and on anyone else. (ii) Submitting an idea to a prospective cooperation partner must be considered a binding offer. It cannot be taken back. How could it? This is at the heart of the information paradox. The offer needs to be firm. With submission the inventor agrees to certain predefined terms – terms he/she will be judged against in the feedback (see below on the rights and obligation of the partner). (iii) Sequence matters. Simultaneous submissions would be highly problematic. Submitting to multiple, especially if potential competitors, would be counterproductive. Only if it is rejected or cooperation is concluded can it be offered to someone else. On the other hand, it is not clear where the demarcation line has to be. Can it be offered to various partners if active in different sectors? Can it be confined to some sector?

What are the commitments during cooperation? Implementing an idea will bring many additional challenges. Who has to work on what? How much? What does that mean for ownership? How much input can be expected from the inventor to refine and modify the idea? How much input to operationalize it? Pragmatic rules need to be established – that may vary considerably by sector, but should be fairly circumstances robust and not too complex – on what typical claims should be. Clear remuneration frameworks may go a long way to address many such issues and sustain adequate incentive structures to support cooperative behaviour and comply with common notions of fairness.

What kind of feedback is needed? The feedback mechanism is key to foster cooperative innovation. Feedback needs to be expected, or defaults may set in. Simultaneity and reasonable timeframes of feedback provision can be expected. But also the content matters, matters even more. What elements is feedback needed on? Who determines what is important? The variance of responses could seriously undermine a working feedback mechanism. Clear structures on what aspects are relevant and what is fair and acceptable feedback is needed. Disgruntlement over inadequate success is different to stealing an idea. Failure can have various forms, and even more explanations. What is constructive feedback, what useless, what slander, what hear-say, and what distortive?

(continued)

Box 11.3 (continued)
Partners: How to Handle Ideas
The one receiving ideas as cooperating partner faces even more difficulties in how to handle ideas. Due to its very nature ideas are special and require special treatment. Normal rules of conduct often do not apply. Many additional considerations need to be included to establish a common understanding of what might be deemed fair behaviour of handling ideas.

When does cooperation start? Anyone open to ideas can be approached. This does not mean they will have to cooperate. They can reject cooperation – due to a lack of time, interest, confidence in the partner, etc. The question is on what basis they need to decide? Ideas are not shared without clear engagement in cooperation. Only a certain level of detail can be asked. Cooperation is entered into before the idea is revealed, or the feedback mechanism does not apply. If too much information is revealed, and deliberately asked without entering into cooperation, the idea often becomes too obvious. A certain clearly stated conduct should be established to shape this process. It should offer sufficient information to engage, and insufficient to the extent of inadvertently revealing too much of the idea. Once cooperation is entered into, the feedback mechanism applies – with all the expectations of fairness.

What happens if an idea is rejected? It is the prerogative of any partner to reject an idea and seize cooperation. This need not be anything bad or harmful to reputation. The manner of how it is done matters greatly though. Rejecting an idea implies a lack of interest. These ideas cannot be taken up at a later stage. What though if the value only becomes apparent in the future or in connection with a different aspect? Once the idea is revealed it is out there. It cannot be taken back. It needs to be clear what obligation to safeguard ideas and act on them come with rejecting an idea. This is utmost tricky. Does a certain timeframe apply? To what degree does a recombination count as a new idea or a reconfiguration of the old? What obligation to the original source of ideas still exists? How will the feedback apply even after the cooperation has been completed and initial feedback been provided?

How to cooperate fairly? No stealing of ideas. This is the baseline. But this falls short of the many other facets and nuances of fair cooperation. Cooperation has many aspects to it. The basic obligation is to honour the commitments made. Any cooperation will have different aspects to consider. Different partners, different ideas, different circumstances, etc. all call for different forms of cooperation. Up front, once a decision to actively cooperate in realising the idea is made, these obligations should be clearly spelled out. Who does what? In what form? Who is remunerated for what? And many, many more such details. Here guidance, especially checklists of what to consider, can help to address these issues. But also some rules of thumb are needed to provide a reference baseline for what could typically be considered as 'fair' in these terms.

(continued)

Box 11.3 (continued)

How to handle ideas in connection with other or own ideas? The simplest form of cooperation would be a one-on-one at a time: a clear idea by the inventor, straightforward implementation by the partner. This, likely, will be more the exception than the rule, especially in the long run. Good and reliable partners, especially big firms, will be offered multiple ideas. Often, even though the initial idea may not be that useful, it may instigate a better still. Often during implementation more ideas are needed to overcome unexpected challenges, or refinements to improve the idea and greatly increase its value. Whose idea then is it? How to treat such situations? What if several ideas are the same? What if they build on each other? Etc. These questions are at the core of the problem of trading ideas. They define our understanding of fairness – and the answers can differ greatly as this territory is utterly unchartered and many diverging claims exist. To provide some more structure, it may help to split the problem at least in two broader components: parallel ideas (two or more separate ideas) and successive ideas (one idea builds on another). *(i) Parallel Ideas:* Often ideas, even most profound and revolutionary ideas, come in multiple versions – often independent of one another and at the same time. New discoveries or new principle enable new applications or products for anyone with some imagination. They replace other processes, or complement each other to service new markets. This in itself is a normal and desirable process. It fosters competition and the evolutionary process of innovation in general. However, it becomes a problem when someone is offered such multiple ideas. Especially large firms, brokers or specialised experts are most prone to be selected by several people with similar, complementary or substitutive ideas. How to handle such ideas? Does the recipient bare the risk, or does the one submitting it? Will it be first-come-first serve, or some form of splitting of the proceeds? If the incentives are skewed towards one or the other, it may be problematic. If anyone submitting bares the risk, and accordingly his share in the return will diminish or even be void, this is problematic. Equally adverse would it be the other way around. An incentive structure is necessary to balance both sides and avert adverse incentive to game the mechanism. *(ii) Successive Ideas:* 'Standing on the shoulders of giants'. Without the magnificent contribution of such 'giants' we would not be able to build on them. One idea is often the basis of the next. If Newton would not have postulated the classical physical principles would Einstein have been able to postulate his? This easily leads to a practical dilemma. If someone has an idea it may be the basis and inspiration for someone else to have a better idea, an idea that may amend, improve or even replace the former. But without the former he/she is unlikely to have come up with the idea in the first place. Hence the first must be credited with the inspiration. The second, however, for the final contribution. What if the first idea in itself is useless, but the refined version is magnificent. Would it be

(continued)

> **Box 11.3** (continued)
> considered only fair to reward also the first? Just the first? Only the second? This is a serious problem in handling ideas. Especially if someone comes up with an idea and approaches an expert. He/she may spot faults, or shortcomings, or be able to go beyond the original idea, add, modify, refine etc. and make the idea better than before. If no one would approach him/her he/she may never have had such an inspiration. How then to attribute the reward?
>
> *How to provide feedback, and on what and when?* At any point during the cooperation process either partner can end the cooperation and initiate the feedback procedure. The feedback mechanism goes both ways. Not only the implementer is considered, but also the inventor. He/she too will receive feedback to capture all relevant information valuable to any following partner to inform them of the trustworthiness and quality of cooperation. Clear scope and format will be needed to capture and administer this. A clear code of conduct on how to provide feedback, in what format, in what tone, and possible remedial actions is essential to make it work (see also feedback from inventors above)

The challenge is to find an acceptable middle ground. While it strengthens some claims, it also restricts others. This is tricky. It is less a problem of inherent differences in the perception of fairness, but of general acceptance. Many aspects can be mediated by establishing a common ground. Sometimes this requires elaborate discussions to establish a sense of being on the same page at least. The residual can often be split down the middle. The bigger issues are often misunderstandings due to perspective, not fairness per se. Hence with a code of conduct comes the demand for solid reasoning. The two can be separated to some extent. If sufficient trust exists in the code of conduct, if it is credible, and reasonable to believe that it is indeed based on solid understanding and profound discussion, this may suffice. If a crisis of confidence arises this can seriously threaten its usage, and cause a divergence in claims. This is a risk. It must therefore be a transparent and open discussion. And it must be an on-going one. Only then can its legitimacy be sustained.

Initially a simple yardstick will suffice. It should instigate further discussion, solidifying its claim. Surely it will need to be amended, augmented, refined, and altered where necessary. None of it is ever set in stone. But something is needed for the beginning. Partners may agree on different Code of Conduct, may propose their own. For simplicity, to reduce interaction costs, to reduce friction potential – you do not want to haggle with everyone, and it is tedious to compare potential partners according to yet another variable that might be critical – a simple common standard would be desirable. But also individual standards that better address sectoral, thematic, procedural particularities may be useful. In a last instance, experience will help to sort out malpractice and quickly promote good practice examples and rules. It is expected to evolve. Indeed it must evolve. The challenge for now still is to take a daring stab at it, and propose an initial outline.

11.4 Guidance

Calling for the provision of guidance may sound quaint, superfluous, and misplaced – a marginal contribution to the challenges facing cooperative innovation. Far from it! It is hugely important. Innovation is daunting when you consider the immense challenges and obstacles from inception to an established market realisation. Adding complexities of cooperation makes it worse still – differentiating cooperation partners, establishing complex feedback mechanisms, etc. Much can be done to make such mechanism work and many aspects need to be considered. Cooperation is not simple. Details matter. Many details. Many aspects need to be addressed in practice. This adds complexity. And this is what makes it less and less accessible, unpractical, and unusable. At the same time it is a novel approach. It needs to be kept simple to be accessible. Later, it may become more complex. First it needs to work. For it to work it requires participation on a larger scale. It must be accessible. Again, a trade-off exists. This time between making it more complex and making the reputation mechanism simpler, making it fool proof to use. More complexity makes the approach more robust, makes it potentially more effective, even fairer. It also makes it less intuitive, less manageable, and more costly. This is where guidance comes in. It can allow for more complexity. Guidance can help reach a level of complexity that is fair enough to be considered reasonably fair, and reasonably reliable, while allowing for broad based participation without being overwhelmed. This can help sustain the mechanism and approach. Indispensable! Clear-cut guidance, transparency, and conveyance of the mechanisms at work are crucial. The better the understanding, the more effective the system will be, and the more demanding it can be designed. It can evolve faster to incorporate more complexities to improve it further. In more applied terms: when designing a novel gadget, say, an iPod you can make it very simple for anyone to use. More can easily use it. But its functionality is limited. If you make it too complex, too many buttons, too many multiple functions per button, etc. it may be able to do more, but it will be much harder to get the hang of it. It may overwhelm many people, deterring them from using it. Somewhere in-between seems best. But where? Without any guidance the device would have to be much simpler than if you can easily explain its additional features. Once it becomes commonplace, with a high uptake, and people become more aquatinted, you can make it more and more complex. Users will learn to appreciate additional functionality. Think of a remote control, think of your car navigation system, think of an email program, a search program, or an operating system. Complexity matters – but it must be accessible. This, to some extent, can be explained. Guidance can be provided; complexity can be phased in.

Many steps of cooperative innovation are new (see Fig. 11.1). They all raise questions of how they will work in practice. Each can be made simple, yet likely less effective, or complex, yet less accessible. The challenge is to find a way to

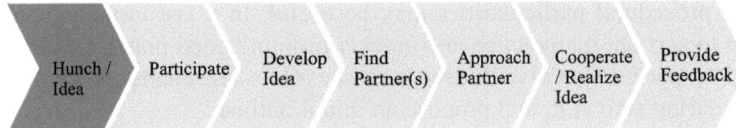

Fig. 11.1 Full cycle: from hunch to feedback

make them reasonably complex and offer guidance to make them more accessible.

The most general aspect: ***participation***. Understanding who can participate, what are their options and chances is important. Clearly the ones with Ideas are essential. So are those able to implement ideas. But who are they? A wide range of possible partners, such as experts, entrepreneurs, firms, and a wide range of those offering services along the way would enrich cooperative innovation. This does not mean the more the merrier – the more appropriate and the better the participants the merrier. The threshold design matters. This can be in form of some kind of deterrent (registration, fee, tediousness, etc.) or in form of appeal and outreach to target groups. It needs to be clear what is expected, what are the chances, what are the advantages and drawbacks of participation. Transparency matters. These are initial questions that may convey some basic notion of how to benefit from participation, and how to do this best. It avoids clutter, and may help to make the entire approach more effective if people make more informed and better choices. Search is more focussed; selection is easier. Particularly if a certain segmentation of participants (firms, entrepreneurs, sectoral experts, service providers, etc.) is introduced this is important.

Once a choice of participation is made, guidance can assist the various aspects of implementation. Foremost: ***Idea development***. From an initial hunch, to a fully defined idea, reflected on, with some initial research, refined, thought through, can be a lengthy process with many steps, and many setbacks. What to look for, what to consider? Providing some clear structures what to do, where to look, and what to look for, can go a long way. Many such concepts exist. They need to be adapted to ideas specifically and made more accessible.[6] This is more than just assistance and circumventing laziness. A real threat of overwhelming the system exists if this is not provided, if this is not used. Hunches are not enough. Cluttering the system, spamming people with half-baked ideas can be a serious problem. As discouraging as it must sound: many people are smart, many people have considered many possible aspects. If you have an idea, to put it bluntly, it is more likely that someone has considered it, than not. It may be labelled differently; it may have been discussed a long time ago. Somewhat likely though it was proposed, was discussed in some form or another. This should not be discouraging. It should be enthralling. It can provide vital inputs, confirm your initial hunch, provide needed input, can inspire more and better ideas yet. But what it means is that you must do your

---

[6] For example, businessmodelgeneration.com/canvas (02.03.2013).

11.4 Guidance

homework, you must spend time researching, providing a better basis for your idea. 'Idea Plans' need to be developed. This need not go far, but requires some initial investigation into the possibilities, the potential, and the prospects of the idea. The methods/product/concept needs to be refined as well as the marketability. Some initial costs estimates, patentability considerations, etc. would be helpful. Knowing what to ask and having access to good information would be helpful. Much guidance is already provided. Better guidance is needed. Services may be created to assist and support the development of ideas. But introducing clear guidance can substantially self-filter, help to make ideas better, help the cooperation process by making it leaner, more structured. There is of course the risk it remains too generic, it may not apply to many ideas. It is considerably different developing a new iPad App, than improving a production process, or developing a revolutionary new gadget. It involves very different questions, very different considerations, very different stakeholders, options, etc. Eventually such more specific guidance will merge, by sector, product, kind, etc. further refining such approaches, assisting hunches to become promising ideas, guiding ideas to its success. Initially a general framework can help spur such development.

Matching: Finding assistance, more importantly, finding potential cooperation partners is tricky. Identifying chances, weighing options, assessing opportunities can be facilitated by providing guidance. Finding the optimal partner, trading off different skills aware of the implications, knowing the shortfalls, the risks and the possibilities helps to make better informed decision, potentially the more effective choices. Here many, many pitfalls arise, often hampering the success of ideas, and the success of cooperation. This is wasteful. Guidance can help. Many dimensions play a role. Explicating some behaviours and considerations may help expose pitfalls, and may thus help avoid them. To illustrate this, consider two, seemingly benign aspects: ambiguity and control. You would think ambiguity is bad. Details are better. You would be right. But this is often not the way people choose. When given limited information, initial impressions can be overly positive. Take an example from online-dating: If someone says he/she likes music, you may conclude, so do I. This would be a good match. Would he/she have said instead they like Krautrock, you might be disinclined to choose such a partner as dissimilarity to your preference for classical music.[7] The same in principle holds for many relevant aspects of matching partners when it comes to cooperative innovation. The more detail, the more apparent dissimilarity becomes. The more ambiguity, the more positive interpretation and association may occur. Guidance may help expose such issues, make aware of such tendencies, etc. and thus facilitate matching. Control is another example. Many people would agree in principle, "It is often better to get a small piece of a very large pie, than to get a large piece of a very small pie." In a way this is a silly saying, yet it bears much truth especially for the current situation. It is a trade-off. If you realise your idea on your own you may retain control and

---

[7] Experiments with online dating have demonstrated a negative relationship between knowledge and liking. For a more elaborate and highly interesting discussion, see Norton et al. (2007).

most of the profits would be yours. Yet, you may not be the one best able to realise the idea. You may fail, or not reach the full potential of the idea. If you would cooperate with an expert or entrepreneur, or even trade the idea it could become much, much more. Even though you would need to share the profits with others, you may still be better off altogether. Which strategy is best depends on the individual case, the idea itself, and your abilities. Think of the pie. It is normally (hopefully!) not the biggest slice but the most pie you want. A desire for control may often hamper that goal. These are but two simple examples, one on the nitty-gritty of matching behaviour, one on the overall considerations of choosing a, and if a, what kind of partner. Many more exist. Many can be serious constraints to success. Minimising such pitfalls can make a cooperative approach more successful. Not everyone has to fall into them, not everyone has to learn on their own, not everyone has to reinvent the wheel. Surely not all aspects of choosing appropriate cooperation can ever be taken into account. Cooperation is much too complex and individual, but where guidance can help, it can help make the process leaner and more effective.

Approaching a potential partner poses yet another complication. How to initiate cooperation? What process, what procedure? When, whom and how can ideas be shared for the mechanism to work? In what format can and should they be shared? These are trivial questions in the grand scheme of things, but none the less important here. Much can go wrong. It very much depends on the design of the feedback mechanism. Can one initiate cooperation, or do both have to agree? When does it start? What commitments does this imply? Depending on the code of conduct, once the idea is submitted, some limitations would apply. No further sharing for a while? No use without consent? Practice will reveal more details, and surely they will evolve. But up front clarity is needed or cooperation, worse the whole approach of a feedback mechanism will fail. Here it can get messy. Spamming and hyperactivity by those submitting ideas can clutter the system. Expectations can get out of hand. On the side of the recipients fine-print will be rampant, complexities become daunting for both. Drawing out the basic principles, conveying the most serious implication makes it fairer to both, the one offering and the one looking for ideas, and easier to engage with either. It needs to be made clear up front what to be judged against, in what way. What are the rights, what the obligations. And what happens if things go wrong. What options exist, what are the implications (see Code of Conduct).

Also during *cooperation* many things can go wrong. Cooperation can be highly diverse, and there likely is no standard format, no blueprint, or success formula. But, there are lessons learned, there are issues to consider. Not everyone needs to make the same mistakes. Some generic guidance can help. It may not guarantee success, but it can foster it. The code of conduct may help here. It can provide a reference point, provide some flagging of potential pitfalls, issues and warning signs of impending failure. But guidance should go beyond. Should address potential issues and propose solutions, should point out areas of conflict and should identify certain warning signs of impending problems. This may help reduce uncertainty and offer help.

Providing feedback can be challenging. The reference should be a clear code of conduct. Providing additional guidance, of what to consider, how to use it, how to weigh different aspects. Different considerations may apply whether cooperation ends amicably or not. Depending on the level of detail of a code of conduct and the flexibility to adapt to individual circumstances guidance is needed. The more rigid, the more guidance on how to apply it; the more flexible, the more guidance on how to use it is needed. This means having a very clear understanding about the mechanism (who provides feedback, who receives it, who will it be available to, when will the other partner see it (before or after their rating), etc.) and the criteria of feedback. Scales need to be clearly explained. Guidance on what information is important, and what is counterproductive can help produce more constructive feedback, useful to others, and reduce slander. A clear understanding of possible repercussions and potential mediation processes can further help make providing feedback more useful and therefore the feedback mechanism more effective.

All four aspects have to be addressed jointly. All have to be addressed thoroughly. Together only can they make, yet any part can break the approach. The devil is in the detail. Explicating some of the fundamental challenges and the intricacies of the various aspects is absolutely necessary to assess the viability and point to possible ways forward. This means delving into technicalities, and details that on the outset seem dull and dreary, but will provide the building blocks of things to come. These are issues to be addressed in order for the *Idea Economy* to evolve and to prosper.

References

Bengtsson, O., & Hsu, D. H. (2010). *How do venture capital partners match with startup founders?* Knut Wicksell Working Paper No 2013:8, Lund University, http://www.lusem.lu.se/media/kwc/working-papers/kwc-wp-2013-8.pdf Accessed 20 August 2013.

Bolton, G. E., Katok, E., & Ockenfels, A. (2004). Trust among internet traders: A behavioral economics approach. *Analyse und Kritik Zeitschrift fuer Sozialtheorie, 26*(1), 185–202.

Bolton, G. E., & Ockenfels, A. (2009). The limits of trust in economic transactions: Investigations of perfect reputation systems. In K. S. Cook, C. Snijders, V. Buskens, & C. Cheshire (Eds.), *eTrust: Forming relationships in the online* (pp. 15–36). New York: Russell Sage Foundation.

Boschma, R. (2005). Proximity and innovation: A critical assessment. *Regional Studies, 39*(1), 61–74.

Bowles, S., & Hwang, S. H. (2008). Social preferences and public economics: Mechanism design when social preferences depend on incentives. *Journal of Public Economics, 92*(8–9), 1811–1820.

Cheng, A., & Friedman, E. (2005). Sybilproof reputation mechanisms. *Proceeding of the 3rd Workshop on Economics of Peer-to-Peer Systems (P2PECON)*. http://www.eecs.harvard.edu/cs286r/courses/fall08/files/paper-CheFri.pdf. Accessed 31 March 2013.

de Jong, J., & Freel, M. (2010). *Geographical distance of innovation collaborations*. Scales research reports H201008, EIM Business and Policy Research. www.entrepreneurship-sme.eu/pdf-ez/H201008.pdf. Accessed 02 March 2013.

Friedman, E., & Resnick, P. (2001). The social cost of cheap pseudonyms. *Journal of Economics and Management Strategy, 10*(2), 173–199.

Huston, L. (2007). *Innovation networks: Looking for ideas outside the company. Interview with Larry Huston*, published, 14 Nov , 2007 in Knowledge@Wharton. http://knowledge.wharton.upenn.edu/article.cfm?articleid=1837. Accessed 02 March 2013.

Kollock, P. (1999). The production of trust in online markets. *Advances in Group Processes, 16*, 99–123.

Levine, B. N., Shields, C., & Margolin, N. B. (2006). *A survey of solutions to the sybil attack*. University of Massachusetts Amherst, Amherst. https://gnunet.org/node/1432. Accessed 02 March 2013.

Mohr, J., & Spekman, R. (1994). Characteristics of partnership success: Partnership attributes, communication behavior, and conflict resolution techniques. *Strategic Management Journal, 15*(2), 135–152.

Nissenbaum, H. (1999). The meaning of anonymity in an information age. *The Information Society, 15*(2), 141–144.

Norton, M. I., Frost, J. H., & Ariely, D. (2007). Less is more: The lure of ambiguity, or why familiarity breeds contempt. *Journal of Personality and Social Psychology, 92*(1), 97–105.

Ostrom, E. (2000). Crowding out citizenship. *Scandinavian Political Studies, 23*(1), 3–16.

Shane, S., & Cable, D. (2002). Network ties, reputation, and the financing of new ventures. *Management Science, 48*(3), 364–381.

Sorenson, O., & Stuart, T. E. (2001). Syndication networks and the spatial distribution of venture capital investments. *The American Journal of Sociology, 106*(6), 1546–1588.

Tadelis, S. (1999). What's in a name? Reputation as a tradeable asset. *The American Economic Review, 89*(3), 548–563.

van der Does de Willebois, E., Halter, E. M., Harrison, R. A., Park, J. W., & Sharman, J. C. (2011). *The puppet masters: How the corrupt use legal structures to hide stolen assets and what to do about it*. Washington, DC: World Bank.

Yamagishi, T. (2003). *The role of reputation in open and closed societies: An experimental study of internet auctioning*. Working paper, Hokkaido University. http://citeseerx.ist.psu.edu/viewdoc/download?doi=10.1.1.7.204&rep=rep1&type=pdf. Accessed 02 March 2013.

Yu, H., Kaminsky, M., Gibbons, P. B., & Flaxman, A. D. (2008) SybilGuard: Defending Against Sybil Attacks via Social Networks. IEEE/ACM Transactions on Networking 16(3):576–589.

Cooperative Innovation and Its Future 12

Will cooperative innovation work? With a practical reputation mechanism in place it becomes possible to reasonably bridge the gap between ideas and its realisation and get people to trust one another when working together to realize ideas. Efforts are under way to operationalize this. Working together with firms and businessmen is becoming accessible to anyone with ideas. The Idea Economy is ready to unfold. Anyone can contribute to it. Everyone will benefit from it.

Cooperative innovation is like hitchhiking. You pick up a stranger and hope you can trust him not to mug you or steal your car. While many people are trustworthy, some are not. As some frightening example, horror movies, and urban legends have taught us, better err on the side of caution: Do not pick up stranger alone in your car. The few bad apples have spoiled the innocence and romantics of hitching a ride: you and a backpack; hitting the road; exploring the world. No more. This system of convenient and efficient transport no longer works. Hardly any but the most daring or carefree stop nowadays to pick up a stranger on the side of the road. Hardly anyone expects them to, so almost no one even tries anymore. The same goes for ideas. Letting a stranger in on your promising idea, a stranger of whom little may be known, especially his intentions, is risky a gamble. No one tries, no one offers. There is no market for ideas. Some prominent examples of ideas being stolen have ruined cooperative innovation for all.[1]

But there is hope. Hitchhiking at least, or a version thereof, is becoming fashionable yet again. Some smart approaches and modern communication have made it possible to find a solution to make sharing a ride with a stranger safe and sound again. New services such as zimride.com have sprung up.[2] They match vacant car seats to 'hitchhikers'. Instead of sticking out their thumb along the road, travellers use the index finger to tap on a mouse or smartphone to signal their need for transport from the cosy convenience of their home or a café. Willing drivers put

---

[1] This is an analogy. The underlying notion of Arrow's and other paradoxes hold just the same.

[2] Also avego.com, blablacar.com, carpooling.com, nuride.com, eRideShare.com or ridejoy.com (for more background see New York Times 2012a, b)

online their route and availability. Drivers with a spare seat get paired with 'hitchhikers'. Everyone has a profile that may tell you something about him or her. As the driver you can chose whom to take: someone going all the way, or only some part of the way; man, women; you can of course be even finickier if you have the choice, looking for a certain taste in music, a love for classical literature, or passion for baking, to liven up your journey. Whatever you feel comfortable with. You arrange for a meet, pick up the hitchhikers, and travel the distance together. Costs are shared – no profits allowed. Once your trip is over, each provides a feedback on the other. Each builds up a profile and a reputation. This is what makes the difference to old-school hitchhiking. The reputation mechanism is what makes it work today. Of course you will only let people you trust ride along. You need to trust they behave. And while you probably do not know your future passenger, others do. Others have provided feedback that gives you an idea about who they are and how they behave. Clearly, misbehaviour will be punished. Those with a bad track record, those others have identified as a 'fraud' you will not take along. Who would agree to share a car for a lengthy trip with someone all others deem annoying, rude, and a fraud? Passengers better behave. Once the reputation is ruined, no one will take them along for the ride. They remain stuck or have to travel alone or much more costly. So they will behave fearing no more rides in the future. Reputation is gold. Key to making any such reputation mechanism work it needs to be ensured that everyone is who they claim they are. If you can use a fake identity, why care about a ruined reputation. Simply start anew with a clean sheet, and again freeride, misbehave however you like. To ensure identity, car sharing platforms have been building in additional safeguards. Want-to-be travellers need to ensure their identity by providing personal verifiable details such as legal documentation and social integration, such as ID numbers and Facebook sign-in. Additional background checks are sometimes employed to further weed out and discourage bad behaviour, and ensure you are not picking up villains, but polite and pleasant company. Such measures have earned Zimride the label "non-creepy hitchhiking service".[3] 'Stranger Danger' no longer applies that crudely. Hitchhiking is back. It now is safer and more convenient than ever. Unsurprisingly, it is growing into a considerable industry. From once feisty little start-ups several companies have grown up to become a serious business – with some serious side effects: carpooling.com, one such platform, which by now is moving over one million people each month alone claims to have saved some 500 million litres of petrol and some one million tons of carbon emissions.[4] Hitchhiking is alive and kicking. Virtual thumbing. Real rides. Real impact.

The same principle can and is applied to a range of other areas. It not only holds for hitchhiking. Such digital revival of old neighbourly traits is spreading fast. Reputation mechanisms work just as well for letting strangers stay at your

---

[3] See Curbed (2012)
[4] See carpooling.com/company/about/ accessed 27.11.2012

apartment (airbnb.com), loaning out your home or garden to strangers (eventup.com), borrowing a stranger's car (relayrides.com), having dinner with strangers (grubwithus.com), or entrusting your darling pet to a perfect stranger (pethomestay.com). It applies to innovation just the same. What works for hitchhiking works also for ideas: trusted cooperation with strangers. Arguably it differs in scope and ambition: it has no predecessor, no tradition; it is treading on unchartered territory; it has not yet been tested. But the same mechanisms hold true. Sound reputation mechanisms with a broadly accepted sense of fairness can indeed provide a market for ideas; a market where anyone can cooperate safely with strangers to realize innovation. This may not be an actual "supermarket for ideas" or eBay.[5] It may not even be a market in its strictest sense – that may never emerge – but something that comes quite close; something that is feasible; something that is pragmatic; something that can be implemented, and implemented now;[6] something that allows those with ideas to work with firms and experts to realize their ideas; something that allows inventors to trust strangers with their idea.

Alas, with innovation it is not as easy as with hitchhiking. Neither the mechanism nor code of conduct yet exists. Building them from scratch and making them work in practice is a significant challenge. All may fail if any number of little things go wrong. Details matter. The nitty-gritty matters gravely. All four elements need to be in place, all four elements need to be considered carefully and addressed with a healthy sense of pragmatism: (i) effective matching mechanisms need to be designed that will get the right people together, and pair up those that can work well together; (ii) a solid reputation mechanism is essential. Without it no trusted cooperation can take place. Identities need to be ensured; reputations slowly established; it needs to be transparent, simple and intuitive. It needs to be available; (iii) a broadly acceptable code of conduct is needed that reflects a shared sense of fairness so it is clear what the feedback is given for and what is expected of any partner; (iv) accessible guidance is crucial – precisely because it is unchartered territory. Practical guidance needs to hold people by the hand to venture into the unknown together. Clarifying each step, how it works, what is expected, what can go wrong.

All the while it will have to be a learning process. Anticipating certain challenges can carry it a long way. Yet, nothing can substitute for trial and error.[7]

---

[5] As envisioned by Norman Macrae (see Economist 1982). Also Dushnitsky and Klueter (2011).

[6] For an excellent overview and discussion, see Gans and Stern (2010). A reputation mechanism based approach does not fulfil the criteria set out by Roth (2007), but it shows the most promise. It may not be an optimal solution, but it is a feasible solution. It certainly is not perfect, and especially an efficiency loss due to 'bilateralism' can pose a problem. "Buyers of ideas should be able to consider multiple offers from multiple potential sellers before contracting with a particular seller." (Gans and Stern 2010, p. 808) This does not work with a reputation mechanism. Inefficiencies may arise (though it can be minimised to some extent through guidance, transparency, and services (especially intermediaries)).

[7] A good example is InnoCentive. Innocent at first, it built on a simple idea, employing a simple approach. But the intricacies mattered. They faced challenges, they learned, they refined their approach, they matured. Now refined and thought through options are available that make the process much more robust and reliable. Similarly the concept of open innovation in firms: by the

'The proof of the pudding is in the eating'. Having a vision helps. Theory helps. The true challenges and how successfully it works will only show in practice. Much has to be figured out along the way. It will remain a trade-off between theory and practice, between what is theoretically sound and practically relevant – to use sufficient theory to define the concept and sufficient pragmatism to realize and adapt it. Indeed, the present state is anecdotal, experimental and frail. No mechanisms are yet established. The analysis provides broad cornerstones, and tries to identify many details and possible hurdles to its realisation that need to be overcome. Best to take them into account early on and address them head on. Many more challenges will arise. It will still be a bumpy ride to full unfolding. Trial and error will surely help move it along. No doubt, many brilliant ideas will refine it. Practical refinements to fairness and feedback will emerge. Codes of Conducts will be established. The mechanisms will be refined. They will mature to allow for even better cooperation. Loopholes will be closed. Legal, financial, and accounting innovations will allow it to expand. New services will arise to facilitate it.

Cooperative innovation has so much potential. It can, and has to improve dramatically still. It will. No doubt. New solutions will arise, possibly better ones. This is but a start. It is a promising one. It is a necessary one to move Cooperative Innovation forward. It may be worthwhile to speculate how this will happen. It may be worthwhile to try to gage the evolution of cooperative innovation in practice. Not the broad and grand strokes of socioeconomic evolution and upheaval, but the more granular, the more specific are of interest here. What will happen to innovation processes once such options to cooperate in the innovation process are in place? How will they evolve? Just as a speculative glimpse, treading into the unknown. It may be worth it in order to frame expectations and anticipate and prepare for change and challenges ahead.

Cooperative innovation is a novel approach. The mechanisms it employs are complex. Though highly anticipated, reluctance to participate is to be expected. It looks complicated. No one yet is used to it. Cooperative innovation needs to prove itself and it needs to be better understood. Peer participation is crucial. A network is only as good as its members. It needs a certain quality and/or quantity of members to become attractive to participate. If only very few people participate initially, it will not be attractive for others. People will start to leave again, and the mechanism will fail. If a certain critical mass is reached, such a network will become increasingly attractive. Hence the more join (at least up to a certain practical degree), the more attractive it will become, and the better such a network will operate.[8] Many services will claim this domain, some new ones, and some revamped existing ones.

---

time Chesbrough suggested means of implementing it, it had ripened. Grand claims of general applicability were dropped and focussed much more narrowly on practical IP trading (not least due to Arrow's fundamental paradox of information), offering very pragmatic approaches to spin in and spin out ideas.

[8] Also see discussions in Farrell and Saloner (1985), Katz and Shapiro (1986), Katz and Shapiro (1994), Katz and Shapiro (1988).

Initially broad, soon dominant players will emerge and more and more niches will sprig up, offering more targeted networks and support structures. This is desirable, but without a common platform may be problematic, risking path dependence and network hopping. Similar assumptions can be made about the participants. The most likely scenario will see young inventers and aspiring entrepreneurs participate, eager to prosper, and able and willing to bare risk: students and young professionals will go first. Increasingly segmentation will set in to stress certain qualifications. More targeted expertise and services will arise. Likely retirees and rentiers will participate, as able and pragmatic experts and practitioners, willing to partake in the innovation process (just as often Angle investors do today). More and more services will pop up. Niches will be filled. As services become more readily available, as the mechanisms prove successful and robust, and as the approach becomes more familiar to a broader audience, the participation base will increase. More and more people from any walks of life will participate. Firms may be initially reluctant, but will gradually participate – first possibly through proxies and use of disclosure screens.

The mode of participation will progress along the same line. In the beginning the focus will be on one-on-one cooperation of complementary partners, partners of similar weight. More and more will experts enter and partnerships become more imbalanced, but also more service oriented. Only gradually will actual trading in ideas emerge once a refined code of conduct crystallizes, and many uncertainties have been addressed. Firms will start participating further refining and expanding cooperative approaches.

It would be wrong however to expect this to be a linear process. The magnitude will vary. A backlog of ideas exists, a stock of ideas that will eventually flood in. Many ideas are still dormant, still in some drawer, someone's head, thought through, but unrealised, as the prospects of feasible realization seemed slim. Now, that realistic options of realizing their ideas open up, these will be unearthed and participation will rise and peak. At the same time as the prospects of profiting from ideas are coming within reach, people will become more aware and more attentive, more actively searching for and pursuing ideas. Both elements will influence participation. This effect may be amplified or overlaid by other aspects. With the first successful projects, the increasing reputation and awareness of the new options, more and more people can be expected to try it. The mechanisms and services will become more sophisticated, more efficient, and more attractive. However, with the first negative examples, with the first problems a certain hesitation and with it a certain gestation period may set in. If doubts raised become too severe, it can threaten its sustainability. On the other hand, the refinement of available mechanisms, the crystallisation of best practice, successful track records, etc. will further lower the risk and costs of realising ideas, hence also attract a wider range of ideas. The opportunities are likely too great to deter participation. A residual risk always exists. The question will be how this can and will be handled. Eventually people will become more focussed on the prospects and hopefully the generation of ideas will increase, along with an even higher increase in the realisation ratio.

Such speculations are no flights of fancy. Anticipating change is important to reduce uncertainly and provide some benchmark of what to expect. It will be a process, not a sudden solution. Such approaches will have to evolve. They will evolve.

This rudimentary sketch of an imperfect approach should not deter anyone from trying. Instead, mapping out some possible steps should inspire to engage, to try, to participate and propose better solutions, to do it better. Surely there are many risks involved in moving forward. Risks and rewards! Given the potential, no risk should be too great. You try, you fail, you learn, you do it better. Or: you try, you gain, and possibly gain a lot. The broader implications are so vast, and so alluring. We simply have to move forward.

References

Curbed (2012). *Zimride site goes live; Strong year for condo sales*. By AR Bobson. 1 August, http://dc.curbed.com/archives/2012/08/zimride-site-goes-live-strong-year-for-condo-sales.php. Accessed 02 March 2013.

Dushnitsky, G., & Klueter, T. (2011). Is there an eBay for ideas? Insights from online knowledge marketplaces. *European Management Review, 8*(1), 17–32.

Farrell, J., & Saloner, G. (1985). Standardization, compatibility, and innovation. *The RAND Journal of Economics, 16*(1), 70–83.

Gans, J. S., & Stern, S. (2010). Is there a market for ideas? *Industrial and Corporate Change, 19*(3), 805–837.

Katz, M. L., & Shapiro, C. (1986). Technology adoption in the presence of network externalities. *Journal of Political Economy, 94*(4), 822–841.

Katz, M., & Shapiro, C. (1988). Systems competition and network effects. *Journal of Economic Perspectives, 8*(2), 93–115.

Katz, M., & Shapiro, C. (1994). Network externalities, competition and compatibility. *The American Economic Review, 75*(3), 424–440.

New York Times (2012a). *Car-pooling makes a surge on apps and social media*. 4 July, by Meece, M. www.nytimes.com/2012/07/05/technology/technology-makes-car-pooling-safer-and-easier.html. Accessed 02 March 2013.

New York Times (2012b). *Ride-sharing services grow popular in Europe*. 1 Oct, by Pfanner, E. www.nytimes.com/2012/10/01/technology/ride-sharing-services-grow-popular-in-europe.html. Accessed 02 March 2013.

Roth, A. E. (2007). The art of designing markets. *Harvard Business Review, 85*(10), 118–126.

The Economist (1982). *We're all intrapreneurial now*. 17th April 1982.

Concluding Remarks 13

The broader landscape is shifting. Innovation is evolving. It is on the verge of a new era. The *Idea Economy* is arriving. It is characterized by a division of labour. Innovation becomes more effective, and more democratic. It also becomes faster – and less a leisurely exercise.

On the one hand there are inventors. Now anyone invents. Anyone can come up with an idea, no matter the skills as a businessman, no matter the experience, no matter the employment situation, age, economic standing, or social origin. Ideas are democratic. Anyone can profit from their idea. Realizing an idea no longer is a question of having the right roommate, the family connection, or serendipitous encounter. Implementation is a service. Experts and firms will do it for you. And indeed, on the other hand there are plenty of gifted entrepreneurs and firms desperate for new ideas. They offer to realize the ideas for inventors. They compete to get hold of new ideas. They use their skills and capacity to bring ideas to market. The two together, inventors and implementers, can achieve a lot: Some invent. Others realize. Both benefit. Both profit. Together they will achieve more and better innovations. Together they can realize the idea more effectively and efficiently than on their own either probably ever could.

This can only work if you are able to trust your partner and cooperation is fair and just. You need to be able to trust your partner not to steal your idea, and treat you fairly working together. The way it works is thus: You have an idea. It is rough, but holds vast potential. You work on it to give it more shape. You refine it, think it through, smooth out the rough edges, and explore its potential. When you think it ready, you look for a partner. This could at first be someone to help you refine it, an expert, a skilled tinkerer, someone who knows how to develop ideas further, into sound concepts, into prototypes. Or you start looking for a firm or entrepreneur with experience, a good track record, someone with the means and skills to realize it. Services are emerging that will assist you, that provide a platform to identify suited partners and engage in constructive cooperation. It is up to you to select from a plethora of potential partners the one you deem best suited to realize your specific idea. This still leaves the question of trust: A feedback mechanism can mitigate this risk. It is of course tempting to take a good idea and run – realize the idea and make away with a juicy profit without having to share any of it. You are left with nothing but a sense of betrayal. The feedback mechanism creates an incentive to behave fairly, and not to steal ideas. Should someone steal the idea, the feedback would reveal them as a thief. You would brandish them a thief, and rightfully so. No one would cooperate with them in the future, with someone who steals ideas. No one would cooperate; no one would share any more of their ideas with them. This may not matter much to you – your idea is gone by then. But it matters to the thief. Not only can he not be sure he will succeed with the stolen idea on his own, and the theft would have been in vain, worse, he will forgo any future opportunities to cooperate and make any further profits from realizing ideas. For a firm this is fatal; for an entrepreneur just as much. In an economy where competition is defined by new ideas those that can no longer access ideas, can no longer compete, will perish. And while the temptation of making a quick buck persists, the fallout of foregone future opportunities would be so much higher that fair cooperation will be a much smarter choice than to prosper shortly only to fail. The choice is between little profit now, or lots later. Reputation is worth a lot. Who would kill the goose that lays golden eggs? The feedback mechanism changes the incentive structure from stealing to cooperation. It makes cooperation not only possible, but a dominant strategy. Both parties have a strong incentive to cooperate and cooperate fairly. Both want to make cooperation work. Both will profit from it. This means the mechanism allows you to put faith in your partner to have the same interest to succeed. You can trust your partner with your idea. You can cooperate to realize the idea or even trade it to firms to implement.

Such cooperation offers new opportunities for both firms and inventors. But what seems convenient and appealing to you, from afar looks like a much more substantial shift. The broader innovation framework is changing. More and more effective innovation takes place. More ideas get realized because all can now participate so that fewer ideas wither and are forgotten in someone's drawer, basement, or garage. Innovation becomes more effective as the competition increases and the implementation process becomes more professionalized, better utilizing existing expertise, skills and resources. Skilled experts will realize ideas. They will bet on the best ones, and implement them quickly and effectively. This reduces wasted potential, of those trying their ill-fated luck as entrepreneurs, or failing to contest well-positioned incumbents. Those with the skills and capacity to quickly turn ideas into products will realize ideas. This makes the most of the idea, and they make best use of their time and skills. With cooperation possible more ideas get realized, and realized more efficient and effectively. Faster and better innovation ensues.

The division of labour expands beyond the production of pins and needles. It goes beyond the physical assembly line. Division of labour now applies to innovation. Some invent. Others realize. The process gets split up – to the benefit of all. Those with ideas need not realize them. Those able to realize ideas need not have ideas but do what they do best: realize. Society benefits from more and better innovation. Thomas Edison's original aspiration of an invention factory may become reality – without brilliant inventors having to waste their potential and trying their luck, struggling to implement their ideas in the market. They can invent and profit from their ideas without being a gifted businessman. Thomas Edison would have had a chance to become a tycoon after all. He could have applied his considerable talent at generating more invention – mass-produce inventions – and create fortunes together with brilliant and shrewd contemporary businessmen the likes of George Westinghouse, Charles Coffin, Henry Ford, Henry Towne, Hugh Chisholm or many others that never got a chance to shine for the lack of ideas on their own, but who would have made most brilliant entrepreneurs. Now this is possible. It will be the norm.

With the nature of innovation changing to resemble more and more a division of labour, more far reaching changes abound. It is a highly dynamic process. It would be wrong to see the division of labour as an extension of the current situation, as offering exciting new opportunities for those with ideas. It will further change the innovation culture. The pressure is on. The geeky tinkerer may come up with an idea. While he may have failed before to realize it, now he has a chance. Indeed. Yet what seemed a leisurely challenge for the reclusive and idealistic is coming more and more to the forefront. As ideas are more and more becoming the critical ingredient, and realizing ones ideas becomes feasible for anyone, the pace of innovation will pick up and the pressure to invent will increase. Tinkering in your garage will become mainstream, will become organized and refined, and no longer the domain of a few. It will be more lucrative, and it will be easier to benefit from it. More ideas will surface and the competition for ideas will increase. To prosper you have to invent or be a very good implementer. Both will become tougher and more competitive. The comfort zone to dwell and procrastinate is shrinking fast. Putting a hunch off too long will no longer be an option. A sense of urgency will spread. The search for new ideas will intensify. To prosper you will need to come up with good ideas. Invention becomes less fun, more work. Realization becomes less a question of luck and chance encounter. It will be a full time job. The innovation process is become more professionalized.

Both those looking for ideas and especially those implementing them will have to adapt to this new situation. To prosper in the *Idea Economy*, attitudes will have to change and organizations adjust. All will have to focus more on ideas not to be left behind.

For those with ideas this means: *Be more attentive*. Recall the bright ideas you had, seek new ones. Rummage around in your drawer, garage, and mind, unravel, unearth your concepts and some-time-to-dos. Keep an open mind. Be curious. Look for ideas. Explore your hunches. Appreciate ideas – they are your future.[1] *Be diligent*: Deliberate your options for a while. Reflect a little. Not too long, but long enough to make informed choices to leverage your skills and ideas. Be diligent and realistic in assessing your ideas. Be diligent and realistic in assessing your abilities to realize them. Consider what is best for your idea. Consider where you best fit in, what you

---

[1] Many have tried to spot some patterns, to codify the serendipitous process, the causal link to ideas (see footnote 180). Much more work on this is likely to emerge. Serial innovators may be good examples: e.g. Arthur Fischer or Dean Kamen.

can contribute. Now it is your benefit you are jeopardizing. Seek assistance where needed. Now you can. Trade the idea when its return is better. Think of the pie. That you succeed is not certain. But the chances have increased substantially. Note: this is not the same. You can profit more, and are more likely to. But you may fail. You make the choices. It is still up to you to make the most of it. If innovation is indeed 1 % inspiration and 99 % perspiration, you too have to sweat it. Others will help, others may stem a load, but you have to go beyond just inspiration. Refine the idea. Be sure to do your part. *Be informed*: Understanding is crucial. Knowing your options, understanding the process, to best use, and not be abused, can help you make the most of your idea. You have much to gain, and much to lose. Making the right choices is not always obvious. But the better you understand how cooperative innovation works the better your chances to succeed. *Be active*: Participate! Try it. Don't be discouraged: "True innovation is complex and tumultuous – full of spurts, frustration and sudden insights."[2] It still is worth it. If you wait too long, others surely will have moved ahead. Realize faster. Push for cooperation. It is no longer a question whether to seek assistance, Not a question of whether to cooperate, but of how best to participate.

For the firms the implications are farther reaching still. They need to compete to attract ideas to survive and turn a profit. The former comfort and splendour of simply being the incumbent is fading. Those that fail to adapt will fade. Firms need to adapt to the new innovation structure. *Participate*: The more and/or the better the ideas you have access to, the better you will fare. Until now you had to rely on your own ingenuity, on spending on uncertain R&D, on questionable concepts on generating ideas, buying start-ups. If you already had a hunch, you could try or always get others to work it out. Now you could also crowdsource it. But the inspiring events, when others proposed an idea to you, were rare to non-existent. Now it will be possible and more frequent. Ideas can come to you. But you need to want them. You need to compete for them. You need to be appealing to those with ideas, in absolute – it needs to be worth their while – as well as in relative terms – alternative paths exist, not least by cooperating with your competitors, or pursuing other realization paths. The better you are prepared, the more engaged you are, the better your chances of grasping the moment, the better your chances of spurting ahead. You need not know all, need not be the epicentre of ideas. You need not employ Nobel Laureates, you need not monopolize all the best and the brightest in your field to hope something may come out. You can, of course. Better, the best ideas come to you. For this you need to participate, actively attract ideas.[3] *Position yourself*: figure out where and how you want to participate. What are your strengths, what is your competitive advantage? In what area? Not just in existing markets, but to launch new innovations. You need a clear concept how you want to be perceived, whom you attract, and how. You have to become a trusted partner, delivering on

---

[2] Quinn (2000, p. 22)
[3] Also see the argument in Govindarajan and Trimble (2010).

what you promise. *Adapt:* Have all your ducks in a row. Create the capacity to participate. Refine your knowledge filters. Adapt your attitude, organization structure, your process and capacities as needed. How will you cooperate, how will you absorb ideas? How will you incubate and realize ideas? Be adaptive to the new environment.[4] Establish a clear process, a reasonable manner of cooperation. *Compete*: Surely you are not the only one looking for idea. Keep abreast of what the competition is doing. You will have to compete with them. Stay ahead of the game. Be more appealing, offer better terms and conditions; make it simpler and more lucrative to cooperate with you. Be a first mover. Establish a sound cooperation record. *Engage*: Cooperative innovation is evolving. You can help shape it. Develop new ways to engage with partners, offer new tools to support them, create new ways of attracting ideas. "The new leaders in innovation will be those who figure out the best way to leverage a network of outsiders. [...] The new leaders in innovation will be those who can understand how to design collaboration networks and how to tap their potential."[5]

The pressure is on. Ideas need to be realized faster. They become obsolete quickly. Others have similar or better ideas. Now that anyone can partake, more ideas get realized, the process needs to become more efficient and the pace will increase. First movers no longer have the luxury to explore or mature gradually. Ideas must be realized quickly and effectively with given means and skills – with partners best situated, established, and with the ready capacity to deliver. Anyone can prosper in the *Idea Economy*. But if you do not make use of your ideas, others will. If you do not partake and cooperate, you will lose out. So, dust off your ideas, take them out of your drawer, your mind, your basement; sharpen your pencil, be attentive, come up with new ones; make the most of your ideas, realize them, profit from them! Your opportunities are endless. Now they are within reach.

References

Arora, A., Fosfuri, A., & Gambardella, A. (2001). Markets for technology: The economics of innovation and corporate strategy. *Industrial and Corporate Change, 10*(2), 419–450.
Gambardella, A., Giarratana, M., & Panico, C. (2010). How and when should companies retain their human capital? Contracts, incentives, and human resource implications. *Industrial and Corporate Change, 19*(1), 1–24.
Govindarajan, V., & Trimble, C. (2010). *The other side of innovation: Solving the execution challenge*. Boston: Harvard Business School Publishing.
Grant, R. (1996). Prospering in dynamically-competitive environments: Organizational capability as knowledge integration. *Organization Science, 7*(4), 375–387.
Lakhani, K. R., & Boudreau, K. J. (2009). How to manage outside innovation. *MIT Sloan Management Review, 50*(4), 69–76.

---

[4] Also see Lakhani and Boudreau (2009), Pisano and Verganti (2008); Arora et al. (2001), also Hellmann; or Grant (1996), O'Connor et al. (2009), Gambardella et al. (2010).
[5] Pisano and Verganti (2008), p. 78/p. 86

O'Connor, G. C., Corbett, A., & Pierantozzi, R. (2009). Create three distinct career paths for innovators. *Harvard Business Review, 87*(12), 78–79.

Pisano, G. P., & Verganti, R. (2008). Which kind of collaboration is right for you? *Harvard Business Review, 86*(12), 78–86.

Quinn, J. B. (2000). Outsourcing innovation: The new engine of growth. *Sloan Management Review, 41*(4), 13–28.

References

Abernathy, W. J., & Rosenbloom, R. S. (1969). Parallel strategies in development projects. *Management Science, 15*(10), 486–505.
Abraham, M. (2008). *The flash of genius* [Movie]. Directed by Abraham, M., screenplay by Railsaback, P., Universal Pictures/Spyglass Entertainment/Strike Entertainment.
Abramovitz, M. (1986). Catching up, forging ahead, and falling behind. *The Journal of Economic History, 46*(2), 385–406.
Acemoglu, D. (1997). Technology, unemployment and efficiency. *European Economic Review, 41*(3), 525–533.
Acemoglu, D. (2002). Technical change, inequality, and the labor market. *Journal of Economic Literature, 40*(1), 7–72.
Acs, Z. J., & Audretsch, D. B. (1990). *Innovation and small firms*. Cambridge: MIT Press.
Acs, Z. J., & Audretsch, D. B. (2005). Entrepreneurship, innovation and technological change. *Foundations and Trends in Entrepreneurship, 1*(4), 1–65.
Acs, Z. J., Audretsch, D. B., & Strom, R. (2009). *Entrepreneurship, growth, and public policy*. Cambridge: Cambridge University Press.
Acs, Z. J., Desai, S., & Klapper, L. (2008). What does entrepreneurship data really show. *Small Business Economics, 31*(3), 265–281.
Aghion, P., & Howitt, P. (1998). *Endogenous growth theory*. Cambridge: MIT Press.
Aghion, P., & Howitt, P. (2007). Capital, innovation, and growth accounting. *Oxford Review of Economic Policy, 23*(1), 79–93.
Ahmad, N., & Hoffman, A. (2008). *A framework for addressing and measuring entrepreneurship* (OECD statistics working paper 2008/2). Paris: OECD. doi:10.1787/243160627270.
Ahmad, N., & Seymour, R. G. (2008). *Defining entrepreneurial activity: Definitions supporting frameworks for data collection* (OECD statistics working papers 2008/1). Paris: OECD. doi:10.1787/243164686763.
Aiken, M., Bacharach, S. B., & French, J. J. (1980). Organizational structure, work process, and proposal making in administrative bureaucracies. *Academy of Management Journal, 23*(4), 631–652.
Aldrich, H. E. (1990). Using an ecological perspective to study organizational founding rates. *Entrepreneurship Theory and Practice, 14*(3), 7–24.
Aldrich, H. E., Carter, N. M., & Ruef, M. (2004). Teams. In W. B. Gartner, K. G. Shaver, N. M. Carter, & P. D. Reynolds (Eds.), *The handbook of entrepreneurial dynamics: The process of organization creation* (pp. 229–310). Thousand Oaks: Sage.
Allen, K. (2001). *Entrepreneurship for dummies*. New York: Wiley.
Allison, P. D., & Stewart, K. A. (1974). Productivity differences among scientists. Evidence for accumulative advantage. *American Sociological Review, 39*(4), 596–606.
Amabile, T. M. (1983). The social psychology of creativity: A componential conceptualization. *Journal of Personality and Social Psychology, 45*(2), 357–376.
Amabile, T. M. (1996). *Creativity in context: Update to the social psychology of creativity*. Boulder: Westwood Press.

Amabile, T. M. (1998). How to kill creativity. *Harvard Business Review, 76*(5), 76–87.
Anderson, H. (2004). Why big companies can't invent. *Technology Review, 107*(4), 56–59.
Andreoni, J. (1991). Reasonable doubt and the optimal magnitude of fines: Should the penalty fit the crime? *The RAND Journal of Economics, 22*(3), 385–395.
Andrew, J. P., Manget, J., Michael, D. C., Taylor, A., & Zablit, H. (2010). Innovation 2010: A return to prominence and the emergence of a new world order. *Boston Consulting Group Report.* www.bcg.com/documents/file42620.pdf. Accessed 02 Mar 2013.
Anton, J. J., & Yao, D. A. (1994). Expropriation and inventions: Appropriable rents in the absence of property rights. *The American Economic Review, 84*(1), 190–209.
Anton, J. J., & Yao, D. A. (1995). Start-ups, spin-offs, and internal projects. *Journal of Law, Economics and Organization, 11*(2), 362–378.
Anton, J. J., & Yao, D. A. (2002). The sale of ideas: Strategic disclosure, property rights, and contracting. *The Review of Economic Studies, 69*(3), 513–531.
Anton, J. J., & Yao, D. A. (2004). Little patents and big secrets: Managing intellectual property. *The RAND Journal of Economics, 35*(1), 1–22.
Anton, J. J., & Yao, D. A. (2005). Markets for partially contractible knowledge: Bootstrapping versus bundling. *Journal of the European Economic Association, 3*(2–3), 745–754.
Arenius, P., & Minniti, M. (2005). Perceptual variables and nascent entrepreneurship. *Small Business Economics, 24*(3), 233–247.
Armington, C. (2004). *Development of business data: Tracking firm counts, growth and turnover by size of firms.* U.S. Small Business Administration, Washington, DC. http://archive.sba.gov/advo/research/rs245tot.pdf. Accessed 02 Mar 2013.
Arora, A. (1995). Licensing tacit knowledge: Intellectual property rights and the market for know-how. *Economics of Innovation & New Technology, 4*(1), 41–60.
Arora, A., & Gambardella, A. (1994). The changing technology of technological change: General and abstract knowledge and the division of innovative labour. *Research Policy, 23*(5), 523–532.
Arora, A., & Gambardella, A. (2010). Ideas for rent: An overview of markets for technology. *Industrial and Corporate Change, 19*(3), 775–803.
Arora, A., & Merges, R. P. (2004). Specialized supply firms, property rights and firm boundaries. *Industrial and Corporate Change, 13*(3), 451–475.
Arora, A., Fosfuri, A., & Gambardella, A. (2001). Markets for technology: The economics of innovation and corporate strategy. *Industrial and Corporate Change, 10*(2), 419–450.
Arrow, K. J. (1962). Economic welfare and allocation of resources for invention. In: R. R. Nelson (Ed.) *The rate and direction of inventive activity: Economic and social factors* (pp. 609–626). New Jersey: Princeton University Press. www.nber.org/books/univ62-1. Accessed 02 Mar 2013.
Arrow, K. J. (1972). Gifts and exchanges. *Philosophy and Public Affairs, 1*(4), 343–362.
Arthur, W. B. (2007). The structure of invention. *Research Policy, 36*(2), 274–287.
Astebro, T. (1998). Basic statistics on the success rate and profits for independent inventors. *Entrepreneurship Theory and Practice, 23*(2), 41–48.
Asterbro, T. (2003). The return to independent invention: Evidence of unrealistic optimism, risk seeking or skewness loving? *The Economic Journal, 113*(484), 226–239.
Asterbro, T., Jeffrey, S. A., & Adomdza, G. K. (2007). Inventor perseverance after being told to quit: The role of cognitive biases. *Journal of Behavioral Decision Making, 20*(3), 253–272.
Atanasov, V. A., Ivanov, V. I., & Litvak, K. (2008). The effect of litigation on venture capitalist reputation. *EFA 2009 Bergen meetings paper.* www.efa2009.org/papers/SSRN-id1343981.pdf. Accessed 02 Mar 2013.
Audia, P. G., & Rider, C. I. (2005). A garage and an idea: What more does an entrepreneur need? *California Management Review, 48*(1), 6–28.
Audia, P. G., & Goncalo, J. (2007). Past success and creativity over time: A study of inventors in the hard disk drive industry. *Management Science, 53*(1), 1–15.
Audretsch, D. B. (2003). *The entrepreneurial society.* Oxford: Oxford University Press.

Audretsch, D. B. (2005). The emergence of entrepreneurship policy. In D. Audretsch, H. Grimm, & C. W. Wessner (Eds.), *Local heroes in the global village: Globalization and the new entrepreneurship policies* (pp. 21–43). New York: Springer.

Audretsch, D. B. (2009). The entrepreneurial society. *The Journal of Technology Transfer, 34*(3), 245–254.

Audretsch, D. B., & Thurik, R. (2000). Capitalism and democracy in the 21st century: From the managed to the entrepreneurial economy. *Journal of Evolutionary Economics, 10*, 17–34.

Audretsch, D. B., & Thurik, R. (2001a). What's new about the new economy? Sources of growth in the managed and entrepreneurial economies. *Industrial and Corporate Change, 10*(1), 267–315.

Audretsch, D. B., & Thurik, R. (2001b). *Linking entrepreneurship to growth* (OECD science, technology and industry working papers 2001/2). Paris: OECD. doi:10.1787/736170038056.

Audretsch, D. B., & Thurik, R. (2004). A model of the entrepreneurial economy. *International Journal of Entrepreneurship Education, 2*(2), 143–166.

Aumann, R. J. (1995). Backward induction and common knowledge of rationality. *Games and Economic Behavior, 8*(1), 6–19.

Axelrod, R. (1984). *The evolution of cooperation*. New York: Basic Books.

Axelrod, R. (1997). *The complexity of cooperation*. Princeton: Princeton University Press.

Ba, S., & Pavlou, P. A. (2002). Evidence of the effect of trust building technology in electronic markets: Price premiums and buyer behavior. *Management Information Systems Quarterly, 26*(3), 243–268.

Bagnall, B. (2006). *On the edge: The spectacular rise and fall of commodore*. Winnipeg: Variant Press.

Bakos, Y., & Dellarocas, C. (2005). *Online reputation and litigation as mechanisms for quality assurance: Strategic implications for profits and efficiency*. www.rhsmith.umd.edu/faculty/cdell/papers/replit.pdf. Accessed 02 Mar 2013.

Banerjee, A., & Duflo, E. (2000). Reputation effects and the limits of contracting: A study of the Indian software industry. *Quarterly Journal of Economics, 115*(3), 989–1017.

Barro, R., & Sala-i-Martin, X. (1995). *Economic growth*. New York: McGraw-Hill.

Barroso, J. M. (2008). *Presentation of the priorities of the Slovenian Presidency of the Council of the EU*. Strasbourg, 16 January 2008, SPEECH/08/17. http://europa.eu/rapid/press-release_SPEECH-08-17_en.htm?locale=en. Accessed 02 Mar 2013.

Bauer, L., & Matis, H. (1988). *Geburt der Neuzeit: Vom Feudalsystem zur Marktgesellschaft*. Munich: DTV Deutscher Taschenbuch.

Baumol, W. J. (1968). Entrepreneurship in economic theory. *The American Economic Review, 58*(2), 64–71.

Baumol, W. J. (1993). Formal entrepreneurship theory in economics: Existence and bounds. *Journal of Business Venturing, 8*, 197–210.

Baumol, W. J. (2002). *The free-market innovation machine: Analyzing the growth miracle of capitalism*. Princeton: Princeton University Press.

BBC News (2012). *Facebook's instagram deal: Can one app be worth $1bn?* By Weber, T., 10 April. www.bbc.co.uk/news/business-17666032

Becker, G. S. (1968). Crime and punishment: An economic approach. *Journal of Political Economy, 76*(2), 169–217.

Bendor, J., Kramer, R. M., & Stout, S. (1991). When in doubt... cooperation in a noisy prisoner's dilemma. *Journal of Conflict Resolution, 35*(4), 691–719.

Bengtsson, O., & Hsu, D. H. (2010). *How do venture capital partners match with startup founders?* Knut Wicksell Working Paper No 2013:8, Lund University, http://www.lusem.lu.se/media/kwc/working-papers/kwc-wp-2013-8.pdf. Accessed 20 August 2013.

Benoit, J. P., & Krishna, V. (1985). Finitely repeated games. *Econometrica, 53*(4), 905–922.

Benson, B. L. (1989). The spontaneous evolution of commercial law. *Southern Economic Journal, 55*(3), 644–661.

Berg, J., Dickhaut, J., & McCabe, K. (1995). Trust, reciprocity, and social history. *Games and Economic Behavior, 10*(1), 122–142.

Berkun, S. (2007). *The myths of innovation*. Sebastopol: O'Reilly.

Bernardo, A. E., Cai, H., & Luo, J. (2009). Motivating entrepreneurial activity in a firm. *The Review of Financial Studies, 22*(3), 1089–1118.

Bernhofen, D. M., El-Sahli, Z., & Kneller, R. (2013). *Estimating the effects of the container revolution on world trade*. CESifo working paper series no. 4136. http://www.cesifo-group.de/DocDL/cesifo1_wp4133.pdf. Accessed 31 Mar 2013.

Berry, T., & Parsons, S. (2008). *3 weeks to startup*. Irvine: Entrepreneurship Press.

Besen, S. M., & Raskind, L. J. (1991). An introduction to the law and economics of intellectual property. *The Journal of Economic Perspectives, 5*(1), 3–27.

Bhide, A. V. (1994). How entrepreneurs craft strategies that work. *Harvard Business Review, 72*(2), 150–161.

Bhide, A. V. (2000). *The origin and evolution of new businesses*. New York: Oxford University Press.

Bhide, A. V. (2008). *The venturesome economy: How innovation sustains prosperity in a more connected world*. Princeton: Princeton University Press.

Biggs, T. (2002). *Is small beautiful and worthy of subsidy?* http://rru.worldbank.org/documents/paperslinks/tylerspaperonsmes.pdf. Accessed 02 Mar 2013.

Biais, B., & Perotti, E. (2008). Entrepreneurs and new ideas. *The RAND Journal of Economics, 39*(4), 1105–1125.

Bierhoff, H. W., & Vornefeld, B. (2004). The social psychology of trust with applications in the internet. *Analyse und Kritik Zeitschrift fuer Sozialtheorie, 26*(1), 48–62.

Bird, B. (1988). Implementing entrepreneurial ideas: The case for intention. *Academy of Management Review, 13*(3), 442–453.

Björk, J., Boccardelli, P., & Magnusson, M. (2010). Ideation capabilities for continuous innovation. *Creativity and Innovation Management, 19*(4), 385–396.

Blackford, M. G. (1991). Small business in America: A historiographic survey. *The Business History Review, 65*(1), 1–26.

Blanchflower, D. G. (2004). Self-employment: More may not be better. *Swedish Economic Policy Review, 11*(2), 15–74.

Blaug, M. (1976). Kuhn versus Lakatos or Paradigms versus research programmes in the history of economics. Reprinted In Hausman, D. M. (1994). *The philosophy of economics – an anthology* (2nd ed., pp. 348–375). Cambridge: Cambridge University Press.

Bloomberg. (2011). *Facebook valuation tops Amazon.com, Trailing only Google on web*. 29 January, by Levy, A. www.bloomberg.com/news/2011-01-28/facebook-s-82-9-billion-valuation-tops-amazon-com-update1-.html. Accessed 02 Mar 2013.

Boldrin, M., & Levine, D. K. (2008). Market size and intellectual property protection. *International Economic Review, 50*(3), 855–881.

Bolton, G. E., Katok, E., & Ockenfels, A. (2004). Trust among internet traders: A behavioral economics approach. *Analyse und Kritik Zeitschrift fuer Sozialtheorie, 26*(1), 185–202.

Bolton, G. E., Katok, E., & Ockenfels, A. (2005). Cooperation among strangers with limited information about reputation. *Journal of Public Economics, 89*(8), 1457–1468.

Bolton, G. E., & Ockenfels, A. (2009). The limits of trust in economic transactions: Investigations of perfect reputation systems. In K. S. Cook, C. Snijders, V. Buskens, & C. Cheshire (Eds.), *eTrust: Forming relationships in the online* (pp. 15–36). New York: Russell Sage Foundation.

Bonawitz, E., Shafto, P., Gweon, H., Goodman, N. D., Spelke, E., & Schulz, L. (2011). The double-edged sword of pedagogy: Instruction limits spontaneous exploration and discovery. *Cognition, 120*(3), 322–330.

Boo. (2012). *The celebration issue*. Issue 1, March, by Bugaboo International. http://issuu.com/bugaboo/docs/boo-us?mode=window&backgroundColor=%23222222. Accessed 10 Dec 2012.

Boschma, R. (2005). Proximity and innovation: A critical assessment. *Regional Studies, 39*(1), 61–74.

Boudreau, K. J., Lacetera, N., & Lakhani, K. (2011). Incentives and problem uncertainty in innovation contests: An empirical analysis. *Management Science, 57*(5), 843–863.
Bowles, S., & Hwang, S. H. (2008). Social preferences and public economics: Mechanism design when social preferences depend on incentives. *Journal of Public Economics, 92*(8–9), 1811–1820.
Boyd, N., & Vozikis, G. (1994). The influence of self-efficacy on the development of entrepreneurial intentions and actions. *Entrepreneurship Theory and Practice, 18*(4), 63–77.
Boyle, J. (2003). The second enclosure movement and the construction of the public domain. *Law and Contemporary Problems, 66*, 33–74.
Boyle, J. (2008). *The public domain: Enclosing the commons of the mind*. New Haven: Yale University Press.
Brabham, D. C. (2010). Moving the crowd at threadless. *Information, Communication & Society, 13*(1), 1122–1145.
Braunerhjelm, P., & Svensson, R. (2010). The inventor's role: Was Schumpeter right? *Journal of Evolutionary Economics, 20*(3), 413–444.
Brazeal, D., & Herbert, T. T. (1999). The genesis of entrepreneurship. *Entrepreneurship Theory and Practice, 23*(3), 29–45.
Brennan, G., & Petti, P. (2004). Esteem, identifiability and the internet. *Analyse und Kritik Zeitschrift für Sozialtheorie, 26*(1), 139–157.
Brock, W. A., & Evans, D. S. (1989). Small business economics. *Small Business Economics, 1*, 7–20.
Brunt, L., Lerner, J., & Nicholas, T. (2008). *Inducement prizes and innovation. Centre for economic policy research discussion paper no 6917*. www.cepr.org/pubs/dps/DP6917.asp. Accessed 02 Mar 2013.
Bruyat, C., & Julien, P. A. (2001). Defining the field of research in entrepreneurship. *Journal of Business Venturing, 16*(2), 165–180.
Bughin, J., et al. (2011). *The impact of internet technologies: Search. McKinsey &Company*. http://ssl.gstatic.com/think/docs/the-impact-of-internet-technologies-search_research-studies.pdf. Accessed 31 Mar 2013.
Bull, I., & Willard, G. E. (1993). Towards a theory of entrepreneurship. *Journal of Business Venturing, 8*, 183–195.
Burgelman, R. A. (1983). Corporate entrepreneurship and strategic management: Insights from a process study. *Management Science, 29*(12), 1349–1364.
Burgelman, R. A. (2002). Strategy as vector and the inertia of coevolutionary lock-in. *Administrative Science Quarterly, 47*(2), 325–357.
Burnham, J. B. (2009). Economic growth, entrepreneurship, and the deployment of technology. In N. Aydogan (Ed.), *Innovation policies, business creation and economic development* (pp. 13–35). New York: Springer.
Burnham, T. C. (2007). High-testosterone men reject low ultimatum game offers. *Proceedings of the Royal Society B, 274*(1623), 2327–2330.
Businessweek (2007). *NineSigma: Nurturing 'Open Innovation'*. Special report 12, June 2007, by Scanlon, J. www.businessweek.com/stories/2007-06-12/ninesigma-nurturing-open-innovation businessweek-business-news-stock-market-and-financial-advice. Accessed 02 Mar 2013.
Businessweek (2009). *The next wave of open innovation*. 8, April 2009, by Hagel, J., & Brown, S. J. www.businessweek.com/innovate/content/apr2009/id2009048_360417.htm. Accessed 02 Mar 2013.
Byrne, J. A. (2011). *World changers: 25 entrepreneurs who changed business as we knew it*. New York: Penguin.
California Energy Commission. (2009). *Appliance efficiency regulations*. Docket Number 09-AAER-1C, California Code of Regulators, Sections pp.1601–1608.
Camerer, C. F. (2003). *Behavioral game theory: Experiments on strategic interaction*. Princeton: Princeton University Press.
Camerer, C. F., & Thaler, R. H. (1995). Anomalies: Ultimatums, dictators and manners. *The Journal of Economic Perspectives, 9*(2), 209–219.

Carbonell, P., Rodríguez-Escudero, A. I., & Pujari, D. (2009). Customer involvement in new service development: An examination of antecedents and outcomes. *The Journal of Product Innovation Management, 26*(5), 536–550.

Caree, M., Van Stel, A., Thurik, R., & Wennekers, S. (2002). Economic development and business ownership: An analysis using data of twenty-three OECD countries in the period 1976–1996. *Small Business Economics, 19*, 271–290.

Carland, J. W., Hoy, F., Boulton, W. R., & Carland, J. A. C. (1984). Differentiating entrepreneurs from small business owners: A conceptualization. *The Academy of Management Review, 9*(2), 354–359.

Carlsson, B., Acs, Z. J., Audretsch, D. B., & Braunerhjelm, P. (2009). Knowledge creation, entrepreneurship, and economic growth: A historical review. *Industrial and Corporate Change, 18*(6), 1193–1229.

Carree, M. A., & Thurik, A. R. (2003). The impact of entrepreneurship on economic growth. In Z. J. Acs & D. B. Audretsch (Eds.), *Handbook of entrepreneurship research – An interdisciplinary survey and introduction*. Boston: Kluwer.

Carsrud, A. L., & Brännback, M. (2009). *Understanding the entrepreneurial mind: Opening the black box*. New York: Springer.

Chandler, A. D. (1962). *Strategy and structure: Chapters in the history of the American Industrial Enterprise*. Cambridge: MIT Press.

Chandler, A. D. (1977). *The visible hand: The managerial revolution in American Business*. Cambridge: Belknap Press.

Chandler, A. D. (1990). *Scale and scope: The dynamics of industrial capitalism*. Cambridge: Belknap Press.

Chandler, A. D. (1994). The competitive performance of U.S. industrial enterprises since the Second World War. *The Business History Review, 68*(1), 1–72.

Chandy, R. K., & Tellis, G. J. (2000). The incumbent's curse? Incumbency, size, and radical product innovation. *The Journal of Marketing, 64*(3), 1–17.

Che, Y. K., & Gale, I. (2003). Optimal design of research contests. *The American Economic Review, 93*(3), 646–671.

Chen, Y., Jeon, G. Y., & Kim, Y. M. (2013). *A day without a search engine: An experimental study of online and offline searches*. http://yanchen.people.si.umich.edu/papers/VOS_2013_03.pdf. Accessed 31 Mar 2013.

Cheng, A., & Friedman, E. (2005). Sybilproof reputation mechanisms. *Proceeding of the 3rd Workshop on Economics of Peer-to-Peer Systems (P2PECON)*. http://www.eecs.harvard.edu/cs286r/courses/fall08/files/paper-CheFri.pdf. Accessed 31 Mar 2013.

Cherensky, S. (1993). A penny for their thoughts: Employee-inventors, preinvention assignment agreements, property, and personhood. *California Law Review, 81*(2), 595–669.

Chesbrough, H., & Rosenbloom, R. S. (2002). The role of the business model in capturing value from innovation: evidence from Xerox Corporation's technology spinoff companies. *Industrial and Corporate Change, 11*(3), 529–555.

Chesbrough, H. W. (2001). Assembling the elephant: A review of empirical studies on the impact of technical change upon incumbent firms. In R. A. Burgelman & H. W. Chesbrough (Eds.), *Comparative studies of technological evolution* (pp. 1–36). Amsterdam: Emerald Group Publishing.

Chesbrough, H. W. (2003a). *Open innovation: The new imperative for creating and profiting from technology*. Cambridge: Harvard Business School Press.

Chesbrough, H. W. (2003b). The era of open innovation. *Sloan Management Review, 44*(3), 35–41.

Chesbrough, H. W. (2003c). The governance and performance of Xerox's technology spin-off companies. *Research Policy, 32*(3), 403–421.

Chesbrough, H. W. (2006a). *Open business models: How to thrive in the new innovation landscape*. Cambridge: Harvard Business School Press.

Chesbrough, H. W. (2006b). Open innovation: A new paradigm for understanding industrial innovation. In H. W. Chesbrough, W. Vanhaverbeke, & J. West (Eds.), *Open innovation: Researching a new paradigm* (pp. 1–12). New York: Oxford University Press.

Chesbrough, H. W., & Crowther, A. K. (2006). Beyond high tech: Early adopters of open innovation in other industries. *R&D Management, 36*(3), 229–236.

Chesbrough, H. W., & Garman, A. R. (2009). How open innovation can help you cope in lean times. *Harvard Business Review, 87*(12), 68–76.

Chesbrough, H. W., Vanhaverbeke, W., & West, J. (2006). *Open innovation: Researching a new paradigm*. New York: Oxford University Press.

Chiaroni, D., Chiesa, V., & Frattini, F. (2010). Unravelling the process from closed to open innovation – Evidence from mature, asset-intensive industries. *R&D Management, 40*(3), 222–245.

Chicago Magazine. (2010). *On Groupon and its founder, Andrew Mason*. August 2010, by Coburn, M. F. www.chicagomag.com/Chicago-Magazine/August-2010/On-Groupon-and-its-founder-Andrew-Mason/. Accessed 02 Mar 2013.

Chicago Magazine. (2012). *How Jake Nickell built his threadless empire*. July 2012, by Froelke Coburn, M. www.chicagomag.com/Chicago-Magazine/July-2012/How-Jake-Nickell-Built-His-Threadless-Empire/. Accessed 02 Mar 2013.

Chicago Tribune. (2011). *Google shake-up returns page to CEO post surprise move comes as challenges mount for internet search giant*. 21 January, by Guynn, J. www.chicagotribune.com/business/ct-biz-0121-google-20110121,0,2955817.story. Accessed 02 Mar 2013.

Christensen, C. M. (1997). *The innovator's dilemma: When new technologies cause great firms to fail*. Cambridge: Harvard Business School Press.

Christensen, J. F., Olesen, M. H., & Kjær, J. S. (2005). The industrial dynamics of open innovation: Evidence from the transformation of consumer electronics. *Research Policy, 34*(10), 1533–1549.

Clinton, H. R. (2010). *Closing remarks at the presidential summit on entrepreneurship*. Washington, DC, 27 April , 2010, PRN:2010/522. www.state.gov/secretary/rm/2010/04/140968.htm. Accessed 02 Mar 2013.

Cohen, W. M., & Levinthal, D. A. (1989). Innovation and learning: The two faces of R & D. *The Economic Journal, 99*(397), 569–596.

Cohen, W. M., & Levinthal, D. A. (1990). Absorptive capacity: A new perspective on learning and innovation. *Administrative Science Quarterly, 35*(1), 128–153.

Coile, R. C. (1977). Lotka's frequency distribution of scientific productivity. *Journal of the American Society for Information Science, 28*(6), 366–370.

Cole, A. H. (1946). An approach to the study of entrepreneurship: A Tribute to Edwin F. Gay. *The Journal of Economic History 6*, Supplement: The tasks of economic history, 1–15.

Compte, O. (1998). Communication in repeated games with imperfect private monitoring. *Econometrica, 66*(3), 597–626.

Cook, K. S., Snijders, C., Biskens, V., & Cheshire, C. (2009). *eTrust: Forming relationships in the online world*. New York: Russell Sage Foundation.

Cooper, R., & Edgett, S. (2008). Ideation for product innovation: What are the best methods? *PDMA visions magazine*, March 2008. www.stage-gate.net/downloads/working_papers/wp_29.pdf. Accessed 02 Mar 2013.

Cooper, A. C. (2003). Entrepreneurship – The past, the present. In Z. J. Acs & D. B. Audretsch (Eds.), *Handbook of entrepreneurship research – An interdisciplinary survey and introduction* (pp. 21–34). Boston: Kluwer.

Cooper, A. C., & Bruno, A. V. (1977). Success among high-technology firms. *Business Horizons, 20*(2), 16–22.

Cooper, A. C., & Saral, K. J. (2010). Entrepreneurship and team participation: An experimental study. *Kauffman foundation small research projects research paper*. doi:10.2139/ssrn.1547186.

Council on Competitiveness. (2004). *Innovate America: National innovation initiative summit and report*. Council on competitiveness, Washington, DC. www.compete.org/images/uploads/File/PDF%20Files/NII_Innovate_America.pdf. Accessed 02 Mar 2013.

Crandall, R. W. (1999). From competitiveness to competition: The threat of minimills to large national steel companies. *Resources Policy, 22*(1–2), 107–118.

Cripps, M. W., Mailath, G. J., & Samuelson, L. (2004). Imperfect monitoring and impermanent reputations. *Econometrica, 72*(2), 407–432.

Csikszentmihalyi, M. (1996). *Creativity: Flow and the psychology of discovery and invention*. New York: Harper Perennial.

Curbed. (2012). *Zimride site goes live; Strong year for condo sales*. By Bobson, A. R. 1 August, http://dc.curbed.com/archives/2012/08/zimride-site-goes-live-strong-year-for-condo-sales.php. Accessed 02 Mar 2013.

Daft, R. L., & Lewin, A. Y. (1990). Can organization studies begin to break out of the normal science straitjacket? An editorial essay. *Organization Science, 1*(1), 1–9.

Dahlander, L., & Gann, D. (2010). How open is innovation. *Research Policy, 39*(6), 699–709.

Davidsson, P. (2004). *Researching entrepreneurship*. New York: Springer.

Davis, L., & Davis, J. (2004). How effective are prizes as incentives to innovation? Evidence from three 20th century contests. *Paper to be presented at the DRUID summer conference 2004 on industrial dynamics*. Innovation and Development Elsinore, Denmark. commercialspace. www.druid.dk/uploads/tx_picturedb/ds2004-1343.pdf. Accessed 02 Mar 2013.

Davis, S. J., Haltiwanger, J., & Schuh, S. (1996). Small business and job creation: Dissecting the myth and reassessing the facts. *Small Business Economics, 8*(4), 297–315.

Day, G. S., & Schoemaker, P. J. H. (2005). Scanning the periphery. *Harvard Business Review, 83*(11), 135–148.

de Jong, J., & Freel, M. (2010). *Geographical distance of innovation collaborations*. Scales research reports H201008, EIM Business and Policy Research. www.entrepreneurship-sme.eu/pdf-ez/H201008.pdf. Accessed 02 Mar 2013.

de Laat, E. A. A. (1996). Patents or prizes: Monopolistic R&D and asymmetric information. *International Journal of Industrial Organization, 15*(3), 369–390.

De Laat, E. A. A. (1997). Patents or prizes: Monopolistic R&D and asymmetric information. *International Journal of Industrial Organization, 15*(3), 369–390.

Dellarocas, C. (2003). The digitization of word of mouth: Promise and challenges of online feedback mechanisms. *Management Science, 49*(10), 1407–1424.

Dellarocas, C. (2005). Reputation mechanism design in online trading environments with pure moral hazard. *Information Systems Research, 16*(2), 209–230.

Dellarocas, C. (2006). Strategic manipulation of internet opinion forums: Implications for consumers and firms. *Management Science, 52*(10), 1577–1593.

Dellarocas, C. (2007). Reputation mechanisms. In T. Hendershott (Ed.), *Handbooks in information systems, volume 1: Economics and information systems* (pp. 629–660). Amsterdam: Elsevier.

Dennis, F. (2008). *How to get rich: One of the world's greatest entrepreneurs shares his secrets*. London: Penguin Books.

Dennis, W. J. (1996). More than you think: An inclusive estimate of business entries. *Journal of Business Venturing, 12*(3), 175–196.

Desai, S. (2009). *Measuring entrepreneurship in developing countries*. UNU-WIDER research paper, no 2009/10.

Diener, K., & Piller, F. (2009). *The market for open innovation: Increasing the efficiency and effectiveness of the innovation process* (Open innovation accelerator survey 2009). Aachen: RWTH-TIM Group.

Dow, J., & da Costa Werlang, S. R. (2008). Nash equilibrium under Knightian uncertainty: Breaking down backward induction. *Journal of Economic Theory, 64*(2), 305–324.

Drnovsek, M., Cardon, M. S., & Murnieks, C. Y. (2009). Collective passion in entrepreneurial teams. In A. L. Carsrud & M. Brännback (Eds.), *Understanding the entrepreneurial mind: Opening the black box* (pp. 191–215). New York: Springer.

Drucker, P. F. (1984). Our entrepreneurial economy. *Harvard Business Review, 62*(1), 59–64.
Drucker, P. F. (1986). *Innovation and entrepreneurship*. New York: HarperCollins. Reprint 2006.
Dulleck, U., & Kerschbamer, R. (2006). On doctors, mechanics, and computer specialists: The economics of credence goods. *Journal of Economic Literature, 44*(1), 5–42.
Dushnitsky, G., & Klueter, T. (2011). Is there an eBay for ideas? Insights from online knowledge marketplaces. *European Management Review, 8*(1), 17–32.
Dyer, J. H., Gregersen, H. B., & Christensen, C. M. (2009). The innovator's DNA. *Harvard Business Review, 87*(12), 61–67.
Eisenhardt, K., & Schoonhoven, C. B. (1990). Organizational growth: Linking founding team, strategy, environment, and growth among U.S. semiconductor ventures, 1978–1988. *Administrative Science Quarterly, 35*(3), 504–529.
Elmquist, M., Fredberg, T., & Ollila, S. (2009). Exploring the field of open innovation. *European Journal of Innovation Management, 12*(3), 326–345.
Enkel, E., Grassmann, O., & Chesbrough, H. W. (2009). Open R&D and open innovation: Exploring the phenomenon. *R&D Management, 39*(4), 311–316.
Enkel, E., Kausch, C., & Gassmann, O. (2005). Managing the risk of customer integration. *European Management Journal, 23*(2), 203–213.
Ernst & Young. (2011). *Nature or nurture? Decoding the DNA of the entrepreneur*. www.ey.com/Publication/vwLUAssets/Nature-or-nurture/$FILE/Nature-or-nurture.pdf. Accessed 02 Mar 2013.
Ernst, H., Leptien, C., & Vitt, J. (2000). Inventors are not alike: The distribution of patenting output among industrial R&D personnel. *IEEE Transactions on Engineering Management, 47*(2), 184–199.
European Commission. (2003). *Green paper entrepreneurship in Europe*. (COM(2003) 27 final).
European Commission. (2004). *Action plan – the European agenda for entrepreneurship*. (COM (2004) 70 final).
European Commission. (2006). Decision No 1639/2006/EC of the European parliament and of the council of 24 October 2006 establishing a competitiveness and innovation framework programme (2007 to 2013). *Official Journal L, 310*, 0015–0040.
European Commission. (2009). *Reviewing community innovation policy in a changing world*. (COM(2009) 442 final).
European Commission. (2011). *Special Eurobarometer 359: Attitudes on data protection and electronic identity in the European Union*. http://ec.europa.eu/public_opinion/archives/ebs/ebs_359_en.pdf. Accessed 02 Mar 2013.
Fairlie, R. W. (2011). *Kauffman index of entrepreneurial activity (1996–2010)*. Kauffman Foundation, Kansas City. www.kauffman.org/uploadedFiles/KIEA_2011_report.pdf. Accessed 02 Mar 2013.
Falk, A., & Fischbacher, U. (2006). A theory of reciprocity. *Games and Economic Behavior, 54*(2), 293–315.
Farrell, J., & Saloner, G. (1985). Standardization, compatibility, and innovation. *The RAND Journal of Economics, 16*(1), 70–83.
Farrell, L., & Walker, I. (1999). The welfare effects of lotto: Evidence from the U.K. *Journal of Public Economics, 72*(1), 99–120.
FastCompany. (2002). *He struck gold on the net (Really)*. By Tischler, L., 31 May. www.fastcompany.com/44917/he-struck-gold-net-really. Accessed 02 Mar 2013.
Feeser, H., & Willard, G. (1990a). Founding strategy and performance: A comparison of high and low growth high tech firms. *Strategic Management Journal, 11*(2), 87–98.
Fehr, E., & Schmidt, K. M. (1999). A theory of fairness, competition, and cooperation. *Quarterly Journal of Economics, 114*(3), 817–868.
Fernald, J. G., & Neiman, B. (2011). Growth accounting with misallocation: Or, doing less with more in Singapore. *American Economic Journal Macroeconomics, 3*(2), 29–74.
Feyerabend, P. K. (1975). *Against method*. London: New Left Book.

Financial Times. (2008). *Buggy maker pushes ahead*. By Steen, M., 15 July. www.ft.com/intl/cms/s/0/aa44c280-5288-11dd-9ba7-000077b07658.html#axzz2EjJcHKKK. Accessed 02 Mar 2013.

Fincher, D. (2010). *The social network*. [Movie] Directed by Fincher, D., screenplay by Sorkin, A. Columbia Pictures/Sony Pictures International.

Florida, R. (2002). *The rise of the creative class: And how it's transforming work, leisure community and everyday life*. New York: Basic Books.

Fogel, R. W. (1999). Catching up with the economy. *The American Economic Review, 89*(1), 1–21.

Forbes. (2000). *Gold rush*. By Coffey, B, 3 July. www.forbes.com/forbes/2000/0703/6601126a.html. Accessed 02 Mar 2013.

Forbes. (2010). *Need to build a community? Learn from threadless*. By Burkitt, L., 7 January. http://www.forbes.com/2010/01/06/threadless-t-shirt-community-crowdsourcing-cmo-network-threadless.html. Accessed 02 Mar 2013.

Foster, R. N. (1986). *Innovation: The attacker's advantage*. New York: Summit Books.

Foss, N. J., & Klein, P. G. (2010). Entrepreneurial alertness and opportunity discovery: origins, attributes, critique. In H. Landström & F. Lohrke (Eds.), *The historical foundations of entrepreneurship research* (pp. 98–120). Cheltenham: Edward Elgar.

Franke, N., & von Hippel, E. (2003). Satisfying heterogenous user needs via innovation toolkits: The case of Apache security software. *Research Policy, 32*(7), 1199–1215.

Franke, N., Schreier, M., & Kaiser, U. (2010). The "i designed it myself" effect in mass customization. *Management Science, 56*(1), 125–140.

Franke, N., Keinz, P., & Klausberge, K. (2012). *"Does this sound like a fair deal?" Antecedents and consequences of fairness expectations in the individual's decision to participate in firm innovation*. Forthcoming in Organization Science, doi:10.1287/orsc.1120.0794.

Fredberg, T., Elmquist, M.,& Ollila-Chalmers, S. (2009). Managing open innovation: Present findings and future directions. VINNOVA Report Nr VR 2008:02, http://www.vinnova.se/upload/EPiStorePDF/vr-08-02.pdf, www.vinnova.se/upload/EPiStorePDF/vr-08-02.pdf. Accessed 2 Mar 2013.

Freeman, C. (1974). *The economics of industrial innovation*. London: Penguin.

Freeman, C., Clark, J., & Soete, L. (1982). *Unemployment and technical innovation: A study of long waves and economic development*. Westport: Greenwood Press.

Feeser, H., & Willard, G. (1990b). Founding strategy and performance: A comparison of high and low growth high tech firms. *Strategic Management Journal, 11*(2), 87–98.

Frese, M. (2009). Towards a psychology of entrepreneurship – An action theory perspective. *Foundations and Trends in Entrepreneurship, 5*(6), 435–494.

Freytag, A., & Thurik, A. R. (2007). Entrepreneurship and its determinants in a cross-country setting. *Journal of Evolutionary Economics, 17*(2), 117–131.

Friedman, E., & Resnick, P. (2001). The social cost of cheap pseudonyms. *Journal of Economics and Management Strategy, 10*(2), 173–199.

Friedman, T. L. (2005). *The world is flat. A brief history of the twenty-first century*. New York: Farrar, Straus and Giroux.

Fudenberg, D., Levine, D., & Pesendorfer, W. (1998). When are nonanonymous players negligible. *Journal of Economic Theory, 79*(1), 46–71.

Fudenberg, D., & Maskin, E. (1986). The folk theorem in repeated games with discounting or with incomplete information. *Econometrica, 54*(3), 533–554.

Fueglistaller, U., Klandt, H., Halter, F., & Müller, C. (2009). *An international comparison of entrepreneurship among students. International report of the Global University entrepreneurial spirit students' survey (GUESSS 2008)*. www.guesssurvey.org/PDF/2009/GUESSS-INT_e_2009.pdf. Accessed 02 Mar 2013.

Füller, J. (2006). Why consumers engage in virtual new product developments initiated by producers. *Advances in Consumer Research, 33*(1), 639–646.

Fullerton, R. L., & McAfee, R. P. (1999). Auctioning entry into tournaments. *The American Economic Review, 107*(3), 573–605.

Gaglio, C. M., & Katz, J. (2001). The psychological basis of opportunity identification: Entrepreneurial alertness. *Small Business Economics, 16*(2), 95–111.
Gaglio, C. M., & Winter, S. (2009). Entrepreneurial alertness and opportunity identification: Where are we now? In A. L. Carsrud & M. Brännback (Eds.), *Understanding the entrepreneurial mind, opening the black box* (pp. 305–325). New York: Springer.
Galbraith, J. K. (1952). *American capitalism: The concept of countervailing power*. Cambridge: The Riverside Press.
Galbraith, J. K. (1967). *The new industrial state*. Boston: Houghton Mifflin.
Galenson, D. W. (2010). Understanding creativity. *Journal of Applied Economics, 13*(2), 351–362.
Gallini, N. T. (2002). The economics of patents: Lessons from recent U.S. patent reform. *The Journal of Economic Perspectives, 16*(2), 131–154.
Gallini, N. T., & Kotowitz, Y. (1985). Optimal R and D processes and competition. *Economica, 52*(207), 321–334.
Gallini, N. T., & Scotchmer, S. (2002). Intellectual property: When is it the best incentive system? *Innovation Policy and the Economy, 2*, 51–77.
Gambardella, A. (2005). Patents and the division of inventive labor. *Industrial and Corporate Change, 14*(6), 1223–1233.
Gambardella, A., Giuri, P., & Luzzi, A. (2007). The market for patents in Europe. *Research Policy, 36*(8), 1163–1183.
Gambardella, A., Giarratana, M., & Panico, C. (2010). How and when should companies retain their human capital? Contracts, incentives, and human resource implications. *Industrial and Corporate Change, 19*(1), 1–24.
Gans, J. S., & Stern, S. (2010). Is there a market for ideas? *Industrial and Corporate Change, 19*(3), 805–837.
Garcia, S. M., & Tor, A. (2009). The N-effect: More competitors, less competition. *Psychological Science, 20*(7), 871–877.
Gartner, W. B. (1985). A conceptual framework for describing the phenomenon of new venture creation. *Academy of Management Review, 10*(4), 696–706.
Gartner, W. B. (1988). Who is an entrepreneur? Is the wrong question. *American Journal of Small Business, 12*(4), 11–32.
Gartner, W. B. (1990). What are we talking about when we talk about entrepreneurship? *Journal of Business Venturing, 5*(1), 15–28.
Gartner, W. B., & Carter, N. M. (2003). Entrepreneurial behavior and firm organizing processes. In Z. J. Acs & D. B. Audretsch (Eds.), *Handbook of entrepreneurship research. An interdisciplinary survey and introduction* (pp. 195–221). Boston: Kluwer.
Gartner, W. B., Shaver, K. G., Carter, N. M., & Reynolds, P. D. (2004). *The handbook of entrepreneurial dynamics: The process of organization creation*. Thousand Oaks: Sage.
Gartner, W. B., & Shane, S. A. (1995). Measuring entrepreneurship over time. *Journal of Business Venturing, 10*(4), 283–301.
Gassmann, O., & Enkel, E. (2006). Open innovation: Externe Hebeleffekte in der Innovation erzielen. *Zeitschrift Führung + Organisation, 3*, 132–138.
Gassmann, O., Enkel, E., & Chesbrough, H. (2010). The future of open innovation. *R&D Management, 40*(3), 313–321.
Ghins, M. (2002). Putnam's no-miracle argument: A critique. In S. Clarke & T. D. Lyons (Eds.), *Recent themes in the philosophy of science: Scientific realism and commonsense* (pp. 121–137). Dordrecht: Kluwer Academic.
Girotra, K., Terwiesch, C., & Ulrich, K. T. (2010). Idea generation and the quality of the best idea. *Management Science, 56*(4), 591–605.
Glaeser, E. L. (2009). The death and life of cities. In R. P. Inman (Ed.), *Making cities work – Prospects and policies for Urban America* (pp. 22–62). Princeton: Princeton University Press.
Glaeser, E. L. (2011). *Triumph of the city. How our greatest invention makes us richer, smarter, greener, healthier, and happier*. New York: Penguin Press.

Godin, K., Clemens, J., & Veldhuis, N. (2008). *Measuring entrepreneurship: Conceptual frameworks and empirical indicators. Studies in entrepreneurship and markets 7*. Fraser Institute. www.fraserinstitute.org/research-news/display.aspx?id=13202. Accessed 02 Mar 2013.

Goldsmith, S., Georges, G., & Burke, T. G. (2010). *The power of social innovation: How civic entrepreneurs ignite community networks for good*. New York: Wiley.

Gorham, G. (1994). Mind-body dualism and the Harvey-Descartes controversy. *Journal of the History of Ideas, 55*(2), 211–234.

Govindarajan, V., & Trimble, C. (2010). *The other side of innovation: Solving the execution challenge*. Boston: Harvard Business School.

Granovetter, M. S. (1973). The strength of weak ties. *The American Journal of Sociology, 78*(6), 1360–1380.

Grant, R. (1996). Prospering in dynamically-competitive environments: Organizational capability as knowledge integration. *Organization Science, 7*(4), 375–387.

Granstrand, O. (2000). *The economics and management of intellectual property: Towards intellectual capitalism*. Cheltenham: Edward Elgar.

Greif, A., Milgrom, P., & Weingast, B. R. (1994). Coordination, commitment, and enforcement: The case of the merchant guild. *Journal of Political Economy, 102*(4), 745–776.

Grilo, I., & Thurik, R. (2006). Latent and actual entrepreneurship in Europe and the US: Some recent developments. *Scientific analysis of entrepreneurship and SMEs, SCALES-paper N200514*. http://www.ondernemerschap.nl/pdf-ez/n200514.pdf. Accessed 02 Mar 2013.

Grove, A. S. (1999). *Only the paranoid survive*. New York: Doubleday.

Guerrero, M., Rialp, J., & Urbano, D. (2008). The impact of desirability and feasibility on entrepreneurial intentions: A structural equation model. *International Entrepreneurship and Management Journal, 4*(1), 35–50.

Güth, W. (1995). On ultimatum bargaining experiments – A personal review. *Journal of Economic Behavior and Organization, 27*(3), 329–344.

Güth, W., & Kliemt, H. (2004). The evolution of trust (worthiness) in the net. *Analyse und Kritik Zeitschrift fuer Sozialtheorie, 26*(1), 203–219.

Güth, W., Schmittberger, R., & Schwarze, B. (1982). An experimental analysis of ultimatum bargaining. *Journal of Economic Behavior and Organization, 3*(4), 367–388.

Güth, W., Mengel, F., & Ockenfels, A. (2007). An evolutionary analysis of buyer insurance and seller reputation in online markets. *Theory and Decision, 63*(3), 265–282.

Habourian, H. (1990). Anonymous repeated games with a large number of players and random outcomes. *Journal of Economic Theory, 51*(1), 92–110.

Hagel, J., & Brown, S. J. (2005). Productive friction: How difficult business partnerships can accelerate innovation. *Harvard Business Review, 83*(2), 82–91.

Hagel, J., Brown, J. S., & Davison, L. (2010). *The power of pull: How small moves, smartly made, can set big things in motion*. New York: Basic Books.

Haltiwanger, J. C., Jarmin, R. S., & Miranda, J. (2009). Business dynamics statistics: An overview. *Ewing Marion Kauffman Foundation*. doi:10.2139/ssrn.1456465.

Haltiwanger, J. C., Jarmin, R. S., & Miranda, J. (2010). Who creates jobs? Small vs. Large vs. Young. *NBER working paper no 16300*. www.nber.org/papers/w16300. Accessed 02 Mar 2013.

Hamel, G. (1999). Bringing silicon valley inside. *Harvard Business Review, 77*(5), 70–84.

Hamilton, V. L., & Rytina, S. (1980). Social consensus on norms of justice: Should the punishment fit the crime? *The American Journal of Sociology, 85*(5), 1117–1144.

Handelsblatt. (2006). *60 Jahre deutsche Wirtschaftsgeschichte: SAP: Langsam, aber gewaltig*. Handelsblatt, 07.04.2006. by Nonnast, T.

Hardin, R. (2004). Internet capital. *Analyse und Kritik Zeitschrift für Sozialtheorie, 26*(1), 122–138.

Hargadon, A., & Sutton, R. I. (2000). Building an innovation factory. *Harvard Business Review, 78*(3), 157–166.

Harhoff, D., & Hoisl, K. (2006). *Institutionalized incentives for ingenuity – patent value and the German Employees' Inventions Act*. http://epub.ub.uni-muenchen.de/1262/1/German_Inventor_Compensation_230106_DP_LMU.pdf. Accessed 02 Mar 2013.

Harrison, B. (1994). The myth of small firms as the predominant job generators. *Economic Development Quarterly, 8*(1), 3–18.

Harrison, B. (1997). *Lean and mean: The changing landscape of corporate power in the age of flexibility*. New York: Guilford Press.

Harron, M. (2000). *American psycho*. [Movie] Directed by Harron, M., screenplay by Harron. M., & Turner, G., based on a novel by BE Ellis, Lions Gate Films.

Headd, B., & Saade, R. (2008). *Do business definition decisions distort small business research results?* U.S. Small Business Administration. www.sba.gov/advo/research/rs330tot.pdf. Accessed 02 Mar 2013.

Heertje, A. (1973). Economics and technical change. London: Wiley.

Hellman, T. (2007). When do employees become entrepreneurs? *Management Science, 53*(6), 919–933.

Hellman, T., & Thiele, V. (2011). Incentives and innovation: A multitasking approach. *American Economic Journal: Microeconomics, 3*, 78–128.

Hellmann, T. H., & Perotti, E. P. (2011). The circulation of ideas in firms and markets. *Management Science, 57*(10), 1813–1826.

Henderson, R. M. (1993). Underinvestment and incompetence as responses to radical innovation: Evidence from the photolithographic alignment equipment industry. *The RAND Journal of Economics, 24*(2), 248–270.

Henderson, R. M. (2006). The innovator's dilemma as a problem of organizational competence. *Journal of Product Innovation Management, 23*(1), 5–11.

Henderson, R. M., & Clark, K. B. (1990). Architectural innovation: The reconfiguration of existing product technologies and the failure of established firms. *Administrative Science Quarterly, 35*(1), 9–30.

Henrekson, M., & Johansson, D. (2009). Gazelles as job creators: A survey and interpretation of the evidence. *Journal of Small Business Economics, 35*(2), 227–244.

Henry, C., Hill, F., & Leitch, C. (2005). Entrepreneurship education and training: Can entrepreneurship be taught? Part I. *Education and Training, 47*(2), 98–11.

Hebert, R. F., & Link, A. N. (1988). *The entrepreneur, mainstream views and radical critiques*. New York: Praeger.

Hebert, R. F., & Link, A. N. (1989). In search of the meaning of entrepreneurship. *Small Business Economics, 1*(1), 39–49.

Hidalgo, C. A. (2009). *The dynamics of economic complexity and the product space over a 42 year period. CID working paper no 189*. www.hks.harvard.edu/centers/cid/publications/faculty-working-papers/cid-working-paper-no.-189. Accessed 02 Mar 2013.

Hidalgo, C. A., & Hausmann, R. (2009). The building blocks of economic complexity. *Proceedings of the National Academy of Sciences of the United States of America, 106*(26), 10570–10575.

Hidalgo, C. A., Klinger, B., Barabasi, A. L., & Hausmann, R. (2007). The product space conditions the development of nations. *Science, 317*(5837), 482–487.

Hoffman, E. (2008). Reciprocity in ultimatum and dictator games: An introduction. In C. R. Plott & V. L. Smith (Eds.), *Handbook of experimental economics results* (Vol. 1, pp. 432–453). Amsterdam: North-Holland.

Holmes, T. J., & Schmitz, J. A. (1990). A theory of entrepreneurship and its application to the study of business transfers. *Journal of Political Economy, 98*(2), 265–294.

Horx, M. (2011). *Das Megatrend-Prinzip: Wie die Welt von morgen entsteht*. Muenchen: Deutsche Verlags-Anstalt.

Howe, J. (2006). *Crowdsourcing. Pure, Unadulterated (and Scalable) crowdsourcing*. [Blog post] 15 June. http://crowdsourcing.typepad.com/cs/2006/06/pure_unadultera.html. Accessed 02 Mar 2013.

Howe, J. (2009). *Crowdsourcing: Why the power of the crowd is driving the future of business.* New York: Random House.

Howells, J. (2006). Intermediation and the role of intermediaries in innovation. *Research Policy, 35*(5), 715–728.

Huston, L. (2007). *Innovation networks: Looking for ideas outside the company. Interview with Larry Huston*, published, 14 Nov , 2007 in Knowledge@Wharton. http://knowledge.wharton.upenn.edu/article.cfm?articleid=1837. Accessed 02 Mar 2013.

Huston, L., & Sakkab, N. (2006). Connect and develop: Inside Procter & Gamble's new model for innovation. *Harvard Business Review, 84*(3), 58–66.

InnoCentive. (2010). *InnoCentive investigation of the challenge driven innovation platform at NASA. An evaluation of the open innovation pilot program between NASA and InnoCentive, Inc.* www.nasa.gov/pdf/572344main_InnoCentive_NASA_PublicReport_2011-0422.pdf. Accessed 02 Mar 2013.

Israel, P. (1998). *Edison: A life of invention.* New York: Wiley.

Iversen, J., Jorgensen, R., & Malchow-Moller, N. (2008). Defining and measuring entrepreneurship. *Foundations and Trends in Entrepreneurship, 4*(1), 1–63.

Jackson, T. (1997). *Inside Intel: Andrew Grove and the rise of the world's most powerful chip company.* New York: Dutton Books.

Jaffe, A. B., & Lerner, J. (2006). Innovation and its discontents. *Innovation Policy and the Economy, 6*(1), 27–65.

Jarrell, G., & Peltzman, S. (1985). The impact of product recalls on the wealth of sellers. *Journal of Political Economy, 93*(3), 512–536.

Jeppesen, L. B., & Lakhani, K. R. (2010). Marginality and problem solving effectiveness in broadcast search. *Organization Science, 21*(5), 1016–1033.

Jorgenson, D. W., & Griliches, Z. (1967). The explanation of productivity change. *The Review of Economic Studies, 34*(3), 249–283.

Johnson, S. (2010). *Where good ideas come from: The natural history of innovation.* New York: Penguin.

Kalil, T. (2006). Prizes for technological innovation. Brookings Institution Discussion Paper 2006–08, http://www.brookings.edu/research/papers/2006/12/healthcare-kalil, www.brookings.edu/research/papers/2006/12/healthcare-kalil. Accessed 2 Mar 2013.

Kamiyama, S., Sheehan, J., & Martinez, C. (2006). *Valuation and exploitation of intellectual property. OECD science, technology and industry working papers 2006/5.* Paris: OECD. doi: 10.1787/307034817055.

Kandori, M., & Matsushima, H. (1998). Private observation, communication and collusion. *Econometrica, 66*(3), 627–652.

Kaplan, B. (1958). Further remarks on compensation for ideas in California. *California Law Review, 46*(5), 699–714.

Kaplan, S. N., & Lerner, J. (2010). It ain't broke: The past, present, and future of venture capital. *Journal of Applied Corporate Finance, 22*(2), 36–47.

Karpoff, J. M., & Lott, J. R. (1993). Reputational penalty firms bear from committing criminal fraud. *Journal of Law and Economics, 36*(2), 757–802.

Katz, J. A. (1991). Endowed positions: Entrepreneurship and related fields. *Entrepreneurship Theory Practice, 15*(3), 53–67.

Katz, J. A. (1992). Secondary analysis in entrepreneurship: An introduction to databases and data management. *Journal of Small Business Management, 30*(2), 74–86.

Katz, J. A. (2003). The chronology and intellectual trajectory of American entrepreneurship education 1876–1999. *Journal of Business Venturing, 18*(2), 283–300.

Katz, J. A. (2004). *2004 Survey of endowed positions in entrepreneurship and related fields in the United States.* Kaufmann Foundation, Kansas City. www.kauffman.org/uploadedfiles/survey_endowed_chairs_04.pdf. Accessed 02 Mar 2013.

Katz, M. L., & Shapiro, C. (1984). How to license intangible property. *Quarterly Journal of Economics, 101*(3), 567–590.

Katz, M. L., & Shapiro, C. (1986). Technology adoption in the presence of network externalities. *Journal of Political Economy, 94*(4), 822–841.
Katz, M., & Shapiro, C. (1988). Systems competition and network effects. *Journal of Economic Perspectives, 8*(2), 93–115.
Katz, M., & Shapiro, C. (1994). Network externalities, competition and compatibility. *The American Economic Review, 75*(3), 424–440.
Kauffman Foundation. (2007). *Entrepreneurship in American Higher Education*. Kaufmann Foundation, Kansas City. www.kauffman.org/uploadedfiles/entrep_high_ed_report.pdf. Accessed 02 Mar 2013.
Kaufman, J. C., & Sternberg, R. J. (2006). *The international handbook of creativity*. Cambridge: Cambridge University Press.
Kawasaki, G. (2004). *The art of the start: The time-tested, battle-hardened guide for anyone starting anything*. London: Penguin Books.
Kay, L. (2011). *Managing innovation prizes in government*. Washington, DC: IBM Center for The Business of Government. www.businessofgovernment.org/report/managing-innovation-prizes-government. Accessed 02 Mar 2013.
Kelley, D. J., Singer, S., & Herrington, M. (2012). *The global entrepreneurship monitor*. Wellesley: Babson College. www.gemconsortium.org/docs/download/2409. Accessed 02 Mar 2013.
Kerr, C., Dunlop, J. T., Harbison, F. H., & Myers, C. A. (1960). *Industrialism and industrial man. The problems of labor and management in economic growth*. Cambridge: Harvard University Press.
Kerr, N. L., & Tindale, R. S. (2004). Group performance and decision making. *Annual Review of Psychology, 55*(1), 623–655.
Kim, P. H., Aldrich, H., & Ruef, M. (2005). *Fruits of co-laboring: Effects of entrepreneurial team stability on the organizational founding process*. Frontiers of entrepreneurship research 2005. Wellesley: Babson College. http://fusionmx.babson.edu/entrep/fer/2005fer/chapter_iii/paper_iii1.html. Accessed 02 Mar 2013.
Kim, W. C., & Mauborgne, R. (1997). *Harvard business review on innovation*. Boston: Harvard Business School.
Kitchen, H., & Powells, S. (1991). Lottery expenditures in Canada: A regional analysis of determinants and incidence. *Applied Economics, 23*(12), 1845–1852.
Klein, B., & Leffler, K. B. (1981). The role of market forces in assuring contractual performance. *Journal of Political Economy, 89*(4), 615–664.
Knight, F. H. (1921). *Risk, uncertainty, and profit*. Boston: Hart, Schaffner & Marx.
Koellinger, P. (2008). Why are some entrepreneurs more innovative than others. *Small Business Economic, 31*(1), 21–37.
Koellinger, P., Minniti, M., & Schade, C. (2007). "I think I can, I think I can" – Overconfidence and entrepreneurial behavior. *Journal of Economic Psychology, 28*(4), 502–527.
Kollock, P. (1999). The production of trust in online markets. *Advances in Group Processes, 16*, 99–123.
Konrad, K. A., & Kovenock, D. (2010). Contests with stochastic abilities. *Economic Inquiry, 48*(1), 89–103.
Kornish, L. J., & Ulrich, K. T. (2011). Opportunity spaces in innovation – Empirical analysis of large samples of ideas. *Management Science, 57*(1), 107–128.
Kremer, M. (1993). Population growth and technological change: One million B.C. to 1990. *Quarterly Journal of Economics, 108*(3), 681–716.
Kremer, M. (1998). Patent buyouts: A mechanism for encouraging innovation. *Quarterly Journal of Economics, 113*(4), 1137–1167.
Kremer, M. (2000). Creating markets for new vaccines. Part II: Design issues. *Innovation Policy and the Economy, 1*, 73–118.
Kreps, D. (1990). Corporate culture and economic theory. In J. E. Alt & K. A. Shepsle (Eds.), *Perspectives on positive political economy* (pp. 90–143). Cambridge: Cambridge University Press.

Kreps, D., & Wilson, R. (1982). Reputation and imperfect information. *Journal of Economic Theory, 27*(2), 253–279.
Kreps, D., Milgrom, P., Roberts, J., & Wilson, R. (1982). Rational cooperation in the finitely repeated prisoner's Dilemma. *Econometrica, 27*(2), 245–252.
Krohmal, B. (2007). *Prominent innovation prizes and reward programs.* KEI research note 2007:1, http://www.keionline.org/misc-docs/research_notes/kei_rn_2008_1.pdf. Accessed 02 Mar 2013.
Krueger, N. F., & Brazeal, D. V. (1994). Entrepreneurial potential and potential entrepreneurs. *Entrepreneurship Theory and Practice, 18*(3), 91–104.
Kuhn, T. S. (1970). *The structure of scientific revolutions.* Chicago: University of Chicago Press.
Kuratko, D. F. (2005). The emergence of entrepreneurship education: Development, trends, and challenges. *Entrepreneurship Theory and Practice, 29*(5), 577–597.
Labianca, J. (2004). The ties that blind. *Harvard Business Review, 82*(10), 19–19.
Lakatos, I., & Feyerabend, P. K. (1974). *Kritik und Erkenntnisfortschritt.* Braunschweig: Vieweg.
Lakhani, K. R. (2006). Broadcast search in problem solving: attracting solutions from the periphery. *Technology Management for the Global Future, 4,* 2450–2468. doi:10.1109/PICMET.2006.296842.
Lakhani, K. R., & Boudreau, K. J. (2009). How to manage outside innovation. *MIT Sloan Management Review, 50*(4), 69–76.
Lakhani, K. R., & Panetta, J. A. (2007). The principles of distributed innovation. *Innovations Technology Governance Globalization, 2*(3), 97–112.
Lakhani, K. R., & Jeppesen, L. B. (2007). Getting unusual suspects to solve R&D puzzles. *Harvard Business Review, 85*(5), 30–32.
Landes, D. L. (1998). *The wealth and poverty of nations: Why some are so rich and others so poor.* New York: Norton.
Landes, D. L., Mokyr, J., & Baumol, W. J. (2010). *The invention of enterprise: Entrepreneurship from ancient Mesopotamia to modern times.* Princeton: Princeton University Press.
Landry, C. (2000). *The creative city: A toolkit for urban innovators.* London: Earthscan Publications.
Langlois, R. N. (2003). The vanishing hand: The changing dynamics of industrial capitalism. *Industrial and Corporate Change, 12*(2), 351–385.
Langlois, R. N. (2004). Chandler in a larger frame: Markets, transaction costs, and organizational form in history. *Enterprise and Society, 5*(3), 355–375.
Lechler, T. (2001). Social interaction: A determinant of entrepreneurial team venture success. *Small Business Economics, 16*(4), 263–278.
Lemley, M. A., & Myhrvold, N. (2008). How to make a patent market. *Hofstra Law Review, 36*(2), 257–259.
Leonard-Barton, D. (1992). Core capabilities and core rigidities: A paradox in managing new product development. *Strategic Management Journal, 13*(2), 111–126.
Lerner, J. (2009). *Boulevard of broken dreams: Why public efforts to boost entrepreneurship and venture capital have failed – And what to do about it.* Princeton: Princeton University Press.
Levine, B. N., Shields, C., & Margolin, N. B. (2006). *A survey of solutions to the sybil attack.* Amherst: University of Massachusetts Amherst. https://gnunet.org/node/1432. Accessed 02 Mar 2013.
Lichtenthaler, U., & Ernst, H. (2007). External technology commercialization in large firms: Results of a quantitative benchmarking study. *R&D Management, 37*(5), 383–397.
Lotka, A. J. (1926). The frequency distribution of scientific productivity. *Journal of the Washington Academy of Science, 16*(2), 317–323.
Love, J. (2008). *Selected innovation prizes and reward programs.* KEI Research Note 2008:1, http://www.keionline.org/misc-docs/research_notes/kei_rn_2008_1.pdf. Accessed 02 Mar 2013.
Low, M. (2001). The adolescence of entrepreneurship research: Specification of purpose. *Entrepreneurship Theory and Practice, 25*(4), 17–25.

Lowe, R. A., & Ziedonis, A. A. (2006). Overoptimism and the performance of entrepreneurial firms. *Management Science, 52*(2), 173–186.
Lucas, R. E. (1988). On the mechanics of economic development. *Journal of Monetary Economics, 22*(1), 3–42.
Machlup, F., & Penrose, E. (1950). The patent controversy in the nineteenth century. *The Journal of Economic History, 10*(1), 1–29.
Maddison, A. (2001). *The world economy: A millennial perspective.* Paris: OECD Development Centre. doi:10.1787/9789264189980-en.
Maddison, A. (2007). *Contours of the world economy 1–2030 AD: Essays in macro-economic history.* Oxford: Oxford University Press.
Maddison, A. (2008). Statistics on world population, GDP and per capita GDP, 1-2006 AD. http://www.ggdc.net/maddison/Maddison.htm. Accessed 6 July 2009.
Magee, G. B. (2005). Rethinking invention: Cognition and the economics of technological creativity. *Journal of Economic Behavior and Organization, 57*(1), 29–48.
Magnusson, P. R. (2009). Exploring the contributions of involving ordinary users in ideation of technology-based services. *Journal of Product Innovation Management, 26*(5), 578–593.
Mailath, G. J., & Samuelson, L. (2006). *Repeated games and reputations: Long-run relationships.* New York: Oxford University Press.
Mankiw, N., Romer, D., & Weil, D. (1992). A contribution to the empirics of economic growth. *Quarterly Journal of Economics, 107*(2), 407–438.
Mariotti, S., DeSalvo, D., & Twole, T. (2000). *The young entrepreneur's guide to starting and running a business* (2nd ed.). New York: Random House.
Martin, X., & Mitchell, W. (1998). The influence of local search and performance heuristics on new design introduction in a new product market. *Research Policy, 26*(7–8), 753–771.
Marvel, M. R., & Lumpkin, G. T. (2007). Technology entrepreneurs' human capital and its effects on innovation radicalness. *Entrepreneurship Theory and Practice, 31*(6), 807–828.
Mashelkar, R. A. (2010). Editorial: Irreverence and Indian science. *Science, 328*(5978), 547.
Matheson, C. (1998). Why the no miracles argument fails. *International Studies in the Philosophy of Science, 12*(3), 263–279.
Maurer, S. M., & Scotchmer, S. (2002). The independent invention defence in intellectual property. *Economica, 69*(276), 535–547.
Maurer, S. M., & Scotchmer, S. (2004). Procuring knowledge. *Advances in the Study of Entrepreneurship Innovation and Economic Growth, 15*, 1–31.
McCloskey, D. (2010). *Bourgeois dignity: Why economics can't explain the modern world.* Chicago: University of Chicago Press.
McDaniel, B. A. (2000). A survey on entrepreneurship and innovation. *The Social Science Journal, 37*(2), 277–284.
McGee, J. E., Peterson, M., Mueller, S. L., & Sequeira, J. M. (2009). Entrepreneurial self-efficacy: Refining the measure. *Entrepreneurship Theory and Practice, 33*(4), 965–988.
McGeer, V. (2004). Developing trust on the internet. *Analyse und Kritik Zeitschrift fuer Sozialtheorie, 26*(1), 91–107.
McKinsey. (2009). *And the winner is...Capturing the promise of philanthropic prizes.* www.mckinsey.com/App_Media/Reports/SSO/And_the_winner_is.pdf. Accessed 02 Mar 2013.
Melnik, M. I., & Alm, J. (2002). Does a seller's ecommerce reputation matter? Evidence from eBay auctions. *The Journal of Industrial Economics, 50*(3), 337–349.
Merges, R. P. (1999a). The law and economics of employee inventions. *Harvard Journal of Law & Technology, 13*(1), 2–53.
Merges, R. P. (1999b). As many as six impossible patent before breakfast: Property rights for business concepts and patent system reform. *Berkeley Technology Law Journal, 14*, 577–615.
Merges, R. P., & Reynolds, G. H. (2000). Proper scope of the copyright and patent power. *Harvard Journal on Legislation, 37*, 45–68.
Mezrich, B. (2010). *The accidental billionaires: The founding of Facebook, a tale of sex, money, genius and betrayal.* New York: Doubleday.

Michelacci, C. (2003). Low returns in R&D due to the lack of entrepreneurial skills. *The Economic Journal, 113*(484), 207–225.

Milgrom, P., North, D., & Weingast, B. (1990). The role of institutions in the revival of trade: The law merchant, private judges, and the champagne fairs. *Economics and Politics, 2*(1), 1–23.

Millard, A. J. (1990). *Edison and the business of innovation*. Baltimore: Johns Hopkins University Press.

Mills, C. W. (1951). *White collar: The American middle classes*. New York: Oxford University Press.

Mitchell, M. L., & Mahoney, M. T. (1989). Crisis in the cockpit? The role of market forces in promoting air travel safety. *Journal of Law and Economics, 32*(2), 329–355.

Mohr, J., & Spekman, R. (1994). Characteristics of partnership success: Partnership attributes, communication behavior, and conflict resolution techniques. *Strategic Management Journal, 15*(2), 135–152.

Moldovanu, B., & Sela, A. (2001). The optimal allocation of prizes in contests. *The American Economic Review, 91*(3), 542–558.

Morgan, J., & Wang, R. (2010). Tournaments for ideas. *California Management Review, 52*(2), 77–97.

Mowery, D. C. (1983). Industrial research and firm size, survival, and growth in American manufacturing, 1921–1946: An assessment. *The Journal of Economic History, 43*(4), 953–980.

Mowery, D. C. (1990). The development of industrial research in U.S. manufacturing. *The American Economic Review, 80*(2), 345–349.

Mumford, M. D., & Hunter, A. T. (2005). Innovation in organizations: A multi-level perspective on creativity. *Research in Multi Level Issues, 4*, 9–73.

Murnighan, J. K. (2008). Fairness in ultimatum bargaining. In C. R. Plott & V. L. Smith (Eds.), *Handbook of experimental economics results* (Vol. 1, pp. 411–416). Amsterdam: North-Holland.

Murphy, P. J., Liao, J., & Welsch, H. P. (2006). A conceptual history of entrepreneurial thought. *Journal of Management History, 12*(1), 12–35.

Narin, F., & Breitzman, A. (1995). Inventive productivity. *Research Policy, 24*(4), 507–519.

National Venture Capital Association. (2009). Venture capital investments Q4 and full year 2008. Press release, http://www.nvca.org/index.php?option=com_docman&task=doc_download&gid=404&Itemid=93, www.nvca.org/index.php?option=com_docman&task=doc_download&gid=404&Itemid=93. Accessed 2 Mar 2013.

National Research Council. (2007). *Innovation inducement prizes at the national science foundation*. Washington, DC: National Academies Press.

Nelson, R. R. (1959). The economics of invention: A survey of the literature. *Journal of Business, 32*(2), 101–127.

Nelson, R. R. (1981). Research on productivity growth and productivity differences: Dead ends and new departures. *Journal of Economic Literature, 19*(3), 1029–1064.

Nelson, K. E. (1993). Dow's energy/WRAP contest- A 12-Yr energy and waste reduction success story. *Proceedings from the fifteenth national industrial energy technology conference*, Houston, 24–25, Mar 1993. http://repository.tamu.edu/bitstream/handle/1969.1/92057/ESL-IE-93-03-03.pdf?sequence=1. Accessed 02 Mar 2013.

Neisser, H. P. (1942). "Permanent" technological unemployment: "Demand for commodities is not demand for labor". *The American Economic Review, 32*(1), 50–71.

New York Times. (2006). *The boss – out of Africa, onto the web*. 17 Dec, by Zipkin, A. www.nytimes.com/2006/12/17/jobs/17boss.html. Accessed 02 Mar 2013.

New York Times. (2008). *If you have a problem, ask everyone*. 22 July, by Dean, C. www.nytimes.com/2008/07/22/science/22inno.html. Accessed 02 Mar 2013.

New York Times. (2009). *ConnectU's 'Secret' $65 million settlement with facebook*. 10 Feb, by Stone, B. http://bits.blogs.nytimes.com/2009/02/10/connectus-secret-65-million-settlement-with-facebook/. Accessed 02 Mar 2013.

New York Times. (2012a). *Car-pooling makes a surge on apps and social media*. 4 July, by Meece, M. www.nytimes.com/2012/07/05/technology/technology-makes-car-pooling-safer-and-easier.html. Accessed 02 Mar 2013.

New York Times. (2012b). *Ride-sharing services grow popular in Europe*. 1 Oct, by Pfanner, E. www.nytimes.com/2012/10/01/technology/ride-sharing-services-grow-popular-in-europe.html. Accessed 02 Mar 2013.

Neyer, A. K., Bullinger, A. C., & Moeslein, K. M. (2009). Integrating inside and outside innovators: A sociotechnical systems perspective. *R&D Management, 39*(4), 410–419.

Nissenbaum, H. (1999). The meaning of anonymity in an information age. *The Information Society, 15*(2), 141–144.

Nonaka, I. (1994). A dynamic theory of organizational knowledge creation. *Organization Science, 5*(1), 14–37.

Nonaka, I., & Takeuchi, T. H. (1995). *The knowledge-creating company: How Japanese companies create the dynamics of innovation*. New York: Oxford University Press.

Norberg-Bohm, V. (2000). Creating incentives for environmentally enhancing technological change: Lessons from 30 years of U.S. energy technology policy. *Technological Forecasting and Social Change, 65*(2), 125–148.

Nordhaus, W. D. (1996). *Do real output and real wage measures capture reality? The history of lighting suggests not*. Cowls foundation working paper no 957, http://cowles.econ.yale.edu/P/cp/p09b/p0957.pdf. Accessed 02 Mar 2013.

Nordhaus, W. D. (1997). Traditional productivity estimates are asleep at the (technological) switch. *The Economic Journal, 107*(444), 1548–1559.

Nordhaus, W. D. (2004). *Schumpeterian profits in the American economy: Theory and measurement*. NBER working paper no 10433. www.nber.org/papers/w10433. Accessed 02 Mar 2013.

North, D. C., & Thomas, R. P. (1973). *The rise of the western world: A new economic history*. Cambridge: Cambridge University Press.

Norton, M. I., Frost, J. H., & Ariely, D. (2007). Less is more: The lure of ambiguity, or why familiarity breeds contempt. *Journal of Personality and Social Psychology, 92*(1), 97–105.

Obama, B. (2009). *Remarks by the president: On a new beginning*, Cairo University, Cairo, 4 June, The White House, Office of the Press Secretary. www.whitehouse.gov/the-press-office/remarks-president-cairo-university-6-04-09. Accessed 02 Mar 2013.

Obama, B. (2012). *State of the Union 2012*. Remarks by the president in State of the union address, United States Capitol, Washington, DC, 24 Jan, The White House, Office of the Press Secretary. www.whitehouse.gov/the-press-office/2012/01/24/remarks-president-state-union-address. Accessed 02 Mar 2013.

O'Connor, G. C., Corbett, A., & Pierantozzi, R. (2009). Create three distinct career paths for innovators. *Harvard Business Review, 87*(12), 78–79.

OECD. (2005). *Oslo manual* (3rd ed.). Paris: OECD. doi:10.1787/9789264013100-en.

OECD. (1998). *Fostering entrepreneurship*. Paris: OECD. doi:10.1787/9789264163713-en.

OECD. (2004). *OECD Compendium II on SME and entrepreneurship related activities carried out by International and Regional Bodies*. www.oecd.org/dataoecd/27/20/36402632.pdf. Accessed 02 Mar 2013.

OECD. (2007). *OECD reviews of innovation policy China*. Paris: OECD. doi:10.1787/9789264039643-en.

OECD. (2008). *Open innovation in global network*. Paris: OECD. doi:10.1787/9789264047693-en.

OECD. (2009a). *Measuring entrepreneurship: A collection of indicators*. 2009 edition, OECD-Eurostat Entrepreneurship Indicators Programme (EIP). www.oecd.org/industry/business-stats/44068449.pdf. Accessed 02 Mar 2013.

OECD. (2009b). *OECD Work on innovation – a stocktaking of existing work. STI working paper 2009/2*. Paris: OECD. doi:10.1787/227048273721.

OECD. (2010a). *SMEs, entrepreneurship and innovation*. Paris: OECD. doi:10.1787/9789264080355-en.

OECD. (2010b). *The OECD innovation strategy – Getting a head start on tomorrow*. Paris: OECD. doi:10.1787/9789264083479-en.
O'Neill, J. J. (2007). *Prodigal genius: The life of Nikola Tesla*. San Diego: The Book Tree.
Oster, E. (2004). Are all lotteries regressive? Evidence from the Powerball. *National Tax Journal, 57*(2), 179–187.
Ostrom, E. (2000). Crowding out citizenship. *Scandinavian Political Studies, 23*(1), 3–16.
Ostrom, E., & Walker, J. (2003). *Trust and reciprocity: Interdisciplinary lessons for experimental research*. New York: Russell Sage Foundation.
Page, S. E. (2007). *The difference: How the power of diversity creates better groups, firms, schools, and societies*. Princeton: Princeton University Press.
Palomeras, N. (2007). An analysis of pure-revenue technology licensing. *Journal of Economics and Management Strategy, 16*(4), 971–994.
Pao, M. L. (2007). An empirical examination of Lotka's law. *Journal of the American Society for Information Science, 37*(1), 26–33.
Park, J., Shin, S. K., & Lawrence, G. L. (2007). Impact of international information technology transfer on national productivity. *Information Systems Research, 18*(1), 86–102.
Parker, S. C. (2008). Statistical issues in applied entrepreneurship research: Data, methods and challenges. In E. Congregado (Ed.), *Measuring entrepreneurship: Building a statistical system* (pp. 9–20). New York: Springer.
Parker, R. S., & Udell, G. G. (1996) The new independent inventor: Implications for corporate policy. *Review of Business, 17*(3), 7–13.
Patil, P. D. (2009) *President of India, Smt. Pratibha Devisingh Patil's address to the joint session of 15th Lok Sabha in New Delhi*. 4th June, AKT/AD/HS/LV (Release ID :49043). http://pib.nic.in/newsite/erelease.aspx?relid=49043. Accessed 02 Mar 2013.
Peberdy, M., & Strowel, A. (2009). *Employee's rights to compensation for inventions – A European Perspective*. PLC Life sciences handbook 2009/10. www.cov.com/files/Publication/4ffe8880-deba-493a-8994-2a69f0da78dd/Presentation/PublicationAttachment/3b0e8983-fe2a-41b5-9e96-2fb6c8a3a8c1/Employee%E2%80%99s%20Rights%20to%20Compensation%20for%20Inventions%20-%20A%20European%20Perspective.pdf. Accessed 02 Mar 2013.
Peneder, M. (2009). The meaning of entrepreneurship: Towards a modular concept. *Journal of Industry Competition and Trade, 9*(2), 77–99.
Pettit, P. (2004). Trust, reliance and the internet. *Analyse und Kritik Zeitschrift für Sozialtheorie, 26*(1), 108–121.
Pettit, P., & Sugden, R. (1989). The backward induction paradox. *The Journal of Philosophy, 86*(4), 169–182.
Pianta, M. (2006). Innovation and employment. In J. Fagerberg, D. C. Mowery, & R. R. Nelson (Eds.), *The Oxford handbook of innovation* (pp. 568–598). Oxford: Oxford University Press.
Pink, D. H. (2005). *A whole new mind*. New York: Penguin.
Pisano, G. P., & Verganti, R. (2008). Which kind of collaboration is right for you? *Harvard Business Review, 86*(12), 78–86.
Plehn-Dujowich, J. M., Serfes, K., & Thiele, V. (2010). *Competing for entrepreneurial ideas: Matching and contracting in the venture capital market*. Searle center on law, regulation, and economic growth working paper no 2010–031. Chicago: Northwestern University. www.economie.uqam.ca/pages/docs/Thiele_Veikko.pdf. Accessed 02 Mar 2013.
Poetz, M. K., & Schreier, M. (2012). The value of crowdsourcing: Can users really compete with professionals in generating new product ideas? *Journal of Product Innovation Management, 29*(2), 245–256.
Polanyi, M. (1943). Patent reform. *The Review of Economic Studies, 11*(2), 61–76.
Poot, T., Faems, D., & Vanhaverbeke, W. (2009). Toward a dynamic perspective on open innovation: A longitudinal assessment of the adoption of internal and external innovation strategies in the Netherlands. *International Journal of Innovation Management, 13*(2), 177–200.
Popper, K. R. (1989). *Logik der Forschung* (9th ed.). Tübingen: Mohr.

Postel-Vinay, F. (2002). The dynamics of technological unemployment. *International Economic Review, 43*(3), 737–760.
Prahalad, C. K., & Bettis, R. (1986). The dominant logic: A new linkage between diversity and performance. *Strategic Management Journal, 7*(6), 485–501.
Prahalad, C. K., & Bettis, R. (1995). The dominant logic: Retrospective and extension. *Strategic Management Journal, 16*(1), 5–14.
Prahalad, C. K., & Ramaswamy, V. (2004). Co-creation experiences: The new practice in value creation. *Journal of Interactive Marketing, 18*(3), 5–14.
Prahalad, C. K., & Krishnan, M. S. (2008). *The new age of innovation: Driving co-created value through global networks*. New York: McGraw-Hill.
Prendergast, R. (2010). Accumulation of knowledge and accumulation of capital in early 'theories' of growth and development. *Cambridge Journal of Economics, 34*(3), 413–431.
Pretzer, W. S. (1989). *Working at inventing: Thomas A. Edison and the Menlo Park experience*. Dearborn: Henry Ford Museum & Greenfield Village.
Purrington, C., & Bettcher, K. E. (2001). *From the garage to the boardroom: The entrepreneurial roots of America's Largest Corporations*. Washington, DC: National Commission on Entrepreneurship. doi:10.2139/ssrn.1260383.
Putnam, H. (1975). *Mathematics, matter and method (Philosophical papers I)*. London: Cambridge University Press.
Quinn, J. B. (2000). Outsourcing innovation: The new engine of growth. *Sloan Management Review, 41*(4), 13–28.
Rajan, R., & Zingales, L. (2001). The firm as a dedicated hierarchy. *Quarterly Journal of Economics, 116*(3), 805–851.
Read, L. E. (1958). *I, pencil. My family tree as told to Leonard E. Read*. www.fee.org/the_freeman/detail/i-pencil/#axzz2HPCcc6O2. Accessed 02 Mar 2013.
Rebel, D. (1993). *Handbuch Gewerbliche Schutzrechte – Übersichten und Strategien: Europa, USA, Japan*. Wiesbaden: Gabler.
Reich, R. (1987). Entrepreneurship reconsidered: The team as hero. *Harvard Business Review, 65*(3), 77–83.
Resnick, P., & Zeckhauser, R. (2002). Trust among strangers in internet transactions: Empirical analysis of eBay's reputation system. *Advances in Applied Microeconomics, 11*, 127–157.
Resnick, P., Zeckhauser, R., Swanson, J., & Lockwood, K. (2006). The value of reputation on eBay: A controlled experiment. *Experimental Economics, 9*(2), 79–101.
Reuters. (2011). *Winklevoss twins end appeal of facebook settlement*. 23 June, by Stempel, J. www.reuters.com/article/2011/06/23/us-facebook-winklevoss-idUSTRE75L7NS20110623. Accessed 02 Mar 2013.
Reynolds, P., Bosma, N., Autio, E., Hunt, S., De Bono, N., Servais, I., Lopez-Garcia, P., & Chin, N. (2005). Global entrepreneurship monitor: Data collection design and implementation 1998–2003. *Small Business Economics, 24*(3), 205–231.
Rice, M., O'Connor, P., Colarelli, G., Peters, L. S., & Morone, J. G. (1998). Managing discontinuous innovation. *Research Technology Management, 41*(3), 52–58.
Riesman, D., Glazer, N., & Denney, R. (1953). *The lonely crowd: A study of the changing American character*. Garden City: Doubleday.
Rifkin, J. (1995). *The end of work: The decline of the global labor force and the dawn of the post-market era*. New York: Putnam.
Rigby, D., & Zook, C. (2002). Open market innovation. *Harvard Business Review, 80*(10), 80–89.
Rivette, K. G., & Kline, D. (2000). *Rembrandts in the attic: Unlocking the hidden value of patents*. Boston: Harvard Business School Press.
Roberts, E. B. (1988). Managing invention and innovation. *Research Technology Management, 31*(1), 11–29. reprint 2007 50(1):35–54.
Robertson, A. (1984). Characteristics of the successful inventor: Some notes on the nature of creativity and the creative mind. *Technovation, 2*, 141–145.

Rohrbeck, R., Hoelzle, K., & Gemuenden, H. G. (2009). Opening up for competitive advantage – How Deutsche Telekom creates an open innovation ecosystem. *R&D Management, 39*(4), 420–430.

Romer, P. M. (1986). Increasing returns and long-run growth. *Journal of Political Economy, 94*(5), 1002–1037.

Romer, P. M. (1990). Endogenous technological change. *Journal of Political Economy, 98*(5), 71–102.

Romer, P. M. (1992). Two strategies for economic development: Using ideas and producing ideas. *Proceedings of the World Bank Annual Conference on Development Economics, 1992*, 63–115.

Romer, P. M. (2007). Economic growth. In: D. R. Henderson (Ed.), *The concise encyclopedia of economics* (2nd ed.). Indianapolis: Liberty Fund. www.econlib.org/library/Enc/Economic Growth.html. Accessed 02 Mar 2013.

Root-Bernstein, R. S. (1989). Who discovers and invents. *Research Technology Management, 32*(1), 43–50.

Rosen, S. (1981). The economics of superstars. *The American Economic Review, 71*(5), 845–858.

Rosen, W. (2010). *The most powerful idea in the world: A story of steam, industry, and invention.* New York: Random House.

Rossman, J. (1931). The motives of inventors. *Quarterly Journal of Economics, 45*(3), 522–528.

Roth, A. E. (2007). The art of designing markets. *Harvard Business Review, 85*(10), 118–126.

Ruef, M., & Lounsbury, M. (2007). The sociology of entrepreneurship. *Research in the Sociology of Organizations, 25*, 1–29.

Ruef, M., Aldrich, H. E., & Carter, N. M. (2003). The structure of founding teams: Homophily, strong ties, and isolation among U.S. entrepreneurs. *American Sociological Review, 68*(2), 195–222.

Sanfey, A. G., Rilling, J. K., Aronson, J. A., Nystrom, L. E., & Cohen, J. D. (2003). The neural basis of economic decision-making in the ultimatum game. *Science, 300*(5626), 1755–1758.

Sakkab, N. Y. (2007). Growing through innovation. *Research Technology Management, 50*(6), 59–64.

Schelling, T. C. (1960). *The strategy of conflict.* Cambridge: Harvard University Press. Reprint 2006.

Schmookler, J. (1957). Inventors past and present. *The Review of Economics and Statistics, 39*(3), 321–333.

Schotter, A., Weiss, A., & Zapater, I. (1996). Fairness and survival in ultimatum and dictatorship games. *Journal of Economic Behavior & Organization, 31*(1), 37–56.

Schramm, C. J. (2004). Building entrepreneurial economies. *Foreign Affairs, 83*(4), 104–115.

Schramm, C. J. (2006). *The entrepreneurial imperative – How America's economic miracle will reshape the world (and change your life).* New York: Collins.

Schramm, C. J. (2008). Toward an entrepreneurial society: Why measurement matters. *Innovations Technology Governance Globalization, 3*(1), 3–10.

Schramm, C. J. (2010) *2010 State of entrepreneurship address.* Washington, DC: National Press Club, 19 Jan, 2010. www.kauffman.org/uploadedfiles/state_of_entrepreneurship_2010.pdf. Accessed 02 Mar 2013.

Shulman, S. (1999). *Owning the future: Staking claims on the knowledge frontier.* Boston: Houghton Mifflin.

Schumpeter, J. A. (1931). *Theorie der wirtschaftlichen Entwicklung* (3rd ed.). München: Duncker und Humbolt.

Schumpeter, J. A. (1934). *Theory of economic development.* New Brunswick: Transaction Publishers. Reprint 2004.

Schumpeter, J. A. (1942). *Capitalism, socialism and democracy* (6th ed.). London: Routledge. Reprint 2006.

Schumpeter, J. A. (1947). The creative response in economic history. *The Journal of Economic History, 7*(2), 149–159.

Schutjens, V. A. J. M., & Wever, E. (2000). Determinants of new firm success. *Papers in Regionla Science, 79*(2), 135–159.

Scotchmer, S. (2004). *Innovation and incentives*. Cambridge: MIT Press.

Selten, R., & Stocker, R. (1986). End behavior in sequences of finite prisoner's Dilemma supergames a learning theory approach. *Journal of Economic Behavior and Organization, 7*(1), 47–70.

Servan-Schreiber, J. J. (1968). *The American challenge*. (With a foreword by Arthur Schlesinger, Jr., Trans. from the French by Ronald Steel). New York: Atheneum.

Shane, S. (2000). Prior knowledge and the discovery of entrepreneurial opportunities. *Organization Science, 11*(4), 448–469.

Shane, S. (2008). *The illusion of entrepreneurship: The costly myth that entrepreneurs, investors, and policy makers live by*. New Haven: Yale University Press.

Shane, S., & Venkataraman, S. (2000). The promise of entrepreneurship as a field of research. *Academy of Management Review, 25*(1), 217–226.

Shane, S., & Cable, D. (2002). Network ties, reputation, and the financing of new ventures. *Management Science, 48*(3), 364–381.

Shane, S., & Eckhardt, J. (2003). Opportunities and entrepreneurship. *Journal of Management, 29*(3), 333–349.

Shane, S., Locke, E. L., & Collins, C. J. (2003). Entrepreneurial motivation. *Human Resource Management Review, 13*(2), 257–279.

Shapiro, C. (1983). Premiums for high-quality products as returns to reputations. *Quarterly Journal of Economics, 98*(4), 659–680.

Shavell, S., & van Ypersele, T. (2001). Rewards versus intellectual property rights. *Journal of Law and Economics, 44*(2), 525–547.

Shaver, K. G., & Scott, L. R. (1991). Person, process, choice: The psychology of new venture creation. *Entrepreneurship Theory & Practice, 16*(2), 23–45.

Sheehan, K. B. (2002). Toward a typology of internet users and online privacy concerns. *The Information Society, 18*(1), 21–32.

Sheehan, K. B., & Hoy, M. G. (2002). Dimensions of privacy concern among online consumers. *Journal of Public Policy and Marketing, 19*(1), 62–73.

Shurkin, J. N. (2006). *Broken genius: The rise and fall of William Shockley, creator of the electronic age*. New York: Palgrave Macmillan.

Silvera, R., & Wright, R. (2010). Search and the market for ideas. *Journal of Economic Theory, 145*(4), 1550–1573.

Smale, S. (1980). The prisoner's dilemma and dynamical systems associated to non-cooperative games. *Econometrica, 48*(7), 1617–1634.

Smart, J. J. C. (1963). *Philosophy and scientific realism*. New York: Humanities Press.

Smil, V. (1994). *Energy in world history*. Boulder: Westview Press.

Smil, V. (2005). *Creating the twentieth century: Technical innovations of 1867–1914 and their lasting impact (Technical revolutions and their lasting impact)*. New York: Oxford University Press.

Smil, V. (2006). *Transforming the twentieth century: Technical innovations and their consequences*. New York: Oxford University Press.

Smith, D. K., & Alexander, R. C. (1999). *Fumbling the future: How Xerox invented, then ignored, the first personal computer*. New York: toExcel.

Sobel, D. (1995). *Longitude: The true story of a lone genius who solved the greatest scientific problem of His time*. New York: Walker and Company.

Sohl, J. (2011). *The angel investor market in 2010: A market on the rebound*. Center for venture research, 12 April, 2011. www.unh.edu/news/docs/2010angelanalysis.pdf. Accessed 02 Mar 2013.

Solnick, S., & Schweitzer, M. (1999). The influence of physical appearance and gender on ultimatum game decisions. *Organizational Behavior and Human Decision Processes, 79*(3), 199–215.

Solomon, G. T., & Fernald, L. W. (1991). Trends in small business and entrepreneurship education in the United States. *Entrepreneurship Theory Practice, 15*(3), 25–40.

Solomon, G. T., Weaver, K. M., & Fernald, L. W. (1994). A historical examination of small business management and entrepreneurship pedagogy. *Simulation Gaming, 25*(3), 338–352.

Solow, R. M. (1956). A Contribution to the theory of economic growth. *Quarterly Journal of Economics, 70*(1), 65–94.

Solow, R. M. (1957). Technical change and the aggregate production function. *The Review of Economics and Statistics, 39*(3), 312–320.

Sommer, S. C., & Loch, C. H. (2004). Selectionism and learning in projects with complexity and unforeseeable uncertainty. *Management Science, 50*(10), 1334–1347.

Sørensen, M. (2007). How smart is smart money? A two-sided matching model of venture capital. *Journal of Finance, 62*(6), 2725–2762.

Sorenson, O., & Stuart, T. E. (2001). Syndication networks and the spatial distribution of venture capital investments. *The American Journal of Sociology, 106*(6), 1546–1588.

Standing, G. (1984). The notion of technological unemployment. *International Labour Review, 123*(2), 127–147.

Stangler, D., & Kedrosky, P. (2010). Firm formation and economic growth exploring firm formation: Why is the number of new firms constant? Kauffman Foundation, Kansas City, http://www.kauffman.org/uploadedFiles/exploring_firm_formation_1-13-10.pdf, www.kauffman.org/uploadedFiles/exploring_firm_formation_1-13-10.pdf. Accessed 2 Mar 2013.

Sternberg, R. J. (1988). *The nature of creativity: Contemporary psychological perspectives.* Cambridge: Cambridge University Press.

Sternberg, R. J., & Lubart, T. (1995). *Defying the crowd: Cultivating creativity in a culture of conformity.* New York: The Free Press.

Stevenson, H. H. (2004). Intellectual foundations of entrepreneurship. In H. P. Welsch (Ed.), *Entrepreneurship: The way ahead* (pp. 3–1). New York: Routledge.

Stine, D. D. (2009). *Federally funded innovation inducement prizes.* CRS report for congress, Congressional Research Service, Washington, DC. www.fas.org/sgp/crs/misc/R40677.pdf. Accessed 02 Mar 2013.

Summers, L. H. (2010). *Remarks of Lawrence H. Summers at the presidential summit on entrepreneurship.* Ronald Reagan Building, Washington, DC, 27 April, The White House, Office of the Press Secretary. www.whitehouse.gov/the-press-office/remarks-lawrence-h-summers-presidential-summit-entrepreneurship. Accessed 02 Mar 2013.

Sutton, J. (1997). Gibrat's legacy. *Journal of Economic Literature, 35*(1), 40–59.

Tadelis, S. (1999). What's in a name? Reputation as a tradeable asset. *The American Economic Review, 89*(3), 548–563.

Tandon, P. (1983). Rivalry and the excessive allocation of resources to research. *Bell Journal of Economics, 14*(1), 152–1650.

Tang, J. (2008). Environmental munificence for entrepreneurs: Entrepreneurial alertness and commitment. *International Journal of Entrepreneurial Behaviour & Research, 14*(3), 128–151.

Tapscott, D., & Williams, A. D. (2006). *Wikinomics: How mass collaboration changes everything.* New York: Penguin.

Taussig, F. W. (1915). *Inventors and money-makers.* New York: Macmillan.

Taylor, C. R. (1995). Digging for golden carrots: An analysis of research tournaments. *The American Economic Review, 85*(4), 872–890.

Terwiesch, C., & Xu, Y. (2008). Innovation contests, open innovation, and multiagent problem solving. *Management Science, 54*(9), 1529–1543.

Terwiesch, C., & Ulrich, K. T. (2009). *Innovation tournaments: Creating and selecting exceptional opportunities.* Cambridge: Harvard Business Press.

Tesla, N. (1919). *My inventions.* Originally published in the electrical experimenter magazine. Available at www.teslaplay.com/auto.htm. Accessed 02 Mar 2013.

Thaler, R. H. (1988). Anomalies: The ultimatum game. *The Journal of Economic Perspectives,* 2(4), 195–206.
The Daily Beast. (2011). *Facebook busted in clumsy smear on Google.* 11 May, by Lyons, D. www.thedailybeast.com/articles/2011/05/12/facebook-busted-in-clumsy-smear-attempt-on-google.html. Accessed 02 Mar 2013.
The Economist. (1982). *We're all intrapreneurial now.* 17th April 1982.
The Economist (2001). The future of the company: A matter of choice. 20 Dec 2001.
The Economist. (2009a). *Special report on entrepreneurship.* An idea whose time has come: Entrepreneurialism has become cool. 12 Mar 2009.
The Economist. (2009b). *Special report on entrepreneurship: Magic formula. The secrets of entrepreneurial success.* 12 Mar 2009.
The Economist. (2009c). *Trolls demanding tolls: Intellectual property comes of age as an alternative investment.* 10 Sept 2009.
The Economist. (2009d). *Filth. The joy of dirt: Why cleanliness may be going out of fashion.* 17 Dec 2009.
The Economist. (2010a). *Innovation prizes. And the winner is...: Offering a cash prize to encourage innovation is all the rage. Sometimes it works rather well.* 5 Aug 2010.
The Economist. (2010b). *Social innovation. Let's hear those ideas: In America and Britain governments hope that a partnership with "social entrepreneurs" can solve some of society's most intractable problems.* 12 Aug 2010.
The Economist. (2010c). *Animal and human behaviour. Manager's best friend: Dogs improve office productivity.* 12 Aug 2010.
The Economist. (2010d). *Promoting innovation. Growth on the cheap: The OECD tells governments how to unleash business's creative potential.* 27 May 2010.
The Economist. (2010e). *Schumpeter: The innovation machine.* 26 Aug 2010.
The Economist. (2011a). *Internet businesses. Another digital gold rush: Internet companies are booming again. Does that mean it is time to buy or to sell.* 12 May 2011.
The Economist. (2011b). *Rig on a roll. Transport: Computer modelling is being used to improve the airflow around big trucks and reduce their fuel consumption.* 2 Jun 2011.
The Economist. (2011c). *Brain scan. Seer of the mirror world: David Gelernter, a pioneering computer scientist, foresaw the modern internet but thinks computers are still too hard to use. Technology Quarterly.* 3 Dec 2011.
The Economist. (2011d). *Big and clever: Why large firms are often more inventive than small ones.* 17 Dec 2011.
The Economist. (2012a). *Technological change. The last Kodak moment? Kodak is at death's door; Fujifilm, its old rival, is thriving. Why?* 14 Jan 2012.
The Economist. (2012b). *Brain scan. Taking the long view: Jeff Bezos, the founder and chief executive of Amazon, owes much of his success to his ability to look beyond the short-term view of things. Technology Quarterly.* 3 Mar 2012.
The Economist. (2012c). *Flushed with pride. Technology and development: Each year 1.5m children die from diarrhoea. Better toilets could reduce the death toll.* Technology Quarterly, 1 Sept 2012.
The Economist. (2012d). *Six degrees of mobilisation. Technology and society: To what extent can social networking make it easier to find people and solve real world problems?* Technology Quarterly, 1 Sept 2012
The New Yorker. (2011). *Creation myth: Xerox PARC, Apple, and the truth about innovation.* 16 May, by Gladwell, M. www.newyorker.com/reporting/2011/05/16/110516fa_fact_gladwell. Accessed 02 Mar 2013.
The New Yorker. (1993). *The flash of genius: Bob Kearns and his patented windshield wiper have been winning millions of dollars in settlements from the auto industry, and forcing the issue of who owns an idea.* 11 Jan, by Seabrook, J. www.newyorker.com/archive/1993/01/11/1993_01_11_038_TNY_CARDS_000363341. Accessed 02 Mar 2013.

Thomke, S., & von Hippel, E. (2002). Customers as innovators: A new way to create value. *Harvard Business Review, 80*(4), 74–81.
Thornton, P. (1999). The sociology of entrepreneurship. *Annual Review of Sociology, 25*, 19–46.
Thurik, A. R. (2008). *Entrepreneurship, economic growth and policy in emerging economies. ERIM report series research in management*, No ERS-2008-060-ORG. http://repub.eur.nl/oai_oai:repub.eur.nl/res/pub/13318/. Accessed 02 Mar 2013.
Thurik, A. R. (2009). Entreprenomics: Entrepreneurship, economic growth, and policy. In Z. Acs, D. Audretsch, & R. Strom (Eds.), *Entrepreneurship, growth, and public policy* (pp. 219–249). Cambridge: Cambridge University Press.
Time Magazine. (1986). *Adios, Amiga?* 24 Feb, by Henry, G. M., Hackman, W. & McCarroll, T. www.time.com/time/magazine/article/0,9171,960694,00.html. Accessed 02 Mar 2013.
Time Magazine. (2010). *The making of America: Thomas Edison*. The inventor. 23 June, by Stengel, R. www.time.com/time/specials/packages/article/0,28804,1999143_1999496_1999498,00.html. Accessed 02 Mar 2013.
Titmuss, R. (1971). The gift of blood. *Society, 8*(3), 18–26.
Troy, I., & Werle, R. (2008). *Uncertainty and the market for patents*. Max-Planck-Institut für Gesellschaftsforschung working paper 08/2. www.mpifg.de/pu/workpap/wp08-2.pdf. Accessed 02 Mar 2013.
Troy, C., Szymanski, M., & Varadarajan, P. (2001). Generating new product ideas: An initial investigation of the role of market information and organizational characteristics. *Journal of the Academy of Marketing Science, 29*(1), 89–101.
Udell, G. G. (1990). It's still caveat, inventor. *Journal of Product Innovation Management, 7*(3), 230–243.
Udell, G. G., Bottin, R., & Glass, D. D. (1993). The Wal-Mart innovation network: An experiment in stimulating American innovation. *Journal of Product Innovation Management, 10*(1), 23–34.
Ueda, M. (2004). Banks versus venture capital: Project evaluation, screening, and expropriation. *Journal of Finance, 59*(2), 601–621.
UK Department of Trade and Industry. (1998). *DTI white paper: Our competitive future: Building the knowledge driven economy*. Report presented to Parliament by the United Kingdom Secretary of State for Trade and Industry, December 1998. http://webarchive.nationalarchives.gov.uk/+/http://www.dti.gov.uk/comp/competitive/. Accessed 02 Mar 2013.
USA Today. (2006). *'Charismatic' founder keeps netflix adapting*. 23 April, by Hopkins, J. http://usatoday30.usatoday.com/tech/products/services/2006-04-23-netflix-ceo_x.htm. Accessed 02 Mar 2013.
Utterback, J. M., & Brown, J. W. (1972). Profiles of the future monitoring for technological opportunities. *Business Horizons, 15*(5), 5–15.
van den Ende, J., & Dolfsma, W. (2005). Technology-push, demand-pull and the shaping of technological paradigms: Patterns in the development of computing technology. *Journal of Evolutionary Economics, 15*(1), 83–99.
van der Does de Willebois, E., Halter, E. M., Harrison, R. A., Park, J. W., & Sharman, J. C. (2011). *The puppet masters: How the corrupt use legal structures to hide stolen assets and what to do about it*. Washington, DC: World Bank.
van Praag, C. M., & Versloot, P. H. (2008). The economic benefits and costs of entrepreneurship: A review of the research. *Foundations and Trends in Entrepreneurship, 4*(2), 65–154.
van Stel, A. (2005). COMPENDIA: Harmonizing business ownership data across countries and over time. *International Entrepreneurship and Management Journal, 1*(1), 105–123.
van Stel, A. (2006). *Empirical analysis of entrepreneurship and economic growth*. New York: Springer.
VanityFair. (2011). *The accidental activist: Part I – twitter was act one*. April, by Kirkpatrick, D. www.vanityfair.com/business/features/2011/04/jack-dorsey-201104
Vesper, K. H. (1990). *New venture strategies* (2nd ed.). Englewood Cliffs: Prentice Hall.
Vesper, K. H., & Gartner, W. B. (1997). Measuring progress in entrepreneurship education. *Journal of Business Venturing, 12*(5), 403–421.

Vivarelli, M. (1995). *The economics of technology and employment: Theory and empirical evidence*. Cheltenham: Elgar.
Vivarelli, M., & Pianta, M. (Eds.). (2000). *The employment impact of innovation: Evidence and policy*. London: Routledge.
von Haberler, G. (1937). *Prosperity and depression*. Geneva: League of Nations.
von Hippel, E. (1982). Get new products from customers. *Harvard Business Review, 60*(2), 117–122.
von Hippel, E. (1988). *The sources of innovation*. New York: Oxford University Press.
von Hippel, E. (2001). Perspective: User toolkits for innovation. *Journal of Product Innovation Management, 18*, 247–257.
von Hippel, E. (2005). *Democratizing innovation*. Cambridge: MIT Press.
von Hippel, E., Thomke, S., & Sonnack, M. (1999). Creating breakthroughs at 3M. *Harvard Business Review, 77*(5), 47–57.
von Hippel, E., & Katz, R. (2002). Shifting innovation to users via toolkits. *Management Science, 48*(7), 821–833.
von Hippel, E., & Jin, C. (2009). The major shift towards user-centered innovation: Implications for China's innovation policymaking. *Journal of Knowledge-based Innovation in China, 1*(1), 16–27.
von Zedtwitz, M., & Gassmann, O. (2002). Market versus technology driven in R&D internationalisation: Four different patterns of managing research and development. *Research Policy, 31*(4), 569–588.
Wagner Weick, C., & Eakin, C. F. (2005). Independent inventors and innovation: An empirical study. *International Journal of Entrepreneurship and Innovation, 6*(1), 5–15.
Wallance, B., Cesarini, D., Lichtenstein, P., & Johannesson, M. (2007). Heritability of ultimatum game responder behavior. *Proceedings of the National Academy of Sciences of the United States of America, 104*(40), 15631–1563.
Wall Street Journal. (2010). *You call that innovation? Companies love to say they innovate, but the term has begun to lose meaning*. 23 May, by Kwoh, L. http://online.wsj.com/article/SB10001424052702304791704577418250902309914.html. Accessed 02 Mar 2013.
Weber, M. (1950). *The protestant ethic and the spirit of capitalism*. (trans: Talcott, P.). London: Scribner. http://ia700306.us.archive.org/5/items/protestantethics00webe/protestantethics00webe.pdf. Accessed 02 Mar 2013.
Weber, M. (1949). The methodology of social sciences. (trans: Shilz, E.S. & Finch, H.A.). Excerpts reprinted in Hausman, D. M. (1994). *The philosophy of economics – an anthology* (2nd ed., pp. 69–82). Cambridge: Cambridge University Press.
Wennekers, A. R. M., & Thurik, A. R. (1999). Linking entrepreneurship and economic growth. *Small Business Economics, 13*(1), 27–55.
Wennekers, S. (2006). *Entrepreneurship at country level: Economic and non-economic determinants*. Erasmus Research Institute of Management Ph.D. Series Research in Management 81. http://repub.eur.nl/res/pub/7982/EPS2006081ORG9058921158Wennekers.pdf. Accessed 02 Mar 2013.
Whyte, W. H. (1956). *The organization man*. New York: Simon and Schuster.
Wilson, R. K., & Eckel, C. C. (2006). Judging a book by its cover: Beauty and expectations in the trust game. *Political Research Quarterly, 59*(2), 189–202.
Wilson, S. (1955). *The man in the gray flannel suit*. New York: Simon and Schuster.
Wingham, D. W. (2004). Entrepreneurship through the ages. In H. P. Welsch (Ed.), *Entrepreneurship: The way ahead* (pp. 27–42). New York: Routledge.
Wired Magazine. (2006). *The rise of crowdsourcing*. 14 June, by Howe, J. www.wired.com/wired/archive/14.06/crowds.html. Accessed 02 Mar 2013.
Woirol, G. R. (1996). *The technological unemployment and structural unemployment debates*. Westport: Greenwood Press.
Wong, P. K., Ho, Y. P., & Autio, E. (2005). Entrepreneurship, innovation, and economic growth: Evidence from GEM data. *Small Business Economics, 24*(3), 335–350.

Woodman, R. W., Sawyer, J. E., & Griffin, R. W. (1993). Toward a theory of organizational creativity. *The Academy of Management Review, 18*(2), 293–321.

World Economic Forum. (2009). *Educating the next wave of entrepreneurs.* REF:150409. www3.weforum.org/docs/WEF_GEI_EducatingNextEntrepreneurs_ExecutiveSummary_2009.pdf. Accessed 02 Mar 2013.

Wright, B. D. (1983). The economics of invention incentives: Patents, prizes, and research contracts. *The American Economic Review, 73*(4), 691–707.

Wu, W. Y., & Sukoco, B. M. (2010). Why should I share? Examining consumer motives and trust on knowledge sharing. *The Journal of Computer Information Systems, 50*(4), 11–19.

Yanagisawa, T., & Guellec, D. (2009). *The emerging patent marketplace.* Science, Technology and Industry Working Papers 2009/9. Paris: OECD. doi:10.1787/218413152254.

Yamagishi, T. (2003). *The role of reputation in open and closed societies: An experimental study of internet auctioning.* Working paper, Hokkaido University. http://citeseerx.ist.psu.edu/viewdoc/download?doi=10.1.1.7.204&rep=rep1&type=pdf. Accessed 02 Mar 2013.

Yosha, O. (1995). Information disclosure costs and the choice of financing source. *Journal of Financial Intermediation, 4*(1), 3–20.

Yu, T. F. L. (2001). Entrepreneurial alertness and discovery. *The Review of Austrian Economics, 14*(1), 47–63.

Yu, H., Kaminsky, M., Gibbons, P. B., & Flaxman, A. D. (2008). SybilGuard: Defending against Sybil attacks via social networks. IEEE/ACM Transactions on Networking, *16*(3), 576–589.

Yusuf, S. (2009). From creativity to innovation. *Technology in Society, 31*(1), 1–8.

Zahra, S. A. (2006). Entrepreneurship and disciplinary scholarship: Return to the fountainhead. In S. A. Alvarez, R. Agarwal, & O. Sorenson (Eds.), *Handbook of entrepreneurship research: Interdisciplinary perspectives* (pp. 253–268). New York: Springer.

Zeckhauser, R. (1996). The challenge of contracting for technological information. *Proceedings of the National Academy of Sciences of the United States of America (PNAS), 93,* 12743–12748.

Ziens, J. D. (2010). *Guidance on the use of challenges and prizes to promote open government.* Executive Office of the President: Office of Management and Budget, Washington, DC. www.whitehouse.gov/sites/default/files/omb/assets/memoranda_2010/m10-11.pdf. Accessed 02 Mar 2013.

Zikmund, W., & D'Amico, M. (1989). *Marketing* (3rd ed.). New York: Wiley.

Zivney, T. L., & Marcus, R. D. (1989). The day the United States defaulted on treasury bills. *The Financial Review, 24*(3), 475–489.

Index

A
Abuse potential, 179
367 Addison Avenue, 49
Adverse selection, 158
Altshuler, Alex, 130
Amazon, 63, 153, 155, 173
Angle investor, 199
Anonymity, 135, 151, 152, 158
Ansari X Prize, 126
Appert, Nicolas, 126
Apple, 7, 9, 39, 46, 49, 64, 106, 151
Appropriability, 11, 47, 62, 63, 66, 79, 90, 91, 118, 120, 121, 132
Arrow, Kenneth, 119, 121
Arrow's Fundamental Paradox of Information, 14, 115, 119–122, 126, 128, 133, 134, 149, 150, 160
Attracting ideas, 205

B
Barenbrug, Max, 99
Bargaining position, 165
Bill and Melinda Gates Foundation, 131
Birthplace of Silicon Valley, 50
Bloomberg, Michael, 103
Bodrov, Yuri, 129
Boiler, Kari, 100
Bugaboo, 99
Business theory, 43

C
C64. *See* Commodore Business Machines
Carpooling, 195, 196
Celluloid, 134
Cheap pseudonyms, 178
Chesbrough, Henry, 14, 64
Chief Innovation Officer, 61

Chrysler, 117
Closed innovation, 36, 40, 59
Clusters, 80
Code of conduct, 177, 183, 184, 192, 193, 197, 198
Coffee cup sleeve, 101
Commoditization of ideas, 83, 123
Commodore Business Machines, 6
COMPETES Act, 28
Competition for ideas, 203
Complementary ideas, 121
Conflict
    of interest, 174
    potential, 82, 127, 128, 143, 162, 177
Connect and Develop, 144–146, 148
Contestable market, 43
Cooperation
    incentive, 155, 157, 160, 161, 165, 202
    threat to, 158
Cooperation mechanism, 148
Cooperative innovation, 12, 13, 73, 75, 77, 82, 84, 85, 87, 88, 93, 94, 106, 109, 116, 123, 143, 145, 146, 148, 151, 153, 162, 163, 166, 171–173, 175, 177, 183, 186, 189–191, 195, 198, 204, 205
Corporate innovation, 13, 59
Corporate R&D, 31, 36, 47, 60, 65, 109, 128, 204
Couchsurfing, 163
Cragin, Bruce, 129
Creativity, 27, 80, 81
Credit
    history, 152, 153
    rating, 153
    score, 162
Cross-pollination, 61, 64, 79, 81, 90, 104
Crowdsourcing, 64, 83, 105, 128, 204

D
Davis, John, 130
Decentralised commissioned research, 128
Demand for information, 121
Democratising innovation, 85, 102, 201
Deterrence, 137, 153, 154, 159, 173, 174, 177, 180
Division of labour, 13, 31, 42, 43, 73–75, 77, 79, 84, 88, 93, 99, 105, 107–109, 119, 133, 144, 172, 177, 201–203
Dorsey, Jack. *See* Twitter
Drucker, Peter, 14, 36, 41, 44
Dyson, James, 118

E
Eastman, George. *See* Kodak, Eastman
Economic evolution, 35
Economics of superstars, 84
Economies of scale, 38, 42, 43
The Economist, 7, 13, 26, 28, 29, 37, 48, 50, 62, 63, 91, 131, 132, 138, 139
Edison, Thomas, 3, 202
Entrepreneurial society, 14, 31, 41, 46, 50, 51, 59, 60, 67, 68, 73, 75, 89, 94, 116
Entrepreneurship
 rise of, 46
 training, 108
Environmental costs, 83
European Commission, 48
European Union, 28
European Year of Creativity and Innovation, 27
Experimental economics, 163
Exxon Valdez, 130

F
Facebook, 101, 117, 152
Fairchild, 5, 62, 118
Fairness, 67, 77, 82, 120, 143, 145, 155, 161, 163, 165, 166, 171, 173, 180, 182–184, 186–188, 197, 198, 201
 inherent notion of, 161, 162, 164, 165
Fair remuneration, 162, 165
Feedback
 evaluation, 158
 mechanism, 89, 153, 173, 177, 179, 180, 183, 184, 186, 187, 192, 193, 201, 202
 abuse, 182
 system, 179
Flash of Genius. *See* Kearns, Robert

Ford Motor Company, 11, 117
Freeriding, 150, 151, 177–179, 196

G
Galilei, Galileo, 125
General Motors, 38, 117
General norms, 162
Goldcorp Challenge, 138
Google, 9, 46, 49, 60, 106, 152
Graphical User Interface (GUI), 39
Groupon, 101

H
Harrison, John, 126, 128
HarvardConnection, 101, 117
Hastings, Reed. *See* Netflix
Henderson, Mike, 91
Hitchhiking, 195–197
Hoffman, Michael, 131
Hyatt, John Wesley, 134

I
IBM, 7, 37, 39, 42, 60, 63, 64, 85, 106
Idea clusters, 103
Idea development, 190
Idea Economy, 12, 13, 31, 73–85, 87–89, 94, 109, 116, 119, 121, 123, 144, 163, 171, 193, 195, 201, 203, 205
Idea plans, 191
Idea theft, 11, 148, 155, 187, 202
Idea trading, 85, 121, 123, 148, 153, 171, 176
Identity, 177
Incumbent, 6, 35, 44, 46, 59, 60, 68, 75, 85, 100, 108, 133, 202, 204
Indian Decade of Innovation, 28
Information and communication technologies, 43, 152
InnoCentive, 126, 129–131, 135–137, 172, 197
Innovation
 culture, 203
 cycle, 108
 horizontal segmentation, 74, 94
 importance of, 27
 incentives, 39
 intermediaries, 135
 internal, 40
 more with less, 27
 nurture, 30
 paths, 13, 66, 73, 74
 open, 67

prizes, 127, 132, 133, 160
vertical segmentation, 74, 94
Intel, 5, 42, 62, 64, 118
Intellectual property, 43, 119, 122, 135, 143, 147
  law, 119, 144
  management, 62
  market, 134
  trading, 62, 64, 67
Intermittent windshield wiper, 11, 117
Internet, 20, 117, 118, 152

K
Kaufman Foundation, 51
Kearns, Robert, 11, 117
Knowledge filter, 38, 61, 74, 84, 144, 205
Kodak, Eastman, 7
Kremer Prize, 126

L
Land of ideas, 28
Lead users, 64, 103
Leatherman, Timothy, 101
Lifetime employment, 42
Light bulb, 3, 20, 21
Lighting, 20
Lindberg, Charles, 126
Linear innovation, 91
Longitude Prize, 125
Lotto-mentality, 82

M
Managed economy, 30, 31, 36, 38, 41, 45, 46, 59, 67, 93
Management science, 44
Market for ideas, 11, 12, 85, 115, 120, 122, 134, 144, 145, 154, 195, 197
Market for reputation, 182
Mason, Andrew, 101
Matching, 172–174, 191
Matching mechanism, 197
Mechanism design, 177–179, 181, 182
Mediation, 182, 183, 193
Medium-to Long-Term Strategic Plan for the Development of Science and Technology, 28
Mehregany, Mehran, 137
Microsoft, 39, 42, 46, 49, 60, 106
Misconduct, 177

N
Narendra, Divya, 117
NASA, 106, 126, 129–131
Nature of innovation, 203
Negative feedback, 155, 177, 181–183
Netflix, 5, 101, 139
  1 Mio Dollar Prize, 139
Network hopping, 179, 199
Newton, Isaac, 125
NineSigma, 105, 137, 138, 172
Nobel Prize, 5, 121
Non-natural persons, 178
Nonoptimal allocation of information, 122
Nordhaus, William, 21

O
Obama, Barack, 28
OECD, 24, 27–29, 36, 46–48
Oil Spill Recover Institute, 130
Online-dating, 191
Online feedback, 152
Online rating platform, 181
Online reputation, 155
Open innovation, 14, 31, 64–68, 74, 75, 84, 87, 89, 92–94, 129, 132, 137, 139, 147, 151, 154, 185
Orteig Prize, 126

P
Parallel ideas, 188
Path dependence, 199
Play-Doh, 91, 106
Procter and Gamble, 64, 144–146, 149, 151

Q
Quality of cooperation, 189

R
Rating agency, 152
Rational incentive problem, 165
Realisation gap, 116, 143, 144
Realisation path, 46, 67, 75, 77, 85, 88, 90, 94, 108, 127, 174
  alternative, 204
  entrepreneurial, 93
Reciprocal feedback, 179
Reinventing The Toilet Challenge, 131
Rembrandts in the Attic, 62

Remedial actions, 189
Remuneration framework, 186
Repeated cooperation, 180
Repeated games and reputation, 150, 157
Reputation mechanism, 143, 150–155, 157, 160, 162, 166, 171, 173, 177–179, 189, 195–197
Request for Proposal, 130
Romer, Paul, 26

S
SAP, 63, 106
Schumpeter, Joseph, 38, 106
Sectoral shift, 84
Seekers, 136
Self-selection bias, 181
Serial entrepreneurship, 108
Shockley Semiconductor, 5
Shockley, William, 5, 62
Sikorsky Prize, 126
SmartTruck, 91
The Social Network, 11, 117; *see also* Facebook
Socioeconomic change, 59
Socioeconomic environment, 30, 35, 40, 47, 75, 81
Solver, 136
Sony, 151
Sorensen, Jay, 101
Spam, 174, 190, 192
Stereotypes, 30
Strategic abuse, 185
Substitutive ideas, 188
Successive ideas, 188
Supermarket for ideas, 197
Sybil attacks, 178

T
Tann Corporation, 117
Tesla, Nikola, 4
Theory of cooperation, 150, 155
Thick market externalities, 80

Threadless, 160
Track record, 26, 153, 154, 158, 159, 196, 201
Transparency, 151, 152, 158, 172, 174, 189, 190, 197
Trial and error, 197
Trust, 148
Twitter, 5

U
Ultimatum games, 163
Unemployment, 84
United States, 48
User innovation, 102

V
Venture capital, 14, 35, 42, 47, 50, 63, 79, 143–145, 149, 151
Venturesome consumers, 82
Vertical integration, 13, 42, 43, 80
Very long run, 20
    GDP, 23–24
    growth, 25
Virtuous reputation loop, 153
von Hippel, Eric, 14

W
Windows-on-a-Mac Prize, 126
Winklevoss, Tyler and Cameron, 11, 117

X
Xerox, 7, 12, 37, 39, 106
    PARC, 61

Z
Zanen, Eduard, 99
Zimride, 195, 196
Zuckerberg, Mark, 11, 117

MIX
Papier aus verantwortungsvollen Quellen
Paper from responsible sources
FSC® C105338

If you have any concerns about our products,
you can contact us on
ProductSafety@springernature.com

In case Publisher is established outside the EU,
the EU authorized representative is:
**Springer Nature Customer Service Center GmbH
Europaplatz 3, 69115 Heidelberg, Germany**

Printed by Libri Plureos GmbH
in Hamburg, Germany